Transgenic Crop Plants

Chittaranjan Kole · Charles H. Michler ·
Albert G. Abbott · Timothy C. Hall
Editors

Transgenic Crop Plants

Volume 1: Principles and Development

 Springer

Editors

Prof. Chittaranjan Kole
Department of Genetics & Biochemistry
Clemson University
Clemson, SC 29634, USA
ckole@clemson.edu

Prof. Albert G. Abbott
Department of Genetics & Biochemistry
Clemson University
Clemson, SC 29634, USA
aalbert@clemson.edu

Prof. Charles H. Michler
Director
Hardwood Tree Improvement and
Regeneration Center at Purdue University
NSF I/UCRC Center for Tree Genetics
West Lafayette, IN 47907, USA
michler@purdue.edu

Prof. Timothy C. Hall
Institute of Developmental & Molecular
Biology
Department of Biology
Texas A&M University
College Station, TX 77843, USA
tim@idmb.tamu.edu

ISBN: 978-3-642-04808-1   e-ISBN: 978-3-642-04809-8
DOI 10.1007/978-3-642-04809-8
Springer Heidelberg Dordrecht London New York

Library of Congress Control Number: 2009939123

Cover design: WMXDesign GmbH, Heidelberg, Germany

Printed on acid-free paper

Springer is part of Springer Science+Business Media (www.springer.com)

Preface

Transgenic Plants – known also as Biotech Plants, Genetically Engineered Plants, or Genetically Modified Plants – have emerged amazingly fast as a boon for science and society. They have already played and will continue to play a significant role in agriculture, medicine, ecology, and environment. The increasing demands for food, feed, fuel, fiber, furniture, perfumes, minerals, vitamins, antibiotics, narcotics, and many health-related drugs and chemicals necessitate development and cultivation of transgenic plants with augmented or suppressed trait(s). From a single transgenic plant (Flavr Savr™ tomato with a longer shelf-life) introduced for commercialization in 1994, we have now 13 transgenic crops covering 800 million hectares in 25 countries of six continents. Interestingly, the 13.3 million farmers growing transgenic crops globally include 12.3 million (90%) small and resource-poor farmers from 12 developing countries. Increasing popularity of transgenic plants is well evidenced from an annual increase of about 10% measured in hectares but actually of 15% in "trait hectares." Considering the urgent requirement of transgenic plants and wide acceptance by the farmers, research on transgenic plants is now being conducted on 57 crops in 63 countries. Transgenic plants have been developed in over 100 plant species and they are going to cover the fields, orchards, plantations, forests, and even the seas in the near future. These plants have been tailored with incorporation of useful alien genes for several desirable traits including many with "stacked traits" and also with silencing of genes controlling some undesirable traits.

Development, applications, and socio-political implications of transgenic plants are immensely important fields now in education, research, and industries. Plant transgenics has deservedly been included in the course curricula in most, if not all, leading universities and academic institutes all over the world, and therefore reference books on transgenic plants with a classroom approach are essential for teaching, research, and extension. There are some elegant reviews on the transgenic plants or plant groups (including a 10-volume series "Compendium of Transgenic Crop Plants" edited by two of the present team of editors C. Kole and T.C. Hall published by Wiley-Blackwell in 2008) and on many individual tools and techniques of genetic transformation in plants. All these reviews could surely serve well the purpose for individual crop plants or particular methodologies. Since transgenic plant development and utilization is studied, taught, and practiced by students, teachers, and scientists of over a dozen disciplines under basic science, agriculture, medicine, and humanities at public and private sectors, introductory reference books with lucid deliberations on the concepts, tools, and strategies to develop and utilize transgenic plants and their global impacts could be highly useful for a broad section of readers.

Deployment of transgenic crop plants are discussed, debated, regulated, and sponsored by people of diverse layers of the society including social activists, policy makers, and the staff of regulatory and funding agencies. They also require lucid deliberations on the deployment, regulations, and legal implications of practicing plant transgenics. More importantly, depiction of the positive and realistic picture of the transgenic plants should and could facilitate mitigation of the negative propaganda against transgenic plants and thereby reinforce moral and financial support from all individuals and platforms of the society. Global population is increasing annually by 70 million and is estimated to grow up to 8 billion by 2025. This huge population, particularly its large section from the developing countries, will suffer because of hunger, malnutrition, and chemical pollution unless we produce more and more transgenic plants particularly with stacked traits. Compulsion to meet the requirements of this growing population on earth and the proven innocuous nature of transgenic plants tested and testified for the last 13 years could substantiate the imperative necessity of embracing transgenics.

Traditional and molecular breeding practiced over the last century provided enormous number of improved varieties in economic crops and trees including wheat and rice varieties that fostered the "green revolution." However, these crop improvement tools depend solely on the desirable genes available naturally, creatable by mutation in a particular economic species or their shuffling for desired recombinations. Transgenic breeding opened a novel avenue to incorporate useful alien genes not only from other cross-incompatible species and genera of the plant kingdom, but also from members of the prokaryotes including bacteria, fungi, and viruses, and even from higher animals including mice and humans. An array of plant genetic engineering achievements starting from developing insect resistance transforming with the *cry* genes in cotton from the bacteria *Bacillus thuringiensis* to the present-day molecular pharming expressing the *interferon-α* gene from human in tobacco evidence for this pan-specific gene transfer.

Human and animal safety is another general concern related to transgenic food or feed. However, there is no reliable scientific documentation of these health hazards even after 13 years of cultivation of transgenic plants and consumption of about one trillion meals containing transgenic ingredients. Utilization of transgenic plants has reduced the pesticide applications by 359,000 tons that would otherwise affect human and animal health besides causing air, water, and soil pollution and also mitigated the chance of consumption of dead microbes and insects along with foods or feeds.

Gene flow from transgenic crop species to their cross-compatible wild relatives is a genuine concern and therefore required testing of a transgenic crop plant before deployment followed by a comprehensive survey of the area for the presence of interfertile wild and weedy plants before introduction of a transgenic crop is being seriously conducted.

Addition of novel genotypes with transgenes in the germplasms is increasing the biological diversity rather than depleting it. Use of the genetically engineered plants has also eliminated greenhouse gas emission of 10 million metric tons through fuel savings. In fact, 1.8 billion liters of diesel has been saved because of reduced tillage and plowing owing only to herbicide-resistant transgenic crops. Many transgenics are now being used for soil reclamation. Above all, cultivation of transgenic crops has returned $44 billion of net income to the farmers. Perhaps these are the reasons that 25 Nobel Laureates and 3,000-plus eminent scientists appreciated the merits and safety and also endorsed transgenic crops as a powerful and safe way to improve agriculture and environment besides the safety of genetically modified foods. Many international and national organizations have also endorsed health and environmental safety of transgenic plants; these include Royal Society (UK), National Academy of Sciences (USA), World Health Organization, Food and Agriculture Organization (UN), European Commission, French Academy of Medicine, and American Medical Association, to name a few.

Production, contributions, and socio-political implications of biotech plants are naturally important disciplines now in education, research, and industries and therefore introductory reference books are required for students, scientists, industries, and also for social activists and policy makers. The two book volumes on "Transgenic Crop Plants" will hopefully fill this gap. These two book volumes have several unique features that deserve mention. The outlines of the chapters for these two books are formulated to address the requirements of a broad section of readers. Students and scholars of all levels will obtain a lot of valuable reading material required for their courses and researches. Scientists will get information on concepts, strategies, and clues useful for their researches. Seed companies and industries will get information on potential resources of plant materials, and expertise, and also for their own R&D activities. In brief, the contents of this series have been designed to fulfill the demands of students, teachers, scientists and industry people, for small to large libraries. Students, faculties, or scientists involved in various subjects will be benefited from this series: biotechnology, bioinformatics, molecular biology, molecular genetics, plant breeding, biochemistry, ecology, environmental science, bioengineering, chemical engineering, genetic engineering, biomedical engineering, pharmaceutical science, agronomy, horticulture, forestry, entomology, pathology, nematology, and virology, just to name a few.

It has been our privilege to edit the 23 chapters of these two books, contributed by 71 scientists from 14 countries, and the list of authors includes one of the pioneers of plant transgenics Prof. Timothy C. Hall (one of the editors also); some senior scientists who have themselves edited books on plant transgenics; and many scientists who have written elegant reviews on invitation for quality books and leading journals. We believe that these two books will hopefully serve the purposes of the broad audience: those who are studying, teaching, practicing, supporting, funding, and also those who are debating for or against plant transgenics. The first volume dedicated to "Principles and Development" elucidates the basic concepts, tools, strategies, and methodologies of genetic engineering, while the second volume "Applications and Safety" enumerates the utilization of transgenic crop plants for various purposes of agriculture, industry, ecology and environment, and also genomics research. The second volume also deliberates comprehensively on the legal and regulatory aspects; addresses the major concerns; and finally justifies the compulsion of practicing plant transgenics.

Little more detail on the contents of the volume "Transgenic Crop Plants: Principles and Development" will hopefully substantiate its usefulness. This volume focuses on the methods for constructing gene vectors, introducing these gene vectors into plant tissue, targeting gene insertion to specific tissues, methods for detecting transgene expression, generation of transgenic plants, and types of traits and bioproducts that are targeted for these technologies. The first chapter of this volume presents glimpses on these aspects and also on those related to deployment of transgenic plants.

One important factor that determines successful transgenc insertion is the decision of explant type for use in transformation. A comprehensive review in Chap. 2 is provided from previous research with both herbaceous and woody plants. The use of morphogenic calli for cereals is discussed along with somewhat standardized protocols for each individual woody species. Gene transfer methods have been discussed including use of *Agrobacterium*, biolistics, electroporation, liposomes, microinjections, and bioactive beads in Chap. 3.

Once a transgene is inserted, markers have been used to either score the success of the transgene event or screen for the successful events. Although much work has been done using *npt* and *gus*, recent work has looked at marker gene removal in the final transgene product in order to belie environmental concerns. Further molecular characterization with Southern blot analysis and PCR confirm definitive transgene integration and copy number. Chapter 4 has been devoted to these critical steps.

Stable and regulated transgene expression, as described in Chap. 5, is necessary for the transgenic plant to express the trait of interest for further research or commercial applications. The use of constitutive and tissue-enhanced promoters along with attention to attachment regions within the DNA, introns, RNA integrity factors, and transcription factors will determine transgene expression and the levels thereof. Besides using transgenes for introduction of nucleic acid for novel trait production, transgenes have also been used to silence native genes for applications such as resistance to nematodes, insects, and viruses. As interesting are the applications to reduce production of compounds in some plants, such as caffeine in coffee and sulfur metabolites in onions, that are disagreeable to portions of the human populations. All of these silencing events rely on RNAi technology to degrade native RNA for those traits of interest. Chapter 6 provides a commentary on the employment of RNAi technology and the implications and outcome of expression and gene silencing have been explicated in Chap. 7.

Researchers have also been successful with organelle transfer, which has applications to molecular pharming, as have been enumerated in Chap. 8. This technique can be employed to overcome some transgene expression difficulties. Cell culture biosynthesis and metabolic engineering are the focus of the last two chapters (Chaps. 9 and 10) in this volume and these chapters offer an intriguing look at research into production of high-value energy and medicinal products, secondary metabolites, and plants with attractive esthetic qualities. Because we still have a rudimentary understanding of many biochemical pathways, we are continuing to gain new knowledge and insight into pathway function, but commercial plant systems are still lacking in most desirable traits when economic viability, environmental safety, and sustainability are taken into account.

We thank the 31 scientists from 9 countries for their elegant and lucid contributions to this volume and also for their sustained support through revising, updating, and fine-tuning their chapters. We also acknowledge the recent statistics we have accessed from the web sites of Monsanto Company on "Conversations about Plant Biotechnology" and "International Service for the Acquisition of Agri-Biotech Applications on ISAAA Brief 39-2008: Executive Summary" and used them in this preface and elsewhere in the volume.

We enjoyed a lot our Clemson-Purdue-Texas A&M triangular interaction, constant consultations, and dialogs while editing this book and also working with the editorial staff of Springer, particularly Dr. Sabine Schwarz, who had been supportive since the inception till the publication of this book.

We look forward to suggestions from all corners for the future improvement of the content and approach of this book volume.

Chittaranjan Kole, Clemson, SC
Charles H. Michler, West Lafayette, IN
Albert G. Abbott, Clemson, SC
Timothy C. Hall, College Station, TX

Contents

Contributors

N. Alburquerque Department of Plant Breeding, CEBAS (CSIC), P.O. Box 164, 30100 Espinardo, Murcia, Spain, nalbur@cebas.csic.es

P. Ananda Kumar NRC on Plant Biotechnology, IARI Campus, New Delhi 110012, India, polumetla@hotmail.com

Pudota B. Bhaskar Department of Horticulture, University of Wisconsin-Madison, 1575 Linden Drive, Madison, WI 53706, USA, pudota1@wisc.edu

Rajib Bandopadhyay Department of Biotechnology, Birla Institute of Technology, Mesra, Ranchi 835215, Jharkhand, India, rajib_bandopadhyay@bitmesra.ac.in

Anjanabha Bhattacharya National Environmental Sound Production Agriculture Laboratory, University of Georgia, Tifton, GA 31794, USA, anjanabha.bhattacharya@gmail.com

John E. Carlson The School of Forest Resources, and Huck Institutes for Life Sciences, Pennsylvania State University, 405C Life Sciences Building, University Park, PA 16802, USA, jec16@psu.edu

Ravindra N. Chibbar Department of Plant Sciences, College of Agriculture and Bioresources, University of Saskatchewan, Saskatoon S7N 5A8, Canada, ravi.chibbar@usask.ca

Michael R. Davey Plant and Crop Sciences Division, School of Biosciences, University of Nottingham, Sutton Bonington Campus, Loughborough, LE12 5RD, UK, mike.davey@nottingham.ac.uk

Chandrakanth Emani Institute of Developmental & Molecular Biology, Department of Biology, Texas A&M University, College Station, TX 77843, USA, chandra@idmb.tamu.edu

Kevin M. Folta Horticultural Sciences Department and the Graduate Program in Plant Molecular and Cellular Biology, University of Florida, Gainesville, FL 32611, USA, kfolta@ufl.edu

Seedhabadee Ganeshan Department of Plant Sciences, College of Agriculture and Bioresources, University of Saskatchewan, Saskatoon, S7N 5A8, Canada, pooba.ganeshan@usask.ca

Timothy C. Hall Institute of Developmental & Molecular Biology, Department of Biology, Texas A&M University, College Station, TX 77843, USA, tim@idmb.tamu.edu

Inamul Haque Department of Biotechnology, Birla Institute of Technology, Mesra, Ranchi 835215, Jharkhand, India, inam.Hoque@yahoo.com

Hiroaki Hayashi School of Pharmacy, Iwate Medical University, 2-1-1 Nishito-kuda, Yahaba, Iwate, 028-3694, Japan, hhayashi@iwate-med.ac.jp

Jiming Jiang Department of Horticulture, University of Wisconsin-Madison, 1575 Linden Drive, Madison, WI 53706, USA, jjiang1@wisc.edu

Miloslav Juricek Institute of Experimental Botany, Academy of Sciences of the Czech Republic, Na Pernikare 15, CZ-160 00, Prague 6, Czech Republic, juricek@ueb.cas.cz

Sunee Kertbundit Department of Plant Science, Faculty of Science, Mahidol University, Rama VI Road, Bangkok 10400, Thailand, stskb@mahidol.ac.th

Ajay Kohli Plant Molecular Biology Laboratory, Plant Breeding Genetics and Biotechnology, International Rice Research Institute, DAPO-7777, Metro Manila, Philippines, a.kohli@cgiar.org

Chittaranjan Kole Clemson University, Department of Genetics and Biochemistry, 111 Jordan Hall, Clemson, SC 29634, USA, ckole@clemson.edu

Sofia Kourmpetli Plant and Crop Sciences Division, School of Biosciences, University of Nottingham, Sutton Bonington Campus, Loughborough LE12 5RD, UK, sbxsk2@nottingham.ac.uk

Haiying Liang Department of Genetics & Biochemistry, Clemson University, 108 Biosystem Research Complex, Clemson, SC 29634, USA, hliang@clemson.edu

Berta Miro Institute for Research on Environment and Sustainability, Devonshire Building, Newcastle University, Newcastle NE17RU, UK, berta.miro@ncl.ac.uk

Hajime Mizukami Department of Pharmacognosy, Graduate School of Pharmaceutical Sciences, Nagoya City University, 3-1 Tanabe-dori, Mizuho-ku Nagoya, 467-8603, Japan, hajimem@phar.nagoya-cu.ac.jp

Kunal Mukhopadhyay Department of Biotechnology, Birla Institute of Technology, Mesra, Ranchi 835215, Jharkhand, India, mkunalus@yahoo.com

Madhugiri Nageswara Rao University of Florida, IFAS, Citrus Research & Education Center, 700 Experiment Station Road, Lake Alfred, FL 33850, USA, mnrao@crec.ifas.ufl.edu

Vikrant Nain NRC on Plant Biotechnology, IARI Campus, New Delhi 110012, India, v_nain@rediffmail.com

A. Piqueras Department of Plant Breeding, CEBAS (CSIC), PO Box 164, 30100 Espinardo, Murcia, Spain, piqueras@cebas.csic.es

William A. Powell College of Environmental Science & Forestry, State University of New York, One Forestry Drive, Syracuse, NY 13210, USA, wapowell@esf.edu

Dharmendra Singh Department of Biotechnology, Birla Institute of Technology, Mesra, Ranchi 835215, Jharkhand, India, damfire@gmail.com

Jaya R. Soneji University of Florida, IFAS, Citrus Research & Education Center, 700 Experiment Station Road, Lake Alfred, FL 33850, USA, jrs@crec.ifas.ufl.edu

Richard M. Twyman Department of Biology, University of York, Heslington, York YO10 5DD, UK, richard@writescience.com

Abbreviations

2, 4-D	2, 4-Dichlorophenoxyacetic acid
2-DOG	2-Deoxyglucose
4MI	4-Methylindole
4-MT	4-Methyltryptophan
4-MUGIuc	4-Methylumbelliferyl glucuronide
5MT	5-Methyltrypthopan
7MT	7-Methyl-DL-tryptophan
AA	Arachidonic acid
aadA/addA	Aminoglycoside 3'-adenyl transferase
ABC	*Arabidopsis* ATP-binding cassette
ABW	Aluminium borate whiskers
ACC	Aminoglycoside acetyltransferase
ADC	Arginine decarboxylase
ADH	Alcohol dehydrogenase
ADP	Adenosine diphosphate
AEC	S-Aminoethyl l-cysteine
AG	*Arabidopsis* agamous
AHAS	Acetohydroxyacid synthase
AK	Aspartate kinase
ALA	Aminolaevulinic acid
ALA	α-Linolenic acid
AL-PCR	Adaptor ligation-PCR
ALS	Acetolactate synthase
amiRNA	Artificial micro-RNA
AMV	Alfalfa mosaic virus
ANS	Anthocyanidin synthase
AP2	*APETALA2* gene
APH	Aminoglycoside phosphotransferase
*aro*A (*epsps*)	5-Enolpyruvylshikimate-3-phosphate synthase gene
ARS	Autonomously replicating sequence
*ars*C	Arsenic reductase gene

AS	Amyrin synthase
AS	Anthranilate synthase
ATP	Adenosine tri-phosphate
AtTSB1	*Arabidopsis* tryptophan synthase beta 1 gene
BA	Benzylaminopurine
BADH	Betaine aldehyde dehydrogenase
bar (*pat*)	Phosphinothricin acetyltransferase gene
BMS	Black mexican sweet
Bnx	Bromoxynil nitrilase gene
BOAA	β-*N*-Oxalylamino-L-alanine
BSMV	Barley stripe mosaic virus
BSV	Banana streak virus
Bt	*Bacillus thuringiensis*
BYDV	Barley yellow dwarf virus
C4H	Cinnamate 4-hydroxylase
Cah	Cyanamide hydratase
CaMV	Cauliflower mosaic virus
CAS	Cycloaretenol synthase
CAT	Chloramphenicol acetyltransferase
cDNA	Complementary-DNA
CDPK	Calcium dependent protein kinase
CHI	Chalcone isomerase
CHR	Chalcone reductase
CHS	Chalcone synthase
CKI1	Cytokinin-Independent 1
CMV	Cytomegalovirus
cp4	Glyphosate resistant 5-enolpyruvylshikimate-3-phosphate synthase
*cpt*II	Carnitine palmitoyltransferase II gene
CRC	Chimeric transcription factor
CRY	Cryptochrome
CSIRO	Commonwealth Scientific and Industrial Research Organization
CsVMV	Cassava vein mosaic virus
CUP1	Yeast metallothionein gene
CYP	Cytochrome P450
CYP	Cytochrome P450 gene
DAAO	D-Amino acid oxidase
DEF	Peptide deformylase
DFR	Dihydroflavonol reductase
DHA	Docosahexaenoic acid
DHDPS	Dihydrodipicolinate synthase
DHFR	Dihydrofolate reductase
DHK	Dihydrokaempferol

DHM	Dihydromyricetin
DHPS	Dihydrodipicolinate synthase
DHPS	Dihydropteroate synthase
DHQ	Dihydroquercetin
DIG	Digoxigenin
DQR	Dihydroquercerin-4 reductase
dsRNA	Double stranded-RNA
EBV	Epstein barr virus
ELISA	Enzyme-linked immunosorbent assay
EPA	Eicosapentanoic acid
EPSP	5-Enolpyruvylshikimate-3-phosphate
EPSPS	5-Enol-pyruvyl shikimate-3-phosphate synthase
EREBP	Ethylene response element binding protein
ERF	Ethylene response factor
ESR1	Enhancer of shoot regeneration 1
F3'5'H	Flavonoid 3', 5'-hydroxylase
F3'H	Flavonoid 3'-hydroxylase
F3H	Flavanone 3b-hydroxylase
FAD3	Fatty acid desaturase 3
FAO	Food and Agricultural Organization
FISH	Fluorescence in situ hybridization
FITC	Fluorescein isothiocyanate
FLC	Flowering locus C gene
FMV	Figwort mosaic virus
FNSII	Flavone synthase II
galT	UDP-glucose:galactose-1-phosphate uridyltransferase
GBSS	Granule bound starch synthase gene
GFP	Green fluorescent protein
gfp	Green fluorescent protein gene
GM	Genetically modified
GOI	Gene of interest
GOX	Glyphosate oxidase
gox	Glyphosate oxidoreductase gene
GS	Glutamine synthase
GSA-AT	Glutamate-1-semialdehyde aminotransferase
gshI	Glutathione synthase I gene
GST	Glutathione-S-transferase
gus	β-Glucuronidase gene
GUS	β-Glucuronidase
H6H	Hyoscyamine-6-hydroxylase
HAS	Human serum albumin
HAT	Histoneacetyl transferase
HD	Histone deacetylase
hph (*hpt*, *aph*IV)	Hygromycin phosphotransferase gene

hpRNA	Hairpin RNA
HPT	Hygromycin phosphotransferase
HSP	Heat shock protein
IAA	Indole acetic Acid
IFS	Isoflavone synthase
IM	Imidazolinone
IME	Intron-mediated enhancement
IPT	Isopentyl transferase
KN1	Homeobox gene knotted1
LA	Linoleic acid
LC-PUFA	Long chain polyunsaturated fatty acid
LDC	Lysine decarboxylase
LF	Lachrymatory factor
LFS	Lachrymatory factor synthase
LRE	Light responsive elements
LT	Lysine and threonine
Luc/lux	Luciferase gene
LUS	Lupeol synthase
MAR	Matrix associated/attachment region
MCA	Metabolic control analysis
mer	Mercuric ion reductase gene
MMV	Mirabilis mosaic virus
mRNA	Messenger-RNA
MT1	Metallothionein 1 gene
MT2	Metallothionein 2 gene
Mtx	Methotrexate
NAA	Naphthalenacetic acid
NoGA2ox3	Oleander gibberellic acid 2-oxidase gene 3
NOS	Nopaline synthase
nos	Nopaline synthase gene
nptII	Neomycin phosphotransferase II
*npt*II	Neomycin phosphotransferase II gene
OCS	Octopine synthase
ODC	Ornithine decarboxylase
OMT	L-O-Methylthreonine
OPH	Organophosphate hydrolase
ORF	Open reading frame
OxO	Oxalate oxidase
PAL	Phenylalanine ammonia lyase
PAT	Phosphinothricin acetyltransferase
PCR	Polymerase chain reaction
PDR	Pathogen-derived resistance
PEG	polyethylene glycol
Pflp	Pepper ferredoxin-like protein

PGA	Plant growth activator
PGL34	Plastoglobulin 34
PGT	p-Hydroxybenzoate geranyltransferase
PHB	p-Hydroxybenzoate/Polyhydroxybutyrate
PHY	Phytochrome
PIG	particle in-flow gun
PMI	Phosphomannose isomerase
pmi	Phosphomannose isomerase gene
PPO	Protoporphyrinogen oxidase
PPT	Phosphinothricin
ppt	Phosphinothricin acetyltransferase gene
PTGS	Post-transcriptional gene silencing
PUFA	Polyunsaturated fatty acids
PVX	Potato virus X
PVY	Potato virus Y
RNAi	RNA interference
ROL/rol	Root locus
RT-PCR	Reverse transcription-PCR
SAAT	Sonication-assisted *Agrobacterium*-mediated transformation
SAM	S-adenosylmethionine
SAUR	Small auxin up RNA
ScBV	Sugarcane bacilliform badnavirus
SDA	Stearidonic acid
siRNA	Short interfering RNA
SOP	Standard operating procedures
SPT	Streptomycin phosphotransferase
STK	Seedstick
SU	Sulfonylureas
SV40	Simian virus 40
TAIL-PCR	Thermal asymmetric interlaced-PCR
TD	Threonine deaminase
TDC	Tryptophan decarboxylase
T-DNA	Transferred DNA
Ti plasmid	Tumour-inducing plasmid
TILLING	Target induced local lesions in genomes
TMV	Tobacco mosaic virus
TP	Triazolopyrimidine
TSP	Total soluble protein
TSSR	Tuber-specific and sucrose responsive
T-strand	Transferred strand
UDP	Uridine diphosphate
UFGT	UDP-glucose:flavonoid 3-O- glucosyltransferase
uidA (gusA)	β-glucuronidase gene
USDA	United States Department of Agriculture

UTR	Untranslated region
UV	Ultraviolet
VIGS	Virus induced gene silencing
WIN1	Wax inducer 1
X-Gluc	5-Bromo-4-chloro-3-indolyl-β-D- glucuronide
YAC	Yeast artificial chromosome
ZAT1/ZntA	Zinc transporter gene

Chapter 1
Generation and Deployment of Transgenic Crop Plants: An Overview

Michael R. Davey, Jaya R. Soneji, M. Nageswara Rao, Sofia Kourmpetli, Anjanabha Bhattacharya and Chittaranjan Kole

1.1 Introduction

As biotechnology increasingly affects almost all aspects of human life, it is essential that the science behind this technology is explained in simple terms to the public to eliminate the misconceptions that may inhibit its acceptability. The basic question that is often asked is what is a gene, a promoter and a terminator? Genes are the basic units of heredity, composed of DNA sequences, which are transmitted from parents to offspring and which, independently or in combination with other genes, control specific traits in an organism. These traits may be, for example, plant height, flower color, fruit and seed size together with regulatory processes, such as assimilate partitioning and drought resistance. Genes are the basis for both the similarity and differences that exist among organisms, and are transmitted from one generation to another. Promoters are DNA sequences that are recognized by RNA polymerase in plant cells and that initiate and regulate transcription, the initial and most important step of gene expression. Terminators are those sequences that command or signal the termination of transcription.

It is possible to identify and to isolate genes from plants, animals, and microorganisms, to modify their promoters, structural sequences and terminators, and to introduce and express chimeric genes in the same or other genus, species, or cultivar. Consequently, it is feasible to control or modify physiological processes. Gene manipulation, combined with the ability to induce cultured plant cells to express their totipotency leading to the regeneration of fertile plants, provides a

M.R. Davey (✉) and C. Kole (✉)
Plant and Crop Sciences Division, School of Biosciences, University of Nottingham, Sutton Bonington Campus, Loughborough, LE12 5RD, UK
e-mail: mike.davey@nottingham.ac.uk

Clemson University, Department of Genetics and Biochemistry, 111 Jordan Hall, Clemson, SC 29634, USA
e-mail: ckole@clemson.edu

C. Kole et al. (eds.), *Transgenic Crop Plants*,
DOI 10.1007/978-3-642-04809-8_1, © Springer-Verlag Berlin Heidelberg 2010

unique opportunity to extend the genetic pool available to breeders for crop improvement.

The successful development of transgenic plants necessitates a reliable tissue culture regeneration system, gene construct(s), suitable vector(s) for transformation and efficient procedures to introduce desired genes into target plants. Once transformation has been performed, it is essential to recover and to multiply the transgenic plants. The latter must be characterized at the molecular and genetic levels for stable and efficient gene expression (Sharma et al. 2005). It may also be necessary to transfer the introduced genes to elite cultivars by conventional breeding.

Prime targets for genetic manipulation include modification of plants to enhance their tolerance to the herbicides used to control weeds, and to confer resistance to insects, bacteria, fungi, and viruses, since these agents account for major crop losses. Other targets include the genetic engineering of plants for biosynthesis of health-care products, increased nutritional value, extension of the shelf-life of crops that deteriorate rapidly following harvest, and tolerance to abiotic stress. Similarly, although not essential for human existence, modification of the esthetic appeal of plants has considerable commercial potential.

1.2   Target Cells and Organelles for Genetic Transformation

A reliable tissue culture-based shoot regeneration system is a pre-requisite for plant genetic transformation. The recognition that, under optimum hormonal and nutritional conditions, somatic cells are totipotent and can be stimulated to develop into whole plants in vitro via organogenesis (shoot formation) or somatic embryogenesis, forms the basis of regeneration in tissue culture (Sharma et al. 2005). Genetic transformation without plant regeneration is of limited or no value. Hence, the identification of explants (cells/tissues/organs) that are capable of regenerating into plants is fundamental to any transformation procedure. Isolated protoplasts (Davey et al. 2005), callus and suspension cultured cells (Rachmawati and Anzai 2006), thin cell layers (Soneji et al. 2007a), leaf disks (Li et al. 2007), root sections (Huang and Ma 1992), stem segments (Song et al. 2006), floral tissues (Zale et al. 2008), epicotyls (Soneji et al. 2007b), hypocotyls (Wang and Xu 2008), cotelydonary nodes (Yi and Yu 2006), and axillary buds (Manickavasagam et al. 2004) have been used for genetic transformation. Explants of mature organs have also been used as target material in transformation experiments to overcome juvenility (Cervera et al. 1998). Tissue culture systems for several plants have been summarized (Khachatourians et al. 2002; Curtis 2004; Loyola-Vargas and Vázuez-Flota 2005) together with aspects of gene introduction into target plants using such systems (Birch 1997; Newell 2000; Sharma et al. 2005; Davey et al. 2008).

Shoot regeneration from cultured cells may lead to chromosomal or genetic variation known as "somaclonal variation." This variation may be useful or detrimental. Tissue culture also requires extensive facilities for maintenance and manipulation of axenic explants, which is labor intensive and expensive.

Consequently, approaches have been reported that reduce or eliminate in vitro procedures. For example, genes have been inserted into pollen and the latter used for fertilization to produce transgenic seed (Saunders and Matthews 1995; Häggman et al. 1997), while Clough and Bent (1998) described a "floral dip" procedure that is discussed later.

In most investigations, gene insertion has been directed primarily to the nuclei of recipient plant cells. Additionally, plastid transformation has been established in several laboratories (Heifetz 2000; Daniell et al. 2002; Maliga 2002, 2004). Extension of plastid transformation to more species constitutes a logical step in the development of genetic manipulation technology (Bock and Khan 2004) as plastid transformation has several advantages for the engineering of gene expression in plants. These advantages include 10–50 times greater transgene expression in plastid genomes, compared to nuclear-inserted genes (Liu et al. 2008a). The plastid genome provides readily obtainable high protein concentrations and the possibility of expressing multiple proteins from polycistronic mRNAs from a single promoter (Maliga 2002). Importantly, uniparental plastid gene inheritance in most crop plants prevents pollen transmission of foreign DNA (Heifetz 2000). As transgenes integrate into the plastid genome via homologous recombination, this facilitates targeted gene replacement and precise transgene control, while sequestration of foreign proteins in plastids prevents adverse interactions with the cytoplasmic environment. Maliga (2004) and Verma and Daniell (2007) discussed the design of vectors for plastid transformation and the selection of transplastomic plants. To date, plastid transformation has been reported in cabbage, lettuce, oilseed rape, petunia, poplar, potato, tobacco, and tomato, with transplastomic plants being regenerated by organogenesis in these cases, or by somatic embryogenesis in carrot, cotton, rice, and soybean (Verma et al. 2008). Extension of plastid transformation to other major crop plants still necessitates reproducible explant, cell, or protoplast-to-plant regeneration systems.

1.3 Methods for Introducing Genes into Plants

Transformation of plants involves the stable introduction of DNA sequences usually into the nuclear genome of cells capable of developing into a whole transgenic plant (Sharma et al. 2005). Once a reliable shoot regeneration system is available, foreign DNA can be introduced into cells by either vector-mediated or direct transfer. Although the technology associated with the construction of chimeric genes is becoming more routine and simple, the transformation process itself remains a comparatively rare event. Consequently, the procedure must be robust and combine reproducible culture of recipient plant cells with efficient gene delivery. Gene transfer experiments focus mainly on maximizing the efficiency of recovery of stably transformed plants, and extending the range of species that can be engineered using a specific procedure.

Agrobacterium-mediated gene transfer and direct DNA transfer into cells by microprojectile bombardment (Fig. 1.1) are the most widely exploited methods for introducing genes into plants because of their ability to transform intact, regenerable tissues and organs. Although aspects of the precise molecular events of *Agrobacterium*-mediated gene delivery are still not fully understood, particularly the transfer and integration of the T-DNA (transferred DNA) from the bacterial tumor-inducing (Ti) plasmid of *Agrobacterium* into the nuclei of recipient plants, *Agrobacterium*-mediated gene delivery remains the preferred method of plant transformation in many laboratories. Lacroix et al. (2006a, 2006b) and Tzfira and Citovsky (2006) proposed mechanisms for the process. Knowledge of foreign gene integration into plant genomes is essential for precise gene targeting in the future.

Immersion of totipotent explants in a suspension of *Agrobacterium* is the main procedure for plant transformation. Several parameters affect transformation, including bacterial virulence, incubation temperature, age of the bacterial suspension, and the cocultivation period of the bacteria with the explants (Gelvin 2003; Wu et al. 2003). Sonication promotes gene delivery (Liu et al. 2006). In planta procedures have been developed to simplify the transformation procedure. Thus, the "floral dip" technique (Clough and Bent 1998) involves immersion of developing flowers in suspensions of *Agrobacterium*, followed by growth of the plants to maturity, the harvesting and germination of seeds, and the selection of transformed seedlings. This procedure, used routinely to transform *Arabidopsis thaliana*, has facilitated progress in understanding the genetics of this plant that is exploited extensively as a model in plant genetics and molecular biology. Chung et al. (2000) compared floral spraying with the floral dip procedure and reported comparable results with the two methods, enabling floral spraying to be used for transforming plants which are too large for the floral dip approach. Probably, in planta techniques will assume increasing importance for gene delivery.

Particle (microprojectile) bombardment has also been exploited extensively for plant transformation (Sharma et al. 2005; Davey et al. 2008) with instruments such as the helium driven HE-1000 device, facilitating technology transfer between laboratories. Microprojectile systems involve high-velocity particles penetrating cell walls and introducing DNA into cells, circumventing the host range limitations of *Agrobacterium*. This transformation procedure is versatile, independent of plant cell type and genotype, and has permitted the transformation of some of the most recalcitrant plants, such as cereals and legumes (Altpeter et al. 2005). Importantly, simple gene constructs, comprising only a promoter, the gene coding sequence and a terminator, may be used for transformation. A criticism of particle bombardment and *Agrobacterium*-mediated gene delivery is the complexity of patterns often associated with the integration of genes into recipient plants, especially with particle delivery. This necessitates detailed molecular analyses to select individuals carrying simple integration events, as such transformed plants are more applicable to longer-term breeding programs. Undoubtedly, the two procedures will continue

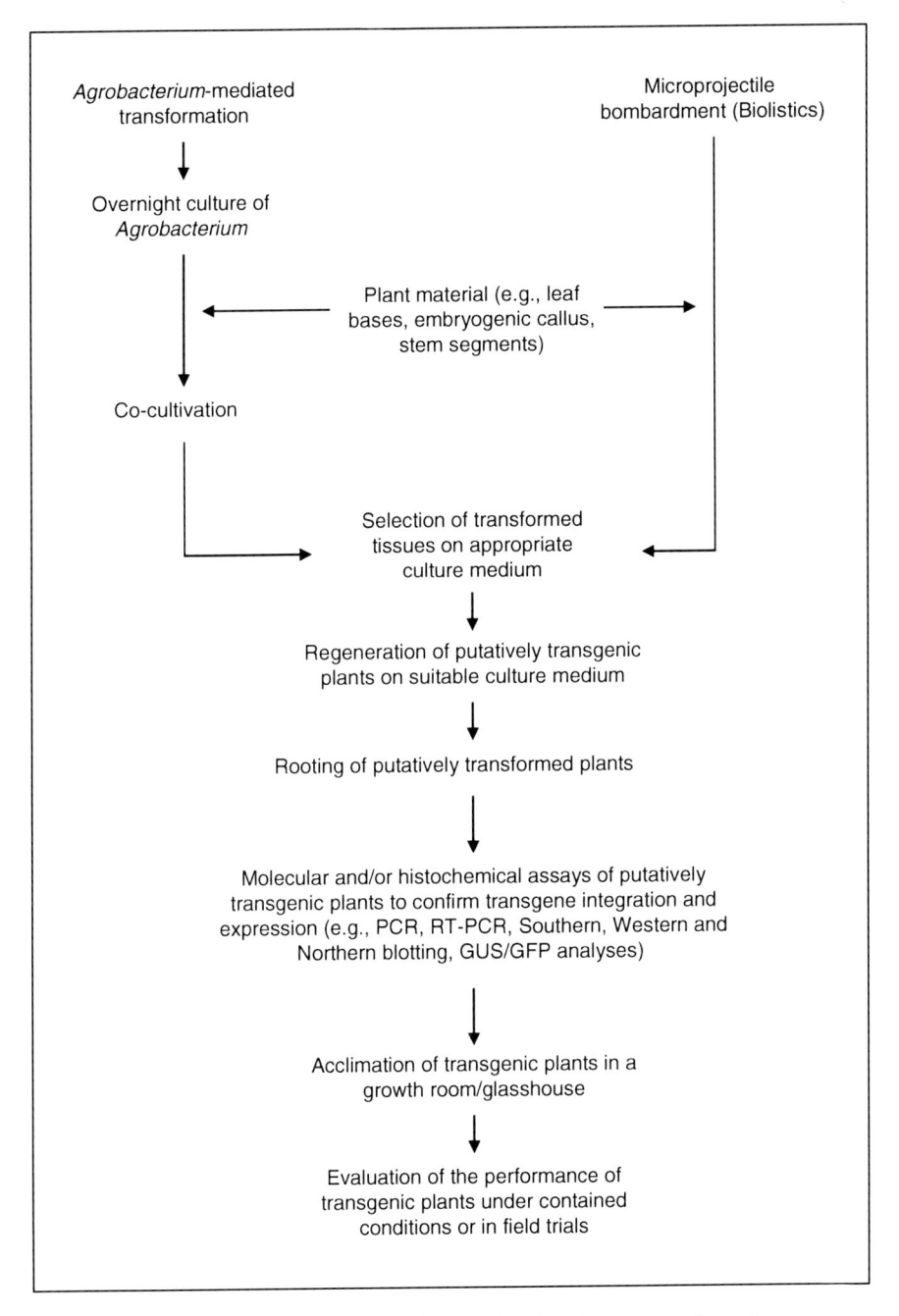

Fig. 1.1 A generalized flow chart depicting the steps involved in plant transformation

to be exploited routinely for gene delivery to plants, the procedure used depending upon the product required and the expertise of the personnel.

Other gene delivery procedures have been exploited, including uptake of DNA into isolated protoplasts, treatment with polyethylene glycol, and/or electroporation inducing DNA uptake. However, the development of robust protoplast-to-plant systems is a labor-intensive, specialized part of the procedure. Several parameters influence transformation, including the stage of the cell cycle of the recipient protoplasts, temperature, pH, and the intensity and duration of the electric field. Even with optimal conditions, the frequency of stable transformation is low and rarely exceeds one transformed cell in every 10^4 treated protoplasts. Protoplast transformation was the procedure of choice for monocotyledons, particularly cereals such as rice, but was superseded by particle bombardment and, more recently, by *Agrobacterium*-mediated gene delivery. Rakoczy-Trojanowska (2002) and Sharma et al. (2005) discussed transformation procedures involving micro- and macroinjection, the use of silicon carbon fibers, and pollen-tube-mediated DNA delivery. Virus-based DNA delivery methods have been reported (Chung et al. 2006). The real success and application of several transformation procedures remains unclear. Specific crops necessitate particular adaptation of techniques to generate transgenic plants, an excellent example being provided by some of the difficulties encountered in applying *Agrobacterium*-mediated gene delivery developed for rice to other cereals (Shrawat and Lörz 2006). However, gene sequencing, as in rice (Matsumoto et al. 2005), and general advances in plant bioinformatics, will facilitate broader application of transformation technology.

1.4 Vector Construction and Genes for Plant Transformation

Vector development has proceeded from the cointegration of foreign genes into the T-DNA region of Ti plasmids, to the construction of disarmed binary and superbinary vectors (Komori et al. 2007; Davey et al. 2008). As Tzfira et al. (2007) explained, although binary vectors were initially revolutionary, subsequent generations of vectors have had more versatility, often being designed for specific transformation purposes (Chung et al. 2005). Some vectors have incorporated recombinase-mediated gene cloning (Karimi et al. 2002). Importantly, advances in vector construction have enabled *Agrobacterium*-mediated transformation to be exploited for gene introduction into monocotyledons (Cheng et al. 2004), as well as dicotyledons. New gene expression technologies developed for nonplant systems rapidly become adapted and exploited in plant biology (Tzfira et al. 2007). This emphasizes the necessity for plant biologists to recognize and exploit developments in fields of research other than their own. A schematic representation of the steps involved in the construction of vectors for plant transformation is shown in Fig. 1.2.

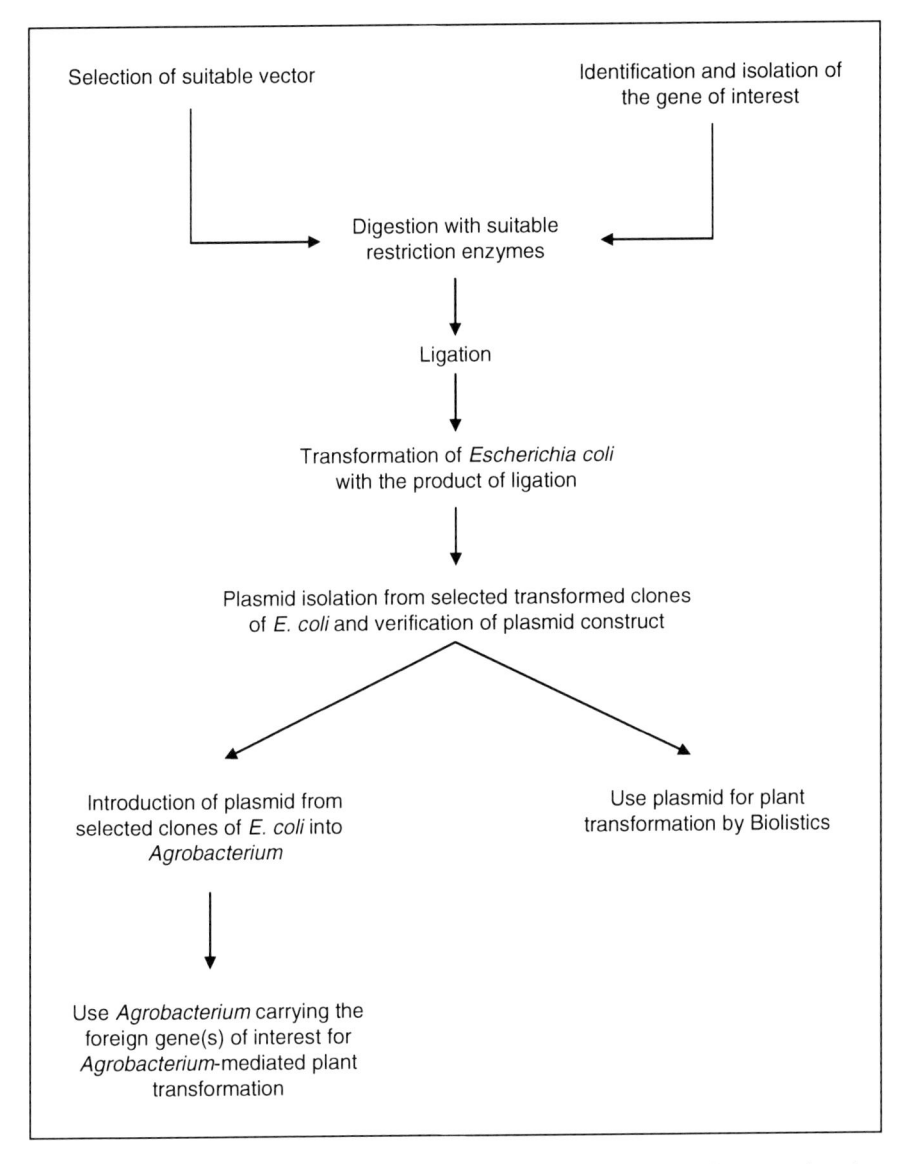

Fig. 1.2 Flow chart depicting the steps involved in the construction of vectors for plant transformation

1.4.1 Promoters for Plant Transformation

Efficient and reliable procedures are essential for constructing vector(s) for plant genetic engineering. Venter (2007) highlighted the importance of focusing attention on promoter construction, because the choice of promoter and its fine-tuning determine constitutive, spatial, and/or temporal transgene expression. Considerable effort has focused on gene promoters. Efficient expression of genes is assured only when they are controlled by plant-derived promoters, or by promoters that are active in plant cells, such as the cauliflower mosaic virus 35S promoter (CaMV 35S). In early transformation assessments, the choice of promoter was governed by promoter availability. The *nos* promoter from the nopaline synthase gene of the T-DNA of the Ti plasmid of *A. tumefaciens* was one of the first to be used in plant genetic engineering, with the 35S promoter from CaMV also featuring in many of the early transformation assessments. Subsequently, other constitutively expressed viral promoters were evaluated, including those from cassava vein mosaic virus (CsVMV), sugarcane bacilliform badnavirus (ScBV), and figwort mosaic virus (Samac et al. 2004; Govindarajulu et al. 2008). The CaMV 35S promoter may have a negative effect on transgene expression in some plants (Yoo et al. 2005). A limitation of the promoters of viral origin is that host plants may recognize and inactivate these sequences (Potenza et al. 2004). However, this may be negated by using promoters of plant origin. Indeed, several promoters including those from *Medicago truncatula* (Xiao et al. 2005), *Vigna radiata* (Cazzonelli et al. 2005), and the tobacco EI1α together with the Cab promoters (Aida et al. 2005) have been evaluated.

Constitutive expression at the incorrect time may have a serious negative effect on plant development, emphasizing the need to refine the promoters for transgene expression. Tissue-specific promoters fulfill this requirement. Examples include a tissue-specific promoter driving a β-1, 3 gluconase gene in pea (Buchner et al. 2002), promoters from fruit-ripening and seed-specific genes (Zakharov et al. 2004) particularly seed storage glutelin genes (Qu et al. 2008) and promoters of glycoproteins in tubers and roots. Flower-specific promoters have application in the genetic manipulation of fruit trees and ornamental plants (Annadana et al. 2002; Sassa et al. 2002). Comparisons of promoter function are important, a cotton α-globulin promoter being evaluated in cotton, *Arabidopsis*, and tobacco (Sunilkumar et al. 2002). Potenza et al. (2004) provided a schematic representation of the sources of many promoters. Tissue-specific promoters have been combined with RNA interference (RNAi) technology to modify flower pigmentation (Nakatsuka et al. 2007a). Modification of promoters may result in changes in tissue and developmental specificities (Kluth et al. 2002). Promoters of considerable potential are those associated with the interaction of plants and microorganisms, such as root-specific promoters involved in nutrient uptake and legume-*Rhizobium* symbiotic associations. These promoters from green tissues confer light-inducible and tissue-specific expression. Cell-type-specific promoters are available, such as those from trichomes, guard cells and stomata, root hairs, phloem (Zhao et al. 2004; Guan and Zhou 2006), and cortical cells (Fruhling et al. 2000). Vectors for plastid transformation normally employ promoters from the plastid genomes of the target plants.

Some plant promoters are induced by biotic and abiotic stress (Pino et al. 2007), wounding (Yevtushenko et al. 2004; Luo et al. 2006), iron deficiency (Kobayashi et al. 2007), and exogenously applied chemicals. The latter include antibiotics, steroids, copper, ethanol (Peebles et al. 2007), inducers of pathogen-related proteins, herbicide safeners and insecticides (Padidam 2003). Synthetic promoters have been assembled, such as a chimeric endosperm-specific promoter for cereal transformation (Oszvald et al. 2008). Liu et al. (2008b) constructed a novel pollen-stigma and carpel-specific promoter, which has potential in controlling pollen and seed-mediated gene flow from genetically manipulated plants. However, some synthetic promoters are unsuitable for plant transformation. For example, the (AocS)(3)AmasPmas promoter driving the *bar* gene for herbicide tolerance inhibited shoot regeneration (Song et al. 2008). Synthetic promoters, with the minimum of sequence similarity, could reduce homology-dependent gene silencing in transgenic plants during gene pyramiding experiments. Indeed, the availability of a broad spectrum of promoters that differ in their ability to regulate temporal and spatial expression patterns of transgenes could increase dramatically the success of transgenic technology (Potenza et al. 2004). Promoter development is still in its infancy. Major advances in transcriptomics, proteomics, and genome sequencing (Yu et al. 2007) will contribute to future development of promoters to drive gene expression in specific cells and tissues.

The correct assembly of constructs for plant transformation is fundamental for maximum gene expression at the correct time in target tissues (Butaye et al. 2005). The merit of bidirectional as well as unidirectional promoters necessitates consideration. Undoubtedly, continued advances in plant genetics, bioinformatics, systems biology, and high through-put gene expression technology will be crucial in predicting coordinated gene expression and the design of synthetic promoters. Terminator sequences must also originate from plant sources or from plant pests such as the CaMV or *Agrobacterium*. Although most investigations are targeted to maximizing gene expression in transgenic plants, the ability to silence genes is equally important in some cases, virus-induced gene silencing (VIGS) being a way of down-regulating expression (Robertson 2004).

1.4.2 *Reporter and Selectable Marker Genes*

Transformation, being a rare event, requires an efficient selection system to distinguish between transformed and nontransformed plant cells. Reporter genes enable cells and tissues to be monitored soon after the transformation procedure to assess the success of a specific construct and/or protocol. Such genes may permit the manual or automated selection of transformed from nontransformed cells, but do not enable transformed cells to outgrow their nontransformed counterparts in culture. In contrast, selectable marker genes provide transformed cells with a competitive advantage, enabling them to outgrow nontransformed cells in vitro, usually in the presence of specific substrates in the culture medium.

Although more than 50 genes have been exploited in nuclear and plastid transformation strategies, only a limited number are used routinely (Miki and McHugh 2004). The *uidA* (*gusA*) gene for β-glucuronidase is a versatile reporter.

In fluorometric and histochemical assays, cleavage of the substrate 5-bromo-4-chloro-3-indolyl-β-D-glucuronide (X-Gluc) by β-glucuronidase results in an indigo compound that is readily observed in transformed cells. A disadvantage of the GUS assay is its destructive nature. Consequently, it has been superseded in many investigations by more versatile, nondestructive assays based on expression of the luciferase (*luc*) gene, or the green fluorescent protein (*gfp*) gene, the latter from the jellyfish, *Aequorea victoria*. Mutant versions of the *gfp* gene that emit blue, cyan, and yellow light are available. Novel proteins from reef coral organisms that fluoresce cyan, red, green, and yellow have also been developed as nondestructive reporters for plant transformation (Wenck et al. 2003). Importantly, significant differences in the excitation and emission wavelengths of some of these proteins permit simultaneous visualization of more than one of these fluorescent proteins in transformed cells. Dixit et al. (2006) emphasized the importance of fluorescent proteins to image dynamic processes within plant cells, highlighting some of the practical issues in exploiting these proteins for live cell imaging. Genes for anthocyanin and carotenoid biosynthesis have also been used to visualize transformed cells prior to their manual selection.

Selection systems have been reported that encourage the growth of transformed cells, although "escapes" may occur, with some nontransformed cells growing in the presence of a selective agent. Commonly used selection systems employ tolerance to antibiotics, particularly kanamycin, encoded by the neomycin phosphotransferase (*npt*II) gene, and to hygromycin through expression of hygromycin phosphotransferase (*hph*, *hpt*, *aph*IV) genes. Phosphinothricin and glyphosate have featured in selection systems based on herbicide resistance, tolerance to phosphothricin being encoded by expression of the *bar* (*pat*) gene, while the *aroA*, *cp4*, and *epsps* and *gox* genes confer tolerance to glyphosate. Streptomycin and spectinomycin have been used to select transplastomic plants. Recently, Pinkerton et al. (2008) introduced resistance based on the enzyme organophosphate hydrogenase, encoded by the bacterial *opd* gene, to generate a new scorable and selectable marker system for transgenic plants. Some investigators have focused on plant genes as selectable markers. For example, Yemets et al. (2008) based selection on a modified plant α-tubulin gene that conferred resistance to dinitroaniline herbicides, with trifluralin as the selective agent. Ogawa et al. (2008) used a mutated rice acetolactate synthase gene to select transgenic plants of wheat. Acetolactate synthase catalyzes the first step in the biosynthesis of the essential branched-chain amino-acids, isoleucine, leucine, and valine, and is a target enzyme for several herbicides. Other procedures have incorporated toxic drugs and metabolite analogs into the culture medium. Genes that stimulate cytokinin biosynthesis stimulated shoot regeneration from transformed cells without the need for selection based on toxic compounds (Zuo et al. 2002). The *Escherichia coli pmi* gene for phosphomannose isomerase converts mannose-6-phosphate, an inhibitor of glycolysis, to fructose-6-phosphate, an intermediate in glycolysis. Expression of *pmi* in plant cells allows transformed cells to grow on medium containing mannose, as in the case of transgenic flax, following *Agrobacterium*-mediated transformation (Lamblin et al. 2007). Future legislation will, almost certainly, demand the elimination of antibiotic resistance genes as selectable markers. Although selectable markers are generally indispensable in plant transformation protocols, they are not required once transgenic plants have been generated. General strategies to eliminate selectable marker genes have been reported. Jia et al. (2006) exploited the Cre/*lox* site-specific recombination system, while Charng et al. (2008) developed an inducible transposon system to terminate selectable marker gene function in transgenic plants. More detailed description of selection systems is presented by Miki and McHugh (2004) and Davey et al. (2008).

1.5 Methods for Screening of Genes Introduced into Putatively Transformed Plants

The strategies used for screening transformed plants usually depend on the type of selectable marker and/or reporter gene used. When an antibiotic resistance gene is employed as a selectable marker, screening is performed by culturing the transformed cells on a medium containing that particular antibiotic (Soneji et al., 2007b, 2007a). In the case of reporter genes, screening is for the distinctive phenotype (Chalfie et al. 1994). However, putative transgenic plants selected by scoring for the presence of selectable marker and/or reporter genes need to be evaluated for the integration and expression of the transgene(s) to minimize escapes. Polymerase chain reaction (PCR)-based screening techniques are used to assess the presence of a specific DNA sequence of the foreign gene of interest, or the selectable marker/ reporter gene by screening putative transgenic plants with primers specific to these gene(s) (Xu et al. 2005; Soneji et al. 2007b, 2007a). Southern hybridization confirms the presence of transgenes and their copy number (Bhat and Srinivasan 2002). Enzyme-linked immunosorbent assay (ELISA) is the preferred method to detect the presence of a specific protein produced by a transgene in a recipient plant. Real-time polymerase chain reaction (RT-PCR) is utilized when more than one gene needs to be analyzed by PCR, along with the detection of the copy number of the desired gene(s) (Yuan et al. 2007).

1.6 Gene Expression in Transgenic Plants

Integration of transgenes into the genomes of plants is a random process, necessitating investigations of their expression in transformed plants. Expression is influenced by several parameters, including the site and pattern of integration, the location of heterochromatic regions, the presence of enhancer elements, the nature of the promoter, gene copy number, truncation, rearrangement, silencing and the presence of any DNA sequences from the vector into which the foreign DNA has been cloned. Although some of these factors can be circumvented by experimental design, it is still necessary to correlate phenotypic differences between transgenic and control plants with transgene expression (Page and Minocha 2005). Thus, transgenic plants require detailed phenotypic, physiological, and molecular analyses to complete their characterization. Techniques such as Western blotting, Northern blotting, ELISA, and quantification and localization of mRNA transcripts are used to analyze transgene expression. These assessments are essential, especially when transgenic material is incorporated into breeding programs.

The use of genetic manipulation in crop improvement also requires transgenes to be expressed either constitutively or in specific cell or tissue types, often at definite stages of plant development (Perret et al. 2003). Although individual transgenic plants within a population may be phenotypically identical, generally they all differ in some subtle way at the molecular level. This emphasizes the requirement to generate as many transgenic plants as possible from an individual experiment and to analyze the maximum number of the regenerants at the phenotypic and molecular levels (Bhat and Srinivasan 2002). Currently, there is no reliable procedure to target foreign genes to specific regions of the genome of plants. It may also be necessary to test individual promoters to establish their expression patterns in different species (Perret et al. 2003). While gene targeting by homologous recombination is potentially extremely important, the development of a routine procedure that incorporates this process remains a major challenge (Cotsaftis and Guiderdoni 2005).

In order to determine the value and application of transformed plants, it is important to understand the inheritance and stability of introduced gene(s). Transmission and segregation analyses of the transgene(s) in subsequent progenies allow insight into transgene inheritance (Yin et al. 2004). *Agrobacterium*-mediated transformation, as well as direct DNA uptake, enables foreign genes to be integrated at a single Mendelian locus, regardless of copy number (Spencer et al. 1992). Stably integrated transgenes are usually inherited in a dominant, Mendelian fashion. However, in subsequent generations, some instability may be observed probably due to rearrangements or methylation of the T-DNA region, and/or to homologous recombination between copies of the transgene inserted into the same nucleus. A non-Mendelian segregation pattern is usually associated with unstable transformation or poor transgene expression (Limanton-Grevet and Jullien 2001).

1.7 Target Genes for Genetic Transformation

Major advances in gene isolation, vector construction, and DNA delivery enable plants to be modified for specific traits, providing an important underpin to conventional breeding. Although genetic engineering reduces the time to integrate desired genes into target plants, it will not replace gene manipulation by sexual hybridization. It has been emphasized that many of the constraints associated with conventional breeding can be overcome by advances at the molecular level (Dalal et al. 2006). Transgenes to be introduced into plants are selected on the basis of their economic/agronomic importance. Recent advances in DNA array technology allow researchers to detect sets of genes that function co-ordinately in the biological processes of interest (Gachon et al. 2005). Several constructs have been developed for use in gene transfer to facilitate the generation of herbicide-, insect-, viral-, fungal-, bacterial-, and nematode-resistant plants (Gubba et al. 2002; Hsieh et al. 2002; Jeanneau et al. 2002; Dasgupta et al. 2003; Grover and Gowthaman 2003; Ranjekar et al. 2003; Prins et al. 2008). Transgenes that may affect quality traits of important crops (Paine et al. 2005), and those for antigens and proteins of pharmaceutical importance, have been introduced into transformation vectors.

Agronomically important genes for biotic and abiotic stresses and quality attributes have been the major focus of research on genetic manipulation, with an extensive range of chimeric genes being introduced into plants (Babu et al. 2003). The majority of transgenes introduced express enzymes that confer novel traits on the respective plants. Proteins lacking enzymatic activity have also been expressed. About 50 important genetically manipulated crops are cultivated in more than 25 countries (Wenzel 2006; James 2008).

1.7.1 Resistance to Biotic and Abiotic Stresses

Biotic and abiotic stresses have a considerable impact on crop growth, development, and productivity throughout the world (Zhao and Zhang 2007). Plant genetic engineering holds the promise of circumventing the problems faced in wide hybridization programs, especially when sources of resistance are not available in taxonomically related species. During the past decade, understanding of the complex molecular events that occur in plant-pathogen interactions has progressed considerably and has provided the opportunity for exploiting the theoretical knowledge and practical skills to generate transgenic plants resistant to pathogens (Grover and Gowthaman 2003). The discovery of abiotic stress-related novel genes, determination of their expression patterns and their roles in adaptation to stress have also provided the foundation for efficient transgenic strategies (Zhao and Zhang 2007).

It is not unexpected that since major crop losses are incited by weeds, insects, viruses, and fungi, increased tolerance to these agents will continue to be a focus of genetic manipulation technology. Transformation of crop plants for increased herbicide tolerance dominated the initial stages of the application of genetic manipulation technology to crop plants. Castle et al. (2006) discussed the ways in which technological advances have been incorporated into agricultural practice and traits introduced into crops such as alfalfa, cotton, maize, oilseed rape, papaya, soybean, and squash, together with the first year of commercialization of the products. Importantly, it is possible to stack transgenes in target plants, conferring tolerance simultaneously to more than one agent.

Behrens et al. (2007) indicated that there has been a rapid increase in the weeds that are tolerant or resistant to the herbicides used with genetically manipulated crops, indicating that such economically important weed management traits may have a finite life. In order to prolong the durability of genetically manipulated herbicide tolerance, these workers developed a nuclear and chloroplast-encoded herbicide balance strategy based on the inexpensive, widely used, and ecologically safe herbicide, dicamba. Similarly, Soberón et al. (2007) discussed the ways in which the evolution of insecticide resistance by insects threatens the application of effective *Bt* toxins from the soil bacterium *Bacillus thuringiensis* that are employed as bacterial sprays, and *Bt* genes that are introduced into genetically manipulated crops. The natural resistance of insects to insecticides will probably necessitate the use of modified *Bt* toxins in the future. Likewise, Gatehouse (2008) stressed the fact that not all pests are targeted adequately by the *Bt* toxins currently in use. *Bt* toxin expression needs to confer adequate protection against target insects, with plastid transformation being superior to nuclear transformation in this respect. Other approaches for maximizing gene expression include the use of novel *Bt* toxins, gene stacking to effect multiple *Bt* toxin expression and protein engineering.

The exploitation of plant defense proteins, such as α-amylase inhibitors and lectins, is also a possibility; novel approaches include the exploitation of new insecticidal proteins such as those from nematodes, the use of bacterial cholesterol oxidase, and the strong insecticidal effect of avidin. Engineering secondary metabolism of plant defense compounds and of the volatiles emitted by plants, and an RNAi approach to generate double-stranded RNAs are also possibilities. Dudareva and Pichersky (2008) discussed the importance of enhancing plant defense by metabolic engineering of volatile compounds, and suggested that priming crops by planting transgenic plants, that constantly emit defense volatiles, among their nontransgenic counterparts, may provide efficient protection. More needs to be known about the properties of specific plant volatiles in terms of their ability to attract or inhibit insect pests.

The status of virus resistance in transgenic plants has advanced considerably since the initial studies involving coat protein-mediated resistance (Prins et al. 2008). The precise mechanism of coat-protein-mediated resistance is not fully understood. It varies with different viruses, but the procedure has been successful in a range of target plants. Other approaches include replicase-mediated resistance and resistance based on movement proteins. RNA-mediated resistance against RNA and DNA viruses is also discussed, as are nonviral sources of resistance using genetic manipulation, particularly an antibody strategy to induce plants to synthesize similar compounds (plantibodies). Transgene-mediated resistances against viroids have been investigated, a promising approach being the expression of recombinant dsRNA-specific RNases by transgenic plants. Several strategies for virus and viroid resistance have been described in the literature, but only a limited number have progressed past the "proof-of-principle" stage, or small-scale field trials (Prins et al. 2008).

In a critique of the deliverables from genetic manipulation technology, Collinge et al. (2008) emphasized the fact that, to date, very few genetically manipulated disease resistant cultivars have been generated, in contrast to plants tolerant to insect pests using a *Bt* approach, and plants that are herbicide tolerant. Indeed, insect- and herbicide-tolerant plants represent more than 90% of all genetically manipulated crops generated to date. Weed control exploiting genetic manipulation technology has been facilitated by understanding the biology of herbicide tolerance and the specificity of synthetic herbicides. Similarly, success in the genetic manipulation of insect resistance was based, at least initially, on knowledge arising from the extensive use of the soil bacterium *B. thuringiensis* as a natural insecticide. Since the organisms that cause disease are taxonomically and physiologically diverse with complex life cycles, Collinge et al. (2008) advocated a balance between classical plant breeding and genetic manipulation to generate disease-resistant plants. They concluded that transgenic fungal and bacterial resistances will probably not be introduced into commercial crops in the near future, although progress in the introduction of a barley class II chitinase gene into wheat to confer resistance to *Fusarium graminearum* represents an advancement in engineering fugal resistance (Shin et al. 2008). Plants experience considerable environmental stresses, with drought posing one of the most important constraints for agriculture on a global scale in the near future (Umezawa et al. 2006; Bhatnagar-Mathur et al. 2008). Tolerance to drought, cold, and salinity are often linked, which may facilitate genetic manipulation to combat these natural agents. Mutasa-Gottgens et al. (2009) showed that genetic modification of gibberellin signaling and metabolism significantly delays bolting in crops such as sugar beet, that are vulnerable to vernalization-induced premature bolting and flowering, reducing crop yield and quality. This approach confirms the potential in genetically modifying plants to minimize yield losses due to unfavorable environmental conditions.

1.7.2 Improvement of Quality

Nutritional value, being one of the most important traits for improvement of crop quality, involves enhancement of the content of amino acids and proteins, micronutrients, vitamins, minerals, dietary fiber, sugars, carbohydrates, starch, lipids and oils, which are essential for a healthy diet (Singh et al. 2008). Staple crops, such as cereals, are low in lysine, while proteins of legumes, roots, tubers, and most vegetables are deficient in sulfur-containing amino acids (Sun 2008). Engineering complex synthetic pathways may not be a simple task, as changing one biosynthetic route may have a detrimental effect on other aspects of metabolism.

Attempts have been made to enhance the essential amino acid and protein content of crops (Sun and Liu 2004). Transgenic technology will continue to be used to biofortify crops to increase vitamins and minerals. Engineering of provitamin A to generate "Golden Rice" and "Golden Rice 2" represents a major technological advance in this respect (Ye et al. 2000; Paine et al. 2005). As vegetables and fruits contribute significantly to human nutrition, they represent another important target for genetic modification in terms of tolerance to abiotic stress, nutritional quality, storage products, aromas and, in certain cases, seedlessness (Fraser et al. 2002; Dalal et al. 2006). Larkin and Harrigan (2007) discussed the attempts made to improve the nutritional value of maize and cotton seed, while others focused on vitamins C (Agius et al. 2003) and E (Chen et al. 2006), particularly on oilseeds (Hunter and Cahoon 2007). Volatiles determine the aromas of fruits, vegetables, and herbs, with genetic engineering being able to ameliorate some of the deficiencies of classical breeding (Dudareva and Pickersky 2008). Tomatoes have been engineered for tolerance to chilling damage (Park et al. 2004), this being of relevance during growth of the plants and during transport of harvested fruit. Delay of fruit ripening and increased shelf-life are also targets for genetic manipulation.

Flavonoids and carotenoids play an important role in human nutrition and health, particularly anticancer activity, and understanding flavonoid and carotenoid biosynthetic pathways has enabled anthocyanins and carotenoids to be up- and down-regulated (Tanaka and Ohmiya 2008). Schijlen et al. (2004) also reviewed the modification of flavonoid biosynthesis in crop plants, while Enfissi et al. (2006) concentrated their attention on the genetic engineering of carotenoids in tomato. Plants have been engineered to produce unusual fatty acids, particularly very long-chain polyunsaturated fatty acids normally found in fish oils and marine organisms (Napier 2007). The longer-term result of engineering complex pathways will be influenced not only by the pathways *per se*, but also by the host plant and physical and chemical parameters. Food allergy is a prevalent medical problem in the western world. Allergen reduction is an important topic for genetic engineering, with RNAi technology being applied to reduce allergens in plants such as apple, peanut, rice, soybean, and tomato (Herman et al. 2003; Gilissen et al. 2005; Le et al. 2006; Chu et al. 2008).

1.7.3 Biopharmaceuticals

Vaccines and antibodies play a major role in human healthcare. The majority of drugs used by humans are derived from plants and have resulted in pharmaceutical companies initiating chemical synthesis of medicinally important compounds (Sharma et al. 1999). However, the full potential of synthesizing compounds has been hampered by production costs and maintaining distribution. The progress in plant transformation has attracted attention in exploiting plants as potential bioreactors or biofactories for the synthesis of immunotherapeutic molecules and recombinant proteins. Plants offer several options for transgene targeting and modification (Warzecha 2008). Indeed, as health care becomes an increasing global issue, the longer-term focus of plant genetic manipulation will be towards the biosynthesis of pharmaceuticals (Zhou and Wu 2006) and other specialty compounds (Fischer et al. 2004, 2007; Yonekura-Sakakibara and Saito 2006). Biofortification of crops with micronutrients is another target for genetic manipulation (Poletti and Sautter 2005). Linked to these goals are issues of biosafety, especially the use of marker genes for antibiotic resistance that are common to many transformation procedures. Davey et al. (2008) presented some of the merits and disadvantages of marker gene technology in the transformation of food crops.

Vaccines such as Hepatitis B surface antigen, Norwalk virus capsid protein, cholera toxin B subunit, Rabies virus glycoprotein, and insulin have been expressed in transgenic plants (Mason et al. 1998; Srinivas et al. 2008), as have immunotherapeutic molecules and industrial proteins, including serum albumin, human α-interferon, human erythropoetin, and murine IgG and IgA immunoglobulins. Oral vaccines synthesized in plants may circumvent some of the limitations of traditional vaccines (Robert and Kirk 2006), especially if vaccines can be synthesized in leafy vegetables that are consumed in the raw state. They will also be cost effective, easy to administer and store, and socioculturally readily acceptable (Lal et al. 2007).

1.7.4 Phytoremediation

Activities, such as intensive mining, agriculture, and military operations, release considerable amounts of toxic heavy metals and organic pollutants, posing a serious threat to living organisms (Cherian and Oliveira 2005). Consequently, there is an urgent requirement to decontaminate polluted environments. Phytoremediation, involving the use of plants and microbes to remove pollutants from contaminated soils, sludge, sediments, groundwater, surface water and waste water, is emerging as a cost-effective and environment-friendly technology compared with conventional methods of remediation (Czako et al. 2006).

Plants harbor highly versatile enzymes such as cytochrome P450 monoxygenases, glutathione S-transferases, glycosyltransferases, laccases, peroxidases, and transporters that detoxify pollutants. Although these enzymes may not completely degrade pollutants, they may form complexes, which can be harvested. In recent years, genetic engineering has been used to introduce key genes to increase the remediation ability of several species. Several genes, such as *merApe9, merB, MT1, MT2, CUP1, gshI, ZAT1, ZntA, arsC* (for heavy metal tolerance), mammalian cytochrome P450 2E1 (*CYP2E1*), *cbn4* (for chlorinated solvents), *CYP1A1, CYP2B6, CYP2C9, CYP2C18, CYP2C19* (for herbicide tolerance), and genes encoding rhamnolipid biosynthesis (for oil degradation), have been overexpressed in transgenic plants (Doty et al. 2000; Dhankher et al. 2002; Lee et al. 2003; Thomas et al. 2003; Cherian and Oliveira 2005; Czako et al. 2006), providing a basis for plant-based phytoremediation.

1.7.5 Floriculture

While food crops will continue to be prime targets for genetic manipulation, ornamentals have featured extensively in genetic manipulation strategies because of the significant contribution of the horticultural industry to the economy of many countries (Tanaka et al. 2005). Ornamentals, especially flower species, are well suited to genetic manipulation. As the end product is not food, it does not necessitate food safety studies, removing major obstacles for commercialization and reducing the cost of production. Chandler and Lu (2005) tabulated the floriculture crops that have been transformed and those with modified characteristics. The latter include disease resistance, herbicide and freezing tolerances (Pennycooke et al. 2003) and, most importantly, modification of pigmentation following manipulation of the genes for pigment biosynthesis (Lu et al. 2003; Tsuda et al. 2004; Suzuki et al. 2007). Attempts have been made to increase the number of flowers produced and extending the life of cut flowers (Shaw et al. 2002). Early and delayed flowering traits have also been introduced (Baker et al. 2002), together with modification of plant architecture (Zheng et al. 2001) and stature (Aswath et al. 2004). The importance of gibberellic acids in controlling plant height in agriculture, horticulture, and silviculture is well recognized (Radi et al. 2006). Dwarf plants may be preferred in amenity planting because of their resistance to unfavorable weather conditions. In this respect, ectopic expression of a gibberellin 2-oxidase from oleander (*NoGA2ox3*) in *Nicotiana tabacum* resulted in dwarf plants (Ubeda-Tomás et al. 2006). Subsequently, Agharkar et al. (2007) demonstrated that genetic manipulation of gibberellin biosynthesis genes can improve the quality of turf grass by increasing the number of vegetative tillers, enhancing turf density under field conditions. Likewise, in order to demonstrate proof of principle and the application of a genetic engineering approach, Dijkstra et al. (2008) overexpressed a gibberellin 2-oxidase gene (*PcGA2ox1*) from *Phaseolus coccineus* to enhance gibberellin inactivation and to induce dwarfism in *Solanum* species. The ability to engineer plant stature through a genetic engineering approach should be of interest to the ornamental industry.

Fragrance will receive more attention (Xiang et al. 2007), since many plants have lost their traditional perfumes through classical breeding. Several approaches have been evaluated to alter scent by genetic modification, as in petunia (Lücker et al. 2001) and carnation (Lavy et al. 2002). However, even though the transgenic plants synthesized more volatiles, the latter could not be detected by humans. In contrast, Zuker et al. (2002) generated carnations with altered floral scent that could be detected by humans, but the resulting plants also had severe alteration in flower color. More recently, Lücker et al. (2004) demonstrated the possibility of modifying the flower fragrance profile by metabolic engineering of tobacco plants using three monoterpene synthases from lemon. These investigators stressed the difficulty of genetically modifying scent because of the need for multigene engineering. Flavonoids and carotenoids are important not only in nutrition and healthcare, as already discussed, but also in flower pigmentation (Nakatsuka et al. 2007b; Tanaka and Ohmiya 2008). Modification of flower color has always been one of the greatest challenges in floricultural plant breeding, since certain colors are difficult to achieve in some species. However, in some cases, genetic manipulation has enabled changes to be made to pigmentation, where classical breeding has failed, by introducing genes from other species and modifying the anthocyanin, carotenoid, or flavonoid biosynthetic pathways. This approach has enabled the generation of purple carnations (Fukui et al. 2003) and blue roses (Potera 2007).

1.8   Risks and Concerns

As with any new technology, there are uncertainties regarding the deployment of genetically engineered plants. There is an increasing concern that insect pests have the capacity to develop resistance against transgenes introduced into plants, or that transgenic properties may be transferred to insects, viruses, and bacteria. Apprehension has also been raised concerning the introgression of transgenes into wild relatives of genetically modified plants and the development of superweeds resulting from introgression of herbicide resistance from transgenic plants to weeds (Sharma et al. 2001, 2002). Transgenic plants may also affect nontarget species and the environment. Food biosafety research has also focused on toxicity and allergenicity of transgenic products.

Although concerns for ecological safety and the human well-being have led to mistrust over the application of genetic manipulation technology, many of these fears appear unsubstantiated or based on misinformation (Stewart et al. 2000). A concerted effort must be made to identify valid concerns and risks, and to provide reliable information to the public. The advent of plant genetic manipulation in vaccine production and quality improvement will increase the emphasis on consumer health benefits, which may facilitate, in turn, acceptance of the use of genetically engineered foods. Active participation of researchers from the fields of biotechnology, ecology, and nutritional sciences may be essential to better determine the biosafety of transgenic plants (Stewart et al. 2000).

1.9   General Conclusions

Modern agricultural biotechnology has been one of the most promising developments in recent years (Sharma et al. 2002). Major advances in understanding gene structure and expression have made significant contributions to the assembly of genes and their regulatory elements for plant genetic engineering. Likewise, progress in DNA delivery technologies has facilitated the introduction of novel genes into a wide range of plants. A common restriction to gene introgression into many crops is the recalcitrance of these plants to express their totipotency in culture. However, the exploitation of procedures that by-pass the requirement for extensive in vitro manipulations should eliminate some of these difficulties. Currently, genetic engineering is not a routine plant breeding tool (Arias et al. 2006), but is an important adjunct to classical breeding (Shewry et al. 2008).

World food supplies will demand more intensive crop production, despite a reduction in available agricultural land because of deterioration of soil quality, drought, climatic change, disease, and political unrest. Farmers will demand more value per unit of agricultural land. Genetic engineering, when used in collaboration with traditional or conventional breeding methods, will be able to increase crop production, increase resistance to major pests and diseases, develop tolerance to adverse weather conditions, improve the nutritional value of some foods, and enhance the durability of products during harvesting or shipping (Sharma et al. 2002). Reduced use of agrochemicals will have less environmental impact. In the future, agriculturally important traits must satisfy not only the requirements of farmers, but also the availability of materials from researchers, governments, distributors, processors, and the opinions of the public (Castle et al. 2006).

Discussions on transgenic crops have placed undue stress on risk assessment, overshadowing potential advantages (Sharma et al. 2002). The issues relating to genetically modified plants, especially food crops, have been analyzed from a scientist's perspective (Lemaux 2008). These issues are not only complex, but are often aggravated by personal opinions, especially by those members of the public who have limited understanding of plant breeding and gene technology. The rapid escalation of increasingly stringent biosafety regulations regarding transgenic plants or food, in the absence of any scientifically proven genetic risk, is most likely to limit application of transgenic research to meet either the production of sustainable staple foods or the alleviation of poverty (Sharma et al. 2002). Moving crop production from one region to another will influence global trade patterns; legislation and the perceived risks of genetically engineered crops will also affect exploitation of these crops (Singh et al. 2006).

What remains clear is that changes in the genetic complement of those plants that contribute to our food supplies are primarily the result, to date, of sexual hybridization. Genetic engineering provides a precise approach to effect genetic modification over a much reduced time-scale. The safety of genetically engineered plants and those generated by conventional breeding needs to be evaluated on a case-by-case basis (Lemaux 2008). Condemning biotechnology for its potential risks without considering the risks associated with prolonging human misery caused by hunger, malnutrition, and infant mortality is unwise and unethical. The global community must endeavor to remain focused on the target of assuring food for all, and cannot afford to be philosophical and elitist regarding any part of a possible solution, including agricultural biotechnology (Sharma et al. 2002).

References

Agius F, Gonzálex-Lamothe R, Caballero JL, Muñoz-Blanco J, Botella MA Valpuesta V (2003) Engineering increased vitamin C levels in plants by over-expression of a D-galacturonic acid reductase. Nat Biotechnol 21:177–181

Agharkar M, Lomba P, Altpeter F, Zhang H, Kenworthy K, Lange T (2007) Stable expression of *AtGA2ox1* in a low-input turfgrass (*Paspalum notatum* Flugge) reduces bioactive gibberellin levels and improves turf quality under field conditions. Plant Biotechnol J 5:791–801

Aida R, Nagaya S, Yoshida K, Kishimoto S, Shibata M, Ohmiya A (2005) Efficient transgene expression in Chrysanthemum, *Chrysanthemum morifolium* Ramat., with the promoters of a gene for Tobacco Elongation Factor 1 α protein. Jpn Agric Res Q 39:269–274

Altpeter F, Baisakh N, Beachy R, Bock R, Capell T, Christou P, Daniell H, Datta K, Datta S, Dix PJ, Fauquet C, Huang N, Kohli A, Mooibroek H, Nicholson L, Nguyen TT, Nugent G, Raemakers K, Romano A, Somers DA, Stoger E, Taylor N, Visser R (2005) Particle bombardment and the genetic enhancement of crops: myths and realities. Mol Breed 15:305–327

Annadana S, Beekwilder MJ, Kuipers G, Visser PB, Outchkourov N, Pereira A, Udayakumar M, De JJ, Jongsma MA (2002) Cloning of the chrysanthemum UEP1 promoter and comparative expression in florets and leaves of *Dendranthema grandiflora*. Transgenic Res 11:437–445

Arias RS, Filichkin SA, Strauss SH (2006) Divide and conquer: development and cell cycle genes in plant transformation. Trends Biotechnol 24:267–273

Aswath CR, Mo SY, Kim S-H, Kim DH (2004) IbMADS4 regulates the vegetative shoot development in transgenic chrysanthemum (*Dendranthema grandiflora* (Ramat.) Kitamura). Plant Sci 166:847–854

Babu RM, Sajeena A, Seetharaman K, Reddy MS (2003) Advances in genetically engineered (transgenic) plants in pest management – an overview. Crop Prot 22:1071–1086

Baker C, Zhang H, Hall G, Scrocki D, Medina A, Dobres MS (2002) The use of the *GAI* and *CO* genes to create novel ornamental plants. In Vitro Cell Dev Biol Plant 38:105

Behrens MR, Mutlu N, Chakraborty S, Dumitru R, Jiang WZ, LaVallee BJ, Herman PL, Clemente TE, Weeks DP (2007) Dicamba resistance: enlarging and preserving biotechnology-based weed management strategies. Science 316:1185–1188

Bhat SR, Srinivasan S (2002) Molecular and genetic analyses of transgenic plants: considerations and approaches. Plant Sci 163:673–681

Bhatnagar-Mathur P, Vadez V, Sharma KK (2008) Transgenic approaches to abiotic stress in plants: retrospect and prospects. Plant Cell Rep 27:411–442

Birch RG (1997) Plant transformation: problems and strategies for practical application. Annu Rev Plant Physiol Plant Mol Biol 48:297–326

Bock R, Khan MS (2004) Taming plastids for a green future. Trends Biotechnol 22:311–318

Buchner P, Rochat C, Wuilleme S, Boutin JP (2002) Characterization of a tissue-specific and developmentally regulated β-1, 3-glucanase gene in pea (Pisum sativum). Plant Mol Biol 49:171–186

Butaye KMJ, Cammue BPA, Delauré SL, De Bolle MFC (2005) Approaches to minimize variation of transgene expression in plants. Mol Breed 16:79–91

Castle L, Wu G, McElroy D (2006) Agricultural input traits: past, present and future. Curr Opin Biotechnol 17:105–112

Cazzonelli C, McCallum E, Lee R, Botella J (2005) Characterization of a strong, constitutive mung bean (*Vigna radiata* L.) promoter with a complex mode of regulation in planta. Transgenic Res 14:941–967

Cervera M, Juarez J, Navarro A, Pina JA, Duran-Villa N, Navarro L, Pena L (1998) Genetic transformation and regeneration of mature tissue of woody fruit plants bypassing the juvenile stage. Transgenic Res 7:51–59

Chalfie M, Tu Y, Euskirchen G, Ward WW, Prasher DC (1994) Green fluorescent protein as a marker for gene expression. Science 263:802–805

Chandler SF, Lu C-Y (2005) Biotechnology in ornamental horticulture. In Vitro Cell Dev Biol-Plant 41:591–601

Charng YC, Li K-T, Tai H-K, Lin N-S, Tu J (2008) An inducible transposon system to terminate the function of a selectable marker in transgenic plants. Mol Breed 21:359–368

Chen S, Li H, Liu G (2006) Progress of vitamin E metabolic engineering in plants. Transgenic Res 15:655–665

Cheng M, Lowe BA, Spencer TM, Ye X, Armstrong CL (2004) Factors influencing *Agrobacterium*-mediated transformation of monocotyledonous species. In Vitro Cell Dev Biol-Plant 40:31–45

Cherian S, Oliveira MM (2005) Transgenic plants in phytoremediation: recent advances and new possibilities. Environ Sci Technol 39:9377–9390

Chu Y, Faustinelli P, Ramos ML, Hajduch M, Thelen JJ, Maleki SJ, Ozias-Akins P (2008) Reduction of IgE binding and non-promotion of *Aspergillus flavus* fungal growth by simultaneously silencing Ara h 2 and Ara h 6 in peanut. J Agric Food Chem 56:11225–11233

Chung M-H, Chen M-K, Pan S-M (2000) Floral spray transformation efficiently generates *Arabidopsis* transgenic plants. Transgenic Res 9:471–486

Chung S-M, Frankman EL, Tzfira T (2005) A versatile vector system for multiple gene expression in plants. Trends Plant Sci 10:357–361

Chung S-M, Vaidya M, Tzfira T (2006) *Agrobacterium* is not alone: gene transfer to plants by viruses and other bacteria. Trends Plant Sci 11:1–4

Clough SJ, Bent AF (1998) Floral dip: a simplified method for *Agrobacterium*-mediated transformation of *Arabidopsis thaliana*. Plant J 16:735–743

Collinge DB, Lund OS, Thordal-Christensen H (2008) What are the prospects for genetically engineered, disease resistant plants? Eur J Plant Pathol 121:217–231

Cotsaftis O, Guiderdoni E (2005) Enhancing gene targeting efficiency in plants: rice is on the move. Transgenic Res 14:1–14

Curtis IS (2004) Transgenic crops of the world. Essential protocols. Kluwer, Dordrecht 454 p

Czako M, Feng X, He Y, Liang D, Pollock R, Marton L (2006) Phytoremediation with transgenic plants. Acta Hortic 725:753–770

Dalal M, Dani RG, Kumar PA (2006) Current trends in the genetic engineering of vegetable crops. Sci Hortic 107:215–225

Daniell H, Khan MS, Allison L (2002) Milestones in chloroplast genetic engineering: an environmentally friendly era in biotechnology. Trends Plant Sci 7:84–91

Dasgupta I, Malathi VG, Mukherjee SK (2003) Genetic engineering for virus resistance. Curr Sci 84:341–354

Davey MR, Anthony P, Power JB, Lowe KC (2005) Plant protoplasts: status and biotechnological perspectives. Biotechnol Adv 23:131–171

Davey MR, Anthony P, Power JB, Lowe KC (2008) Transgenics for genetic improvement of plants. In: Kole C, Abbott AG (eds) Principles and practices of plant genomics. Vol 2: molecular breeding. Science Publication, New Hampshire, pp 434–464

Dhankher OP, Li YJ, Rosen BP, Shi J, Salt D, Senecoff JF, Sashti NA, Meagher RB (2002) Engineering tolerance and hyperaccumulation of arsenic in plants by combining arsenate reductase and gamma-glutamylcysteine synthetase expression. Nat Biotechnol 20:1140–1145

Dijkstra C, Adams E, Bhattacharya A, Page AF, Anthony P, Kourmpetli S, Power JB, Lowe KC, Thomas SG, Hedden P, Phillips AL, Davey MR (2008) Over-expression of a gibberellin 2-oxidase gene from *Phaseolus coccineus* L. enhances gibberellin inactivation and induces dwarfism in *Solanum* species. Plant Cell Rep 27:463–470

Dixit R, Cyr R, Gilroy S (2006) Using intrinsically fluorescent proteins for plant imaging. Plant J 45:599–615

Doty SL, Shang TQ, Wilson AM, Tangen J, Westergreen AD, Newman LA, Strand SE, Gordon MP (2000) Enhanced metabolism of halogenated hydrocarbons in transgenic plants containing mammalian cytochrome P450 2E1. Proc Natl Acad Sci USA 97:6287–6291

Dudareva N, Pichersky E (2008) Metabolic engineering of plant volatiles. Curr Opin Biotechnol 19:181–189

Enfissi EMA, Fraser PD, Bramley PM (2006) Genetic engineering of carotenoid formation in tomato. Phytochem Rev 5:59–65

Fischer R, Stoger E, Schillberg S, Christou P, Twyman RM (2004) Plant-based production of biopharmaceuticals. Curr Opin Plant Biol 7:152–158

Fischer R, Twyman RM, Hellwig S, Drossard J, Schillberg S (2007) Facing the future with pharmaceuticals from plants. In: 11th Annual Conference International Association Plant Tissue Culture Biotechnology, Biotechnology and Sustainable Agriculture 2006 and Beyond. pp. 13–32

Fraser PD, Romer S, Shipton CA, Mills PB, Kiano JW, Misawa N, Drake RG, Schuch W, Bramley PM (2002) Evaluation of transgenic tomato plants expressing an additional phytoene synthase in a fruit specific manner. Proc Natl Acad Sci USA 99:1092–1097

Fruhling M, Schroder G, Hohnjec N, Puhler A, Perlick AM, Kuster H (2000) The promoter of the *Vicia faba* L. gene *VfEnod12* encoding an early nodulin is active in cortical cells and nodule primordia of transgenic hairy roots of *Vicia hirsuta* as well as in the prefixing zone II of mature transgenic *V. hirsuta* root nodules. Plant Sci 160:67–75

Fukui Y, Tanaka Y, Kusumi T, Iwashita T, Nomoto K (2003) A rationale for the shift in colour towards blue in transgenic carnation flowers expressing the flavonoid 3', 5'-hydroxylase gene. Phytochemistry 63:15–23

Gachon CM, Langlois-Meurinne M, Henry Y, Saindrenan P (2005) Transcriptional co-regulation of secondary metabolism enzymes in *Arabidopsis*: functional and evolutionary implications. Plant Mol Biol 58:229–245

Gatehouse JA (2008) Biotechnological prospects for engineering insect-resistant plants. Plant Physiol 146:881–887

Gelvin S (2003) *Agrobacterium*-mediated plant transformation: the biology behind the "Gene-Jockeying" tool. Microbiol Mol Biol Rev 67:16–37

Gilissen LJ, Bolhaar SI, Matos CI, Rouwendal GJ, Boone MJ, Krens FA, Zuidmeer L, Van Leeuwen J, Akkerdaas Hoffmann-Sommergruber K, Knulst AC, Bosch D, Van de Weg WE, Van Ree R (2005) Silencing the major apple allergen Mal d 1 by using the RNA interference approach. J Allergy Clin Immunol 115:364–369

Govindarajulu M, Elmore JM, Fester T, Taylor CG (2008) Evaluation of constitutive viral promoters in transgenic soybean roots and nodules. Mol Plant Microbe Interact 21:1027–1035

Grover A, Gowthaman R (2003) Strategies for development of fungus-resistant transgenic plants. Curr Sci 84:330–340

Guan C, Zhou X (2006) Phloem specific promoter from a satellite associated with a DNA virus. Virus Res 115:150–157

Gubba A, Gonsalves C, Stevens MR, Tricoli DM, Gonsalves D (2002) Combining transgenic and natural resistance to obtain broad resistance to tospovirus infection in tomato (*Lycopersicon esculentum* mill). Mol Breed 9:13–23

Häggman HM, Aronen TS, Nikkanen TO (1997) Gene transfer by particle bombardment to Norway spruce and Scots pine pollen. Can J For Res 27:928–935

Heifetz PB (2000) Genetic engineering of the chloroplast. Biochimie 82:655–666

Herman EM, Helm RM, Jung E, Kinney AJ (2003) Genetic modification removes an immunodominant allergen from soybean. Plant Physiol 132:36–43

Hsieh TH, Lee JT, Yang PT, Chiu LH, Charng YY, Wang YC, Chan MT (2002) Heterology expression of the *Arabidopsis* C-repeat/dehydration response element binding factor 1 gene confers elevated tolerance to chilling and oxidative stresses in transgenic tomato. Plant Physiol 129:1086–1094

Huang H, Ma H (1992) An improved procedure for transforming *Arabidopsis thaliana* (Landsberg *erecta*) root explants. Plant Mol Biol Rep 10:372–383

Hunter SC, Cahoon EB (2007) Enhancing vitamin E in oilseeds: unravelling tocopherol and tocotrienol biosynthesis. Lipids 42:97–108

James C (2008) Global Status of Commercialized Biotech/GC Crops: 2008: http://www.isaaa.org/resources/publications/briefs/39/pressrelease/default.html

Jeanneau M, Gerentes D, Foueillassar X, Zivy M, Vidal J, Toppan A, Perez P (2002) Improvement of drought tolerance in maize: towards the functional validation of the *Zm-Asr1* gene and increase of water use efficiency by over-expressing C4–PEPC. Biochimie 84:1127–1135

Jia H, Pang Y, Chen X, Fang R (2006) Removal of the selectable marker gene from transgenic tobacco plants by expression of Cre recombinase from a Tobacco Mosaic Virus vector through agroinfection. Transgenic Res 15:375–384

Karimi M, Inzé D, Depicker A (2002) GATEWAY vectors for *Agrobacterium*-mediated plant transformation. Trends Plant Sci 7:193–195

Khachatourians GG, McHughen A, Scorza R, Nip W-K, Hui YH (2002) Transgenic plants and crops. Marcel Dekker, New York, USA

Kluth A, Sprunck S, Becker D, Lörz H, Lutticke S (2002) 50 deletion of a gbss1 promoter region leads to changes in tissue and developmental specificities. Plant Mol Biol 49:669–682

Kobayashi T, Yoshihara T, Itai RN, Nakanishi H, Takahashi M, Mori S, Nishizawa NK (2007) Promoter analysis of iron-deficiency-inducible barley *IDS3* gene in *Arabidopsis* and tobacco plants. Plant Physiol Biochem 45:262–269

Komori T, Imayama T, Kato N, Ishida Y, Ueki J, Komari T (2007) Current status of binary and superbinary vectors. Plant Physiol 145:1155–1160

Lacroix B, Li J, Tzfira T, Citovsky V (2006a) Will you let me use your nucleus? How *Agrobacterium* gets its T-DNA expressed in the host plant cell. Can J Physiol Pharmacol 84:333–345

Lacroix B, Tzfira T, Vainstein A, Citovsky V (2006b) A case of promiscuity: *Agrobacterium's* endless hunt for new partners. Trends Genet 22:29–37

Lal P, Ramachandran VG, Goyal R, Sharma R (2007) Edible vaccines: current status and future. Int J Med Microbiol 25:93–102

Lamblin F, Aimé A, Hano C, Roussy I, Domon JM, Van Droogenbroeck B, Lainé E (2007) The use of the phosphomannose isomerase gene as alternative selectable marker for *Agrobacterium*-mediated transformation of flax (*Linum usitatissimum*). Plant Cell Rep 26:765–772

Larkin P, Harrigan GG (2007) Opportunities and surprises in crops modified by transgenic technology: metabolic engineering of benzylisoquinoline alkaloid, gossypol and lysine biosynthetic pathways. Metabolomics 3:371–382

Lavy M, Zucker A, Lewinsohn E, Larkov O, Ravid U, Vainstein A, Weiss D (2002) Linalool and linalool oxide production in transgenic carnation flowers expressing the *Clarkia breweri* linalool synthase gene. Mol Breed 9:103–111

Le LQ, Mahler V, Lorenz Y, Scheurer S, Biemelt S, Vieths S, Sonnewald U (2006) Reduced allergenicity of tomato fruits harvested from Lyc e 1-silenced transgenic tomato plants J Allergy Clin Immunol 118:1176–1183

Lee S, Moon JS, Ko TS, Petros D, Goldsbrough PB, Korban SS (2003) Overexpression of *Arabidopsis* phytochelatin synthase paradoxically leads to hypersensitivity to cadmium stress. Plant Physiol 131:656–663

Lemaux PG (2008) Genetically engineered plants and foods: a scientist's analysis of the issues (Part I). Annu Rev Plant Biol 59:771–812

Li HQ, Xu J, Chen L, Li MR (2007) Establishment of an efficient *Agrobacterium tumefaciens*-mediated leaf disc transformation of *Thellungiella halophila*. Plant Cell Rep 26:1785–1789

Limanton-Grevet A, Jullien M (2001) *Agrobacterium*-mediated transformation of *Asparagus officinalis* L.: molecular and genetic analysis of transgenic plants. Mol Breed 7:141–150

Liu CW, Lin CC, Yiu JC, Chen JJW, Tseng MJ (2008a) Expression of a *Bacillus thuringiensis* toxin (*cryIAb*) gene in cabbage (*Brassica oleracea* L. var. *capitata* L.) chloroplasts confers high insecticidal efficacy against *Plutella xylostella*. Theor Appl Genet 117:75–88

Liu Y, Yang H, Sakanishi A (2006) Ultrasound: mechanical gene transfer into plant cells by sonoporation. Biotechnol Adv 24:1–16

Liu Z, Zhou C, Wu K (2008b) Creation and analysis of a novel chimeric promoter for the complete containment of pollen- and seed-mediated gene flow. Plant Cell Rep 27:995–1004

Loyola-Vargas VM, Vázquez-Flota F (2005) Methods in molecular biology. Plant cell culture protocols, 2nd edn. Humana Press, Totowa, NJ, p 393

Lu C, Chandler SF, Mason JG, Brugliera F (2003) Florigene flowers: from laboratory to market. In: Vasil IK (ed) Plant biotechnology 2002 and beyond. Kluwer, Dordrecht, pp 333–336

Lücker J, Bouwmeester HJ, Schwab W, Blaas J, van der Plas LHW, Verhoeven HA (2001) Expression of *Clarkia S*-linalool synthase in transgenic petunia plants results in the accumulation of *S*-linalyl-β-D-glucopyranoide. Plant J 27:315–324

Lücker J, Schwab W, van Hautum B, Blaas J, van der Plas LHW, Bouwmeester HJ, Verhoeven HA (2004) Increased and altered fragrance of tobacco plants after metabolic engineering using three monoterpene synthases from lemon. Plant Physiol 134:510–519

Luo K, Deng W, Xiao Y, Zheng X, Li Y, Pei Y (2006) Leaf senescence is delayed in tobacco plants expressing the maize *knotted1* gene under the control of a wound-inducible promoter. Plant Cell Rep 25:1246–1254

Maliga P (2002) Engineering the plastid genome of higher plants. Curr Opin Plant Biol 5:164–172

Maliga P (2004) Plastid transformation in higher plants. Annu Rev Plant Biol 55:289–313

Manickavasagam M, Ganapathi A, Anbazhagan VR, Sudhakar B, Selvaraj N, Vasudevan A, Kasthurirengan S (2004) *Agrobacterium*-mediated genetic transformation and development of herbicide-resistant sugarcane (*Saccharum* species hybrids) using axillary buds. Plant Cell Rep 23:134–143

Mason HS, Haq TA, Clements JD, Arntzen CJ (1998) Edible vaccine protects mice against *E. coli* heat-labile enterotoxin (LT): potatoes expressing a synthetic LT-B gene. Vaccine 16:1336–1343

Matsumoto T, Wu J, Kanamori H, Katayose Y, Fujisawa M et al (2005) The map-based sequence of the rice genome. Nature 436:793–800

Miki B, McHugh S (2004) Selectable marker genes in transgenic plants: applications, alternatives and biosafety. J Biotechnol 107:193–232

Mutasa-Gottgens E, Qi A, Mathews A, Thomas S, Phillips AV, Hedden P (2009) Modification of gibberellin signalling (metabolism & signal transduction) in sugar beet: analysis of potential targets for crop improvement. Transgenic Res 18:301–308

Nakatsuka T, Pitaksutheepong C, Yamamura S, Nishihara M (2007a) Induction of differential flower pigmentation patterns by RNAi using promoters with distinct tissue-specific activity. Plant Biotechnol Rep 1:251–257

Nakatsuka T, Abe Y, Kakizaki Y, Yamamura S, Nishihara M (2007b) Production of red-flowered plants by genetic engineering of multiple flavonoid biosynthetic genes. Plant Cell Rep 26:1951–1959

Napier JA (2007) The production of unusual fatty acids in transgenic plants. Annu Rev Plant Biol 58:295–319

Newell CA (2000) Plant transformation technology: developments and applications. Mol Biotechnol 16:53–65

Ogawa T, Kawahigashi H, Toki S, Handa H (2008) Efficient transformation of wheat by using a mutated rice acetolactate synthase gene as a selectable marker. Plant Cell Rep 27:1325–1331

Oszvald M, Gardonyi M, Tamas C, Takacs I, Jenes B, Tamas L (2008) Development and characterization of a chimaeric tissue-specific promoter in wheat and rice endosperm. In Vitro Cell Dev Biol Plant 44:1–7

Padidam M (2003) Chemically regulated gene expression in plants. Curr Opin Plant Biol 6:169–177

Page AF, Minocha SC (2005) Analysis of gene expression in transgenic plants. In: Pena L (ed) Methods in molecular biology, vol 286. Transgenic Plants. Humana Press, Totowa, New Jersey, USA, pp 291–311

Paine JA, Shipton CA, Chaggar S, Howells RM, Kennedy MJ, Vernon G, Wright SY, Hinchliffe E, Adams JL, Silverstone AL, Drake R (2005) Improving the nutritional value of golden rice through increased provitamin a content. Nat Biotechnol 23:482–487

Park E, Jeknic Z, Sakamoto A, DeNoma J, Yuwansiri R, Murata N, Chen THH (2004) Genetic engineering of glycine betaine synthesis in tomato protects seeds, plants, and flowers from chilling damage. Plant J 40:474–487

Peebles CAM, Gibson SI, Shanks JV, San K-Y (2007) Characterization of an ethanol-inducible promoter system in *Catharanthus roseus* hairy roots. Biotechnol Prog 23:1258–1260

Pennycooke JC, Jones ML, Stushnoff C (2003) Down-regulating α-galactosidase enhances freezing tolerance in transgenic Petunia. Plant Physiol 133:901–909

Perret SJ, Valentine J, Leggett JM, Morris P (2003) Integration, expression and inheritance of transgenes in hexaploid oat (*Avena sativa* L.). J Plant Physiol 160:931–943

Pinkerton TS, Howard JA, Wild JR (2008) Genetically engineered resistance to organophosphate herbicides provides a new scoreable and selectable marker system for transgenic plants. Mol Breed 21:27–36

Pino MT, Skinner JS, Park EJ, Jeknic Z, Hayes PM, Thornashow MF, Chen THH (2007) Use of a stress inducible promoter to drive ectopic AtCBF expression improves potato freezing tolerance while minimizing negative effects on tuber yield. Plant Biotechnol J 5:591–604

Poletti S, Sautter C (2005) Biofortification of the crops with micronutrients using plant breeding and/or transgenic strategies. Minerva Biotechnol 17:1–11

Potenza C, Aleman L, Sengupta-Gopalan C (2004) Targetting gene expression in research, agricultural and environmental applications: Promoters used in plant transformation. In Vitro Cell Dev Biol Plant 40:1–22

Potera C (2007) Blooming biotech. Nat Biotechnol 25:963–965

Prins M, Laimer M, Noris E, Schubert J, Wassenegger M, Tepfer M (2008) Strategies for antiviral resistance in transgenic plants. Mol Plant Pathol 9:73–83

Qu LQ, Xing YP, Liu WX, Xu XP, Song YR (2008) Expression pattern and activity of six glutelin gene promoters in transgenic rice. J Exp Bot 59:2417–2424

Rachmawati D, Anzai H (2006) Studies on callus induction, plant regeneration and transformation of Javanica rice cultivars. Plant Biotechnol 23:521–524

Radi A, Lange T, Niki T, Koshioka M, Lange MJP (2006) Ectopic expression of pumpkin gibberellin oxidases alters gibberellin biosynthesis and development of transgenic Arabidopsis plants. Plant Physiol 140:528–553

Rakoczy-Trojanowska M (2002) Alternative methods of plant transformation – a short review. Cell Mol Biol Lett 7:849–858

Ranjekar PK, Patankar A, Gupta V, Bhatnagar R, Bentur J, Ananda Kumar P (2003) Genetic engineering of crop plants for insect resistance. Curr Sci 84:321–329

Robert JS, Kirk DD (2006) Ethics, biotechnology, and global health: the development of vaccines in transgenic plants. Am J Bioeth 6:W29–W41

Robertson D (2004) VIGS vectors for gene silencing: many targets, many tools. Annu Rev Plant Biol 55:495–519

Samac DA, Tesfaye M, Dornbusch M, Saruul P, Temple SJ (2004) A comparison of constitutive promoters for expression of transgenes in alfalfa (*Medicago sativa*). Transgenic Res 13:349–361

Sassa H, Ushijima K, Hirano HA (2002) Pistil-specific thaumatin/*PR5*-like protein gene of Japanese pear (*Pyrus serotina*): sequence and promoter activity of the 50 region in transgenic tobacco. Plant Mol Biol 50:371–377

Saunders JA, Matthews BF (1995) Pollen electrotransformation in tobacco. In: Nickoloff JA (ed) Methods in molecular biology, vol 55: plant cell electroporation and electrofusion protocols. Humana Press, Totowa, NJ, pp 81–88

Schijlen EGWM, de Vos CHR, van Tunen AJ, Bovy AG (2004) Modification of flavonoid biosynthesis in crop plants. Phytochemistry 65:2631–2648

Sharma AK, Mohanty A, Singh Y, Tyagi AK (1999) Transgenic plants for the production of edible vaccines and antibodies for immunotherapy. Curr Sci 77:524–529

Sharma HC, Sharma KK, Seetharama N, Ortiz R (2001) Genetic transformation of crop plants: risks and opportunities for the rural poor. Curr Sci 80:1495–1508

Sharma KK, Sharma HC, Seetharama N, Ortiz R (2002) Development and deployment of transgenic plants: biosafety considerations. In Vitro Cell Dev Biol Plant 38:106–115

Sharma KK, Bhatnagar-Mathur P, Thorpe TA (2005) Genetic transformation technology: status and problems. In Vitro Cell Dev Biol Plant 41:102–112

Shaw JF, Chen HH, Tsai MF, Kuo CI, Huang LC (2002) Extended flower longevity of *Petunia hybrida* plants transformed with *boers*, a mutated ERS gene of *Brassica oleracea*. Mol Breed 9:211–216

Shewry PR, Jones HD, Halford NG (2008) Plant biotechnology: transgenic crops. Adv Biochem Eng Biotechnol 111:149–186

Shrawat AK, Lörz H (2006) *Agrobacterium*-mediated transformation of cereals: a promising approach to crossing barriers. Plant Biotechnol J 4:575–603

Shin S, Mackintosh CA, Lewis J, Heinen SJ, Radmer L, Dill-Macky R, Baldridge GD, Zeyen RJ, Muehlbauer GJ (2008) Transgenic wheat expressing a barley class II chitinase gene has enhanced resistance against *Fusarium graminearum*. J Exp Bot 59:2371–2378

Singh OV, Ghai S, Paul D, Jain RK (2006) Genetically modified crops: success, safety assessment, and public concern. Appl Microbiol Biotechnol 71:598–607

Singh HP, Vasanthaiah HKN, Nageswara Rao M, Soneji JR, Singh B, Lohithaswa HC, Kole C (2008) Molecular mapping and quality breeding. In: Kole C, Abbott AG (eds) Principles and practices of plant genomics, vol 2: molecular breeding. Science Publ, New Hampshire, Jersey, Plymouth, pp 103–142

Soberón M, Pardo-López L, López I, Gómez I, Tabashnik BE, Bravo A (2007) Engineering modified *Bt* toxins to counter insect resistance. Science 318:1640–1643

Soneji JR, Nageswara Rao M, Chen C, Gmitter FG Jr (2007a) Regeneration from transverse thin cell layers of mature stem segments of citrus. In: Plant and Animal Genome XV Conference, San Diego, California, USA

Soneji JR, Chen C, Nageswara Rao M, Huang S, Choi YA, Gmitter FG Jr (2007b) *Agrobacterium*-mediated transformation of citrus using two binary vectors. Acta Hortic 738:261–264

Song J, Lu S, Chen ZZ, Lourenco R, Chiang VL (2006) Genetic transformation of *Populus trichocarpa* genotype Nisqually-1: a functional genomic tool for woody plants. Plant Cell Physiol 47:1582–1589

Song GQ, Sink KC, Callow PW, Baughan R, Hancock JF (2008) Evaluation of a herbicide-resistant trait conferred by the *bar* gene driven by four distinct promoters in transgenic blueberry plants. J Am Soc Hortic Sci 133:605–611

Spencer TM, O'Brien JV, Start WG, Adams TR, Gordon-Kamm WJ, Lemaux PG (1992) Segregation of transgenes in maize. Plant Mol Biol 18:201–210

Srinivas L, Sunil Kumar GB, Ganapathi TR, Revathi CJ, Bapat VA (2008) Transient and stable expression of hepatitis B surface antigen in tomato (*Lycopersicon esculentum* L.). Plant Biotechnol Rep 2:1–6

Stewart CN Jr, Richards HA IV, Halfhill MD (2000) Transgenic plants and biosafety: science, misconceptions and public perceptions. Biotechniques 29:832–843

Sun SSM (2008) Transgenics for new plant products, applications to tropical crops. In: Moore PH, Ming R (eds) Genomics of tropical crop plants. Springer, New York, USA, pp 63–81

Sun SSM, Liu QQ (2004) Transgenic approaches to improve the nutritional quality of plant proteins. In Vitro Cell Dev Biol Plant 40:55–162

Sunilkumar G, Connell JP, Smith CW, Reddy AS, Rathore KS (2002) Cotton α-globulin promoter: isolation and functional characterization in transgenic cotton, *Arabidopsis*, and tobacco. Transgenic Res 11:347–359

Suzuki S, Nishihara M, Nakatsuka T, Misawa N, Ogiwara I, Yamamura S (2007) Flower color alteration in *Lotus japonicus* by modification of the carotenoid biosynthetic pathway. Plant Cell Rep 26:951–959

Tanaka Y, Katsumoto Y, Brugliera F, Mason J (2005) Genetic engineering in floriculture. Plant Cell Tissue Organ Cult 80:1–24

Tanaka Y, Ohmiya A (2008) Seeing is believing: engineering anthocyanin and carotenoid biosynthetic pathways. Curr Opin Biotechnol 19:190–197

Thomas JC, Davies EC, Malick FK, Endreszl C, Williams CR, Abbas M, Petrella S, Swisher K, Perron M, Edwards R, Ostenkowski P, Urbanczyk N, Wiesend WN, Murray KS (2003) Yeast metallothionein in transgenic tobacco promotes copper uptake from contaminated soils. Biotechnol Prog 19:273–280

Tsuda S, Fukui Y, Makamura N, Katsumoto Y, Yonekura-Sakakibara K, Fukuchi-Mizutani M, Ohira K, Ueyama Y, Ohkawa H, Holton TA, Kusumi T, Tanaka Y (2004) Flower color modification of *Petunia hybrida* commercial varieties by metabolic engineering. Plant Biotechnol 21:377–386

Tzfira T, Citovsky V (2006) *Agrobacterium*-mediated genetic transformation of plants: biology and biotechnology. Curr Opin Biotechnol 17:147–154

Tzfira T, Kozlovsky SV, Citovsky V (2007) Advanced expression vector systems: new weapons for plant research and biotechnology. Plant Physiol 145:1087–1089

Ubeda-Tomás S, García-Martínez JL, López-Díaz I (2006) Molecular, biochemical and physiological characterization of gibberellin biosynthesis and catabolism genes from *Nerium oleander*. J Plant Growth Regul 25:52–68

Umezawa T, Fujita M, Fujita Y, Yamaguchi-Shinozaki K, Shinozaki K (2006) Engineering drought tolerance in plants: discovering and tailoring genes to unlock the future. Curr Opin Biotechnol 17:113–122

Venter M (2007) Synthetic promoters: genetic control through *cis* engineering. Trends Plant Sci 12:118–124

Verma D, Daniell H (2007) Chloroplast vector systems for biotechnology applications. Plant Physiol 145:1129–1143

Verma D, Samson NP, Koya V, Daniell H (2008) A protocol for expression of foreign genes in chloroplasts. Nat Protocol 3:739–758

Wang G, Xu Y (2008) Hypocotyl-based *Agrobacterium*-mediated transformation of soybean (*Glycine max*) and application for RNA interference. Plant Cell Rep 27:1177–1188

Warzecha H (2008) Biopharmaceuticals from plants: a multitude of options for posttranslational modifications. Biotechnol Genet Eng 25:315–330

Wenck A, Pugieux C, Turner M, Dunn M, Stacy C, Tiozzo A, Dunder E, van Grinsven E, Khan R, Sigareva M, Wang WC, Reed J, Drayton P, Oliver D, Trafford H, Legris G, Rushton H, Tayab S, Launis K, Chang Y-F, Chen D-F, Melchers L (2003) Reef-coral proteins as visual, non-destructive reporters for plant transformation. Plant Cell Rep 22:244–251

Wenzel G (2006) Molecular plant breeding: achievements in green biotechnology and future perspectives. Appl Microbiol Biotechnol 70:642–650

Wu H, Sparks C, Amoah B, Jones HD (2003) Factors influencing successful *Agrobacterium*-mediated genetic transformation of wheat. Plant Cell Rep 21:659–668

Xiang L, Milc JA, Pecchioni N, Chen L-Q (2007) Genetic aspects of floral fragrance in plants. Biochemistry 72:351–358

Xiao K, Zhang C, Harrison M, Wang ZY (2005) Isolation and characterization of a novel plant promoter that directs strong constitutive expression of transgenes in plants. Mol Breed 15:221–231

Xu C-J, Yang L, Chen KS (2005) Development of a rapid, reliable and simple multiplex PCR assay for early detection of transgenic plant materials. Acta Physiol Plant 27:283–288

Ye X, Al-Babili S, Klöti A, Zhang J, Lucca P Beyer P, Potrykus I (2000) Engineering the provitamin A (beta-carotene) biosynthetic pathway into (carotene-free) rice endosperm. Science 287:303–305

Yemets A, Radchuk V, Bayer O, Bayer G, Pakhomov A, Baird WV, Blume YB (2008) Development of transformation vectors based upon a modified plant α-tubulin gene as the selectable marker. Cell Biol Int 32:566–570

Yevtushenko DP, Sidorov VA, Romero R, Kay WW, Misra S (2004) Wound-inducible promoter from poplar is responsive to fungal infection in transgenic potato. Plant Sci 167:715–724

Yi X, Yu D (2006) Transformation of multiple soybean cultivars by infecting cotyledonary-node with *Agrobacterium tumefaciens*. Afr J Biotechnol 5:1989–1993

Yin Z, Plader W, Malepszy S (2004) Transgene inheritance in plants. J Appl Genet 45:127–144

Yonekura-Sakakibara K, Saito K (2006) Review: genetically modified plants for the promotion of human health. Biotechnol Lett 28:1983–1991

Yoo SY, Bomblies K, Yoo SK, Yang JW, Choi MS, Lee JS, Weigel D, Ahn JH (2005) The 35S promoter used in a selectable marker gene of a plant transformation vector affects the expression of the transgene. Planta 221:523–530

Yu SM, Ko SS, Hong CY, Sun HJ, Hsing YI, Tong CG, Ho THD (2007) Global functional analyses of rice promoters by genomics approaches. Plant Mol Biol 65:417–425

Yuan JS, Burris J, Stewart NR, Mentewab A, Stewart CN Jr (2007) Statistical tools for transgene copy number estimation based on real-time PCR. BMC Bioinformatics 8:S6

Zakharov A, Giersberg M, Hosein F, Melzer M, Müntz K, Saalbach I (2004) Seed-specific promoters direct gene expression in non-seed tissue. J Exp Bot 55:1463–1471

Zale J, Agarwal S, Loar S, Steber CM (2008) Floral transformation of wheat. In: Jones HD, Shewry PR (eds) Methods in molecular biology, transgenic wheat. Barley and oats. Humana Press, New York, USA, pp 105–113

Zhao Y, Liu Q, Davis R (2004) Transgene expression in strawberries driven by a heterologous phloem-specific promoter. Plant Cell Rep 23:224–230

Zhao F, Zhang H (2007) Transgenic rice breeding for abiotic stress tolerance – present and future. Chin J Biotechnol 23:1–6

Zheng ZL, Yang ZB, Jang JC, Metzger JD (2001) Modification of plant architecture in chrysanthemum by ectopic expression of the tobacco phytochrome B1 gene. J Am Soc Hortic Sci 126:19–26

Zhou LG, Wu JY (2006) Development and application of medicinal plant tissue cultures for production of drugs and herbal medicinals in China. Nat Prod Rep 23:789–810

Zuker A, Tzfira T, Ben-Meir H, Ovadis M, Shklarman E, Itzhaki H, Forkmann G, Martens S, Neta-Sharir I, Weiss D, Vainstein A (2002) Modification of flower color and fragrance by antisense suppression of the flavonone 3-hydroxylase gene. Mol Breed 9:33–41

Zuo J, Niu Q-W, Ikeda Y, Chua N-H (2002) Marker-free transformation: increasing transformation frequency by the use of regeneration-promoting genes. Curr Opin Biotechnol 13:173–180

Chapter 2
Explants Used for the Generation of Transgenic Plants

A. Piqueras, N. Alburquerque, and K.M. Folta

2.1 Introduction

The objective of this chapter is to discuss the types of explants more frequently used in the currently published transformation protocols as well as the morphogenic pathways selected for the regeneration of the transgenic plants.

The process of plant genetic transformation can be divided into three phases: (1) foreign DNA transfer into the plant genome, (2) regeneration of the transformed explant into a normal plant, and (3) selection of transgenic plants and confirmation of their transgenic nature. To develop efficient transformation protocols both in herbaceous and woody plants, different explants of distinct morphogenetic potential have been used. For example, hypocotyls, cotyledons, leaves, stems, and roots are all used as starting materials for transformation and regeneration. The following chapter is divided into two principal parts: the first discusses explant selection and utilization in a series of herbaceous crops and the second reviews explant usage in select woody species.

In herbaceous plants, leaf segments are the preferred explants. Organogenesis and somatic embryogenesis, the two more important morphogenetic alternatives in plant tissue culture (Piqueras and Debergh 1999), have been used depending on the recalcitrance of the selected plant. Adventitious shoot regeneration is most frequently used for transgenic shoot regeneration followed by direct somatic embryogenesis from explants or embryogenic cultures.

The following sections present a review of the literature regarding explant selection for transformation and regeneration. These foundations should serve as

K.M. Folta (✉)

Horticultural Sciences Department and the Graduate Program in Plant Molecular and Cellular Biology, University of Florida, Gainesville, FL, USA

e-mail: kfolta@ifas.ufl.edu

C. Kole et al. (eds.), *Transgenic Crop Plants*,

DOI 10.1007/978-3-642-04809-8_2, © Springer-Verlag Berlin Heidelberg 2010

an outstanding platform to test new experimental systems, as well as define new protocols to hasten processes in established models.

2.2   Explants Used for the Transformation of Herbaceous Plants

2.2.1   Cereals

For genetic transformation of cereals, various methods of direct and indirect transfer of foreign DNA are used, and their morphogenic calli or immature embryos (with subsequent stimulation of morphogenic callus formation) are normally used as explants. However, most cereal crops are characterized by a low morphogenic potential and that significantly limits their application for genetic engineering (Danilova 2007).

As a general rule, the morphogenic calli are used as explants for genetic transformation of cereal crops. Alternatively, immature embryos can be used with subsequent initiation of morphogenic callus formation. Regeneration is the next and probably the most important step in genetic transformation of plants. The overall efficiency of transformation greatly depends on the regeneration potential of explants. Much work is being done in many scientific centers throughout the world to increase the regeneration potential of cereal crops. New methods and approaches are tried to widen the range of transformable crops and increase the regeneration potential of calli used. In most cases, the regeneration potential depends on explant type, genotype, and the composition of the cultivation medium (Rout and Lucas 1996; Cheng et al. 2003; Eudes et al. 2003). Traditionally for induction of morphogenesis in vitro, the phytohormones and synthetic hormone-like regulators were used that belong to the auxin family in different combinations with cytokinins (for example BA). Among the most commonly used inducers are 2,4-D-picloram (4-amino- 3,5,6-trichloropicolinic acid) and dicamba (3,6-dichloroanisic acid) (Bahieldin et al. 2000) .

2.2.2   Brassica

Various methods used for Brassica transformation and the factors affecting transformation efficiencies have been reviewed by Poulsen (1996). *Agrobacterium tumefaciens*-mediated transformation is most widely used for Brassica and it is generally quite efficient and practical for most species in the genus. Although seedlings parts, such as hypocotyls, cotyledons, and cotyledonary petioles, are the most common explants, pieces of flowering stalks also regenerate well (Christey and Earle 1991). Flowering stalk explants are less convenient to obtain and more subject to contamination, but have some advantages, particularly when the supply of seeds of a particular genotype is limited (Metz et al. 1994). However, there is still a need for developing efficient transformation methods to overcome genotype dependency (Cardoza and Stewart 2004).

2.2.3 Cassava

The plant tissue types used for transformation of cassava include shoots induced by organogenesis (Siritunga et al. 2004; Puonti-Kaerlas et al. 1997) and germinating somatic embryos. Direct shoot induction from cotyledons of somatic embryos has been used in both biolistics (Zhang and Puonti-kaerlas 2000) and Agroinfection (Msikita et al. 2002). However, plant regeneration efficiency is highly variable (5–70%) and genotype dependent (Zhang and Puonti-kaerlas 2000). As a result of these cultivar-dependent differences, a variety of tissues including axillary buds (Puonti-Kaerlas et al. 1997), apical leaves (Siritunga et al. 2004), and floral meristems (Woodward and Puonti-Kaerlas 2001) have been used, as various groups found differing success with the various explants from different cultivars.

2.2.4 Potato

Many protocols used now are based on a two-stage regeneration and transformation using stem and leaf explants developed by Visser et al. (1989). The relative ease of shoot regeneration from different tissues of potato (e.g., stem section, leaf, petiole, and tuber disk) conditions the development of the systems used for transformation of this species. At the present moment, most protocols use a two-step regeneration procedure with a callus induction stage followed by a shoot outgrowth stage. The callus stage is minimized to prevent the high incidence of somaclonal variation reported in potato (Heeres et al. 2002). This initial stage is facilitated by treating the explant with zeatin or zeatin riboside in combination with low levels of NAA or IAA. The second stage has the zeatin level reduced by 20% and the auxin level reduced by a factor of 10, plus the addition of gibberellin to stimulate shoot elongation. Usually regeneration rates are high and after 10 weeks of culture ten shoots per explant is a common result.

2.2.5 Sugarcane

Both embryogenic calli (Arencibia et al. 1998) and meristematic tissues from micropropagated plants (Enríquez et al. 1998) have been used as explants for the transformation of sugarcane. The production of sugarcane transgenic plants by agroinfection has been achieved by combining several tissue culture procedures, particularly the use of young regenerable material characterized by the presence of actively dividing cells competent for agroinfection, pre-induction of the regeneration capacity and treatments to improve the adhesion of *Agrobacterium* during cocultivation (Arencibia et al. 2000).

2.2.6   *Banana*

Several types of explants have been used for banana transformation (Gómez-Lim and Litz 2004), the most frequently used are embryogenic cultures (Khanna et al. 2004). These cultures are usually induced from immature male (Escalant et al. 1994) and female flowers (Grapin et al. 2000). Wounded meristems of in vitro plantlets have been used for banana transformation as well (May et al 1995). Although transformed plants were regenerated, this procedure has not been widely used because of low transformation rates and chimeras.

2.2.7   *Carnation*

Stem sections were the first explants used for carnation transformation (Lu et al. 1991); thin sections of node explants have been also used for carnation transformation (Nontaswatsri et al. 2004). So far, the most reproducible and efficient explant used for carnation transformation has been the leaf base of micropropagated plantlets (Firoozabady et al. 1995; Van Altvorst et al. 1995; Kinouchi et al. 2006). By using the leaf base from in vitro grown shoots three cultivars representing three major commercial carnation groups have been transformed in what could be considered a proof to the cultivar independence of this method, transformation efficiency was high and fully transformed carnation plants were produced without chimerism (Firoozabady et al. 1995; Van Altvorst et al. 1995). An example of carnation regeneration is presented in Fig. 2.1.

2.2.8   *Tomato*

Tomato (*Solanum lycopersicum*) was one of the first crops for which a genetic transformation system was reported involving regeneration by organogenesis from *Agrobacterium*-transformed explants. Commonly, cotyledons from seeds of different tomato lines have been chosen as explants (Frary and Van Eck 2005; Sun et al. 2006; Qiu et al. 2007). With this procedure several tomato cultivars transformed obtained transformation efficiencies that range from 10 to 14% (Van Eck et al. 2006).

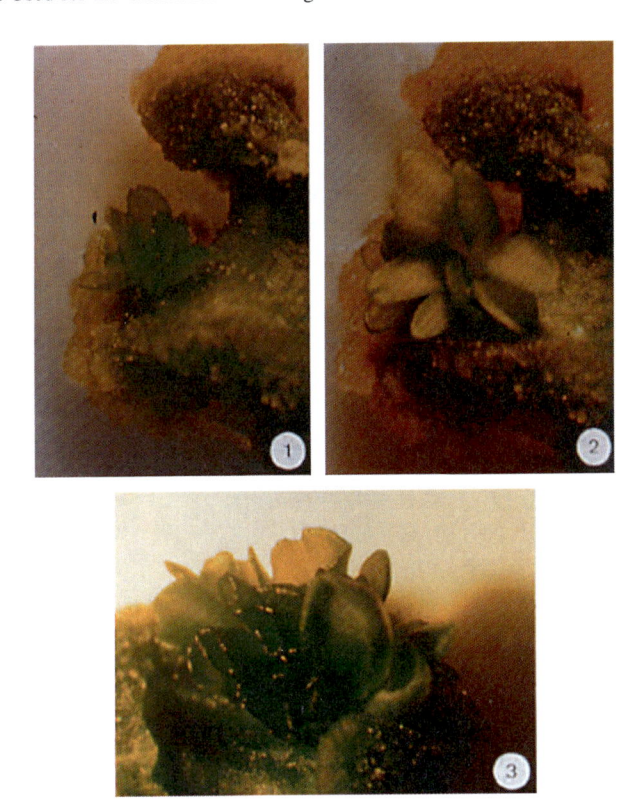

Fig. 2.1 Sequence of morphogenic events leading to adventitious bud regeneration at the leaf base of carnation. (**a**) Emergence of meristemoid, (**b**) Initial shoot cluster formation, (**c**) Developed adventitious shoot

2.2.9 Soybean

Transgenic soybean [*Glycine max* L. Merr] plants have been produced by different methods using explants as embryogenic suspension cultures, proliferating meristems, immature cotyledons, or shoot and axillary meristems from mature cotyledons. Microprojectile bombardment has been used to transform embryogenic suspension cultures and shoot meristems (McCabe et al. 1988; Parrott et al. 1989a; Finner and McMullen 1991; Rech et al. 2008). *Agrobacterium*-mediated transformation methods have been developed with the rest of target tissues (Aragao et al. 2000; Ko et al. 2003, 2004; Liu et al. 2004). The criteria for choosing the type of explants to transform are strongly influenced by the genotype; for instance, the embryogenic response varied with genotype (Parrott et al. 1989b). In a recent work, Cao et al. (2008) have found that there are significant differences among soybean genotypes in their susceptibility to *Agrobacterium rhizogenes* when germinated seedlings have been infected.

2.2.10 Alfalfa

Although different transformed plants of alfalfa (*Medicago sativa* L.) have been obtained by different methods (Samac and Temple 2004), the use of leaf explants of a highly regenerable genotype infected with *A. tumefaciens* followed by induction of somatic embryos have allowed the recovery of transgenic alfalfa plants with extremely high efficiencies (Samac and Austin-Phillips 2006). Samac (1995) reported that transformation at the whole plant level is germplasm dependent, while in tissue culture the bacterial strain used is the critical variable for successful transformation. Other authors (Desgagnes et al. 1995) found a high influence of the genotype, the expression vector and the bacterial stain on the ability to produce stable transgenic material by the method described earlier.

2.2.11 Sunflower

Sunflower (*Helianthus annuus* L.) is considered one of the most difficult species to be genetically transformed because its competent cells for regeneration are not able to be transformed. To overcome this problem, different approaches have been reported such as the origin of the explants, the transformation vectors or systems, and even combinations of approaches. Schrammeijer et al. (1990) developed a routine *A. tumefaciens*-mediated transformation protocol using meristems from late-stage embryonic tissues as efficient explants with low transformation efficiencies. Later, other groups have reported changes in the procedure that have allowed improvement of the efficiency or reduction of chimeral shoot and plant production derived from this kind of explant (Knittel et al. 1994; Rao and Rohini 1999; Molinier et al. 2002).

2.2.12 Cucumber

Cucumber (*Cucumis sativus* L.) transformation has been approached with different kinds of explants using *A. tumefaciens*. Cotyledons (Chee and Slightom 1991; Tabei et al. 1998), hypocotyls (Trulson et al. 1986; Nishibayashi et al. 1996), or petioles (Raharjo et al. 1996) have been used as explants. Embryogenic callus tissues have been bombarded with microprojectiles coated with several plasmid DNAs (Chee and Slightom 1992). One method, which has advantages like fast and efficient plant regeneration from a wide range of genotypes, consisted of producing direct regeneration from leaf microexplants selected on kanamycin-containing medium. The transformation efficiencies varied from 0.8 to 6.5% depending on the genotype and the construct (Yin et al. 2005). Other method involves regeneration from a long-term established embryogenic suspension culture, obtaining higher transformation efficiencies (from 6.4 to 17.9%) (Burza et al. 2006).

2.2.13 Eggplant

Different authors have developed regeneration and transformation procedures to transform eggplant (*Solanum melongena* L.) using seedling explants like hypocotyl, epicotyl, and node segments as well as cotyledon segments (Arpaia et al. 1997; Magioli et al. 1998), leaf disks (Yadav and Rajam 1998), or roots (Franklin and Sita 2003). When an *Agrobacterium*-mediated transformation protocol was used with cotyledons, hypocotyls and leaves from two eggplant genotypes high transformation efficiencies were reported (Van Eck and Snyder 2006).

2.2.14 Melon

In melon species, the transformation frequency is very low due to the production of "escapes" (Guis et al 1998; Galperin et al. 2003). In previous studies, transgenic plants were generated using adventitious shoot organogenesis. To reduce the problem of "escapes," an alternative regeneration system that can enable transformation was developed, and several groups reported the production of somatic embryos from melon cell suspension cultures (Guis et al. 2000). Published protocols for melon genetic engineering use the process of organogenesis (Dong et al. 1991). Although this sometimes leads to problems, such as abnormal embryos, the liquid culture system is considered very useful for the efficient selection of transformed tissues, as whole explants absorb antibiotics more easily when suspended in liquid media than when cultured on solidified media. Embryogenesis is also a useful regeneration system for transgenic research because none of the transgenic plants are chimeric (Asaka-Kennedy et al. 2004). The efficiency of embryogenesis in melons is closely related to the genotype (Oridate et al. 1992).

2.2.15 Strawberry

Strawberry transformation has been reported for many genotypes, although confined to diploid wild species (e.g., *Fragaria vesca*) and cultivated octoploid varieties (*Fragaria* x *ananassa*). Most of the transformation reports generate transgenic plants from *Agrobacterium*-mediated transformation of leaf disks or cut leaves (El Mansouri et al. 1996; Passey et al. 2003; Landi and Mezzetti 2005). In some cases petiole segments are particularly prolific (Folta et al. 2006). New systems in strawberry transformation are being constantly retooled or developed because of strawberry's position as a functional system to test gene activity in the valued Rosaceae family (Shulaev et al. 2008). A demonstration of regeneration from diverse tissues is shown in Fig. 2.2.

Fig. 2.2 Shoot emergence from diploid strawberry leaf explants. Culture conditions have been optimized for prolific regeneration from a number of diploid accessions, providing an outstanding functional genomics resource for rapid elucidation of gene function

2.3 Explants Used for the Transformation of Woody Plants

Woody plant transformation has become a central area of interest, for valuable forest products, both economical and ecological. Studies of lumber-crop transformation are driven by a need for improved lumber, food, paper, fuels, and other materials that are derived from tree crops. In addition, valuable nutritious fruit and nut products are borne from woody perennials. Lumber, biomass, food, and ecologically intended tree crops will benefit from in vitro propagation or development of genetically enhanced germplasm. Interest has been piqued since the unveiling of the *Populus* genome. There is now extensive interest in transformation of tree crops to validate in planta the findings of genome sequencing and functional genomics studies. The generally long juvenile periods and dormancy issues make breeding efforts and crop improvement strategies difficult and arduous. The need for improved woody species is being met in two ways. First, direct engineering and deployment of tree crops featuring genes of interest permit favorable traits to be directly introduced into production scenarios. A few such examples include the notable engineering feats of introducing papaya ringspot virus resistance into papaya (reviewed in Tripathi et al. 2008) and plum pox virus resistance into stone fruits (Ravelonandro et al. 1997). Transformation has also been used to engineer rootstocks to reduce the time before scion reproductive competence. The goal is to introduce genes associated with inducing the floral transition so that traditional breeding techniques could be implemented. Such milestones have been met in poplar (Hsu et al. 2006) and citrus (Nishikawa et al. 2007). The challenge to these processes is the engineering itself, as many tree species are recalcitrant to genetic manipulation.

For generation of transgenic materials, it is necessary to regenerate organs and/ or embryos using fairly standardized protocols. In these cases, explant selection is central to successful transformation and regeneration. The tissue chosen, as well as the health status and developmental state of the donor plant are critical to the success of adventitious shoot production. In just about all cases callusing and shoot production was greatly accelerated by specific growth regulators, viz. thiadiazuron, discussed elsewhere in this volume. This part of the chapter focuses on economically important woody crops, primarily those used in fruit and forest industries.

2.3.1 Almond

The almond (*Prunus dulcus* Mill.) literature presents several complementary studies of transformation or regeneration, but relatively efficient protocols that combine the two have only been recently developed. *Agrobacterium*-mediated transformation (albeit without regeneration) of almond leaf disks was reported in 1995 (Archilletti et al. 1995). Successful transformation and regeneration of almond was reported 4 years later. Miguel and Oliveira (1999) used the four most recent fully expanded leaves from 3-week-old shoots in culture as explants. A subset of the leaves received a pretreatment on the callus induction media for 3–4 days before wounding. These tissues were cut with a scalpel dipped in bacterial suspension. The efficiency of this approach was low, but successful, and could be greatly increased with addition of acetyosyringone to cocultivation medium (Costa et al. 2006). A contemporaneous study carefully examined regeneration conditions for two major cultivars (Ainsley et al. 2000) leading to higher regeneration efficiency. This study also used young leaves from in vitro grown plantlets, cutting them into 5-mm^2 pieces. Additional reports examined the regeneration from embryonic cotyledons under different culture conditions and growth regulators, greatly improving efficiency (Ainsley et al. 2001b). A complementary study from the same group examined transformation protocols (Ainsley et al. 2001a) but did not report regeneration. Efficient transformation and regeneration has been reported using the same explants – the fully expanded leaves from 21- to 28-day-old micropropagated shoots, cut across the midrib (Costa et al. 2006; Ramesh et al. 2006).

2.3.2   Apple

Generation of adventitious shoots from apple (*Malus* x *domestica* Borkh.) leaf explants has been documented since 1983 (Liu et al., 1983a, 1983b). Genetic transformation of apple tissue was reported two decades ago using leaf explants (James et al. 1989; Maheswaran et al. 1992), and stable integration and segregation of transgenes were later reported (James et al. 1994, 1996). In these early studies *Agrobacterium*-mediated transformation was used with leaf explants, and the effects of explant age, orientation, and genotype were explored (Debondt et al. 1994; Yepes and Aldwinckle 1994; Puite and Schaart 1996). Regeneration from embryonic cotyledons and axes was reported (Keulemans and Dewitte 1994). The effect of various growth regulators was also assessed on leaf explants (Yancheva et al. 2003) as were the effects of acetyosyringone and explant pretreatment (Seong and Song 2008). Internodal seedling explants were shown to be especially amenable to transformation, particularly when etiolated (Liu et al. 1998). Another study utilized stem microcuttings and *A. rhizogenes* to create chimeric apple trees that later yielded transgenic plants (Lambert and Tepfer 1992). Leaf explants have also been successfully used for particle bombardment (Gercheva et al. 1994).

2.3.3   Apricot

Apricots (*Prunus armeniaca* L.) are less routinely transformed than other *Prunus* counterparts such as plum (reviewed later in this chapter), but have benefited from careful studies that have defined optimal conditions. Early studies defined the conditions of regeneration, and sometimes transformation, from a variety of explant types. Transfer of genes using *Agrobacterium* has been accomplished using embryonic cotyledons, leaf disks, and somatic embryos as initial culture material (Machado et al. 1994). Efficient regeneration was achieved from young in-vitro-derived leaves (typically the first four from 21-day-old in vitro plants), but was highly dependent on genotypes and cytokinin and gelling agent used (Perez-Tornero et al. 2000; Burgos and Alburquerque 2003), and could be accelerated with ethylene inhibitors (Burgos and Alburquerque 2003). A series of studies have increased the efficiency of transformation and regeneration of the "Helena" cultivar using a series of selection strategies, growth regulators, and culture conditions, but the explant source has remained unchanged (Petri et al., 2004, 2005a, 2005b, 2008a). Emerging shoots from an apricot explant are presented in Fig. 2.3.

2.3.4   Blueberry

Successful transformation of leaf materials from a blueberry hybrid (*Vaccinium corymbosum* x *V. angustifolium*) was first achieved by Graham et al. (1996). Since

Fig. 2.3 Adventitious bud regeneration on a leaf segment base of apricot cv. Helena

then, multiple studies have examined gene transfer to cut leaf sections from multiple *V. corymbosum* genotypes raised in vitro (Cao et al. 1998; 2003; Sink and Song 2004; Song and Sink 2004). One of the factors affecting transformation efficiency is explant age, as those removed from recently transferred source material typically performed better (Cao et al. 1998). Song and Sink (2004) report successful transformation and regeneration using leaf explants when the petiole is removed and the distal third of the leaf blade is discarded. This latter protocol has been successfully implemented in testing various promoters in this valuable, yet recalcitrant crop (Song et al. 2008).

2.3.5 Birch

The first reports of birch transformation were obtained in Japanese white birch (*Betula platyhphylla*). Here leaf disk explants were transformed and regenerated with reasonable efficiency (Mohri et al. 1997). A study in silver birch (*Betula pendula*) tested regeneration in leaf, internodal stem segments, and nodal stem segments, and the results showed that all were capable of producing adventitious shoots at a high (> 90%) rate (Lemmetyinen et al. 1998). Transformation was performed on explants that were precultured and wounded before cocultivation (Lemmetyinen et al. 1998). Cultivated callus and shoots were used for biolistic-assisted transformation of silver birch, producing stable plantlets (Valjakka et al. 2000).

2.3.6 Citrus

There is a desire to transform both the scions and the rootstocks of major citrus cultivars, as a changing spectrum of pests, pathogens, and production challenges forces new and rapid innovation. Many genotypes have been regenerated, namely sour orange (*Citrus aurantium* L.), sweet orange (*Citrus sinensis* L. Osbeck), grapefruit (*Citrus paradisi*), mandarin (*Citrus reshni* Hort. ex Tan.), alemow (*Citrus macrophylla* Wester), and the hybrid Troyer citrange (*Citrus sinensis* [L.] Osbeck) among others. The development of transgenic plants was performed by generating transgenic shoots that could be grafted to seedling rootstocks. These shoots arose from internodal stem segments from 5-week-old seedlings, or on the cut end of epicotyls treated with *Agrobacterium* (Pena et al. 1995). In one particular study epicotyl segments from germinated seedlings were cultured in darkness, horizontally, and robust shooting occurred on the basipetal end, arising from the cambial region (Bordon et al. 2000).

2.3.7 Cherry

Cherry is also considered to be recalcitrant to transformation and regeneration. A number of reports have demonstrated the ability to regenerate shoots of several sweet cherries (*P. avium*) grown in vitro (Grant and Hammatt 2000; Bhagwat and Lane 2004; Feeney et al. 2007; Canli and Tian 2008). Bhagwat and Lane (2004) tested a series of explants in two cultivars, comparing the furled leaves at the apex to the expanding leaves to mature leaves, with and without perpendicular wounds across the midrib. The results showed that regeneration occurred only in the upper expanded leaves that were wounded. An evaluation of several explant types was performed by Feeney et al. (2007). In this study explants arose from orchard trees, demonstrating that organogenic callus could be derived from ex vitro materials. Wounding was also advantageous in regeneration of orchard and in vitro tissues. The study by Feeney et al. (2007) also indicates that callus formation becomes less robust as explants are located near the bottom of the plant, as those proximal to the shoot tip performed better. In sour cherry (*P. cerasus* L.) the cultivar "Montmorency" was used for regeneration and transient expression, again using leaf explants with cuts perpendicular to the midrib, much like in apricot (see earlier). These protocols were the basis of transformation of the "Montmorency" cultivar, where additional culture conditions were evaluated (Song and Sink 2006).

2.3.8 *Eucalyptus*

The first report of transgenic eucalyptus (*Eucalyptus globulus* Labill.) trees describes *Agrobacterium*-mediated transformation of wounded seedlings (Moralejo et al. 1998). The process was also described with careful detail using *Eucalyptus camaldulensis* hypocotyl segments and organogenesis from callus (Ho et al. 1998). Transgenic plants bearing genes for resistance to herbicide and insect larvae were generated from cut hypocotyls and cotyledons of 2-week-old seedlings (Harcourt et al. 2000). Stable transformation was also achieved by sonicating seeds or seedlings in the presence of *Agrobacterium* (Gonzalez et al. 2002), where the most efficient transformation occurred in the intersection of the root and shoot or cotyledons. Particle bombardment of eucalyptus hybrid callus derived from seedling hypocotyls and cotyledons also resulted in successful regeneration of stable transgenic trees (Sartoretto et al. 2002). While the typical goal of a transformation system is to generate stable plants bearing a transgene, other attempts have examined stably engineered tissue to study wood formation (Spokevicius et al. 2005) or cell fate in transformed cambium (Van Beveren et al. 2006).

2.3.9 *Kiwi*

Successful transformation and/or regeneration was/were reported for several species of *Actinidia*. The first reports appear from *Actinidia chinensis* (Suezawa et al. 1988). Callus was produced from field-grown leaves and then regenerated through cell suspension cultures. Later, Rugini et al. (1991) transformed elite kiwi germplasm, inserting the *rol A, B,* and *C* genes from *Agrobacterium rhizongenes*, with the intent of affecting root morphology and rooting ability. Here leaf disks were used to directly generate roots, approaching 100% efficiency. At the same time other groups found great success from cocultivation of kiwi hypocotyls or stem segments (Uematsu et al. 1991). In a separate study, kiwi hypocotyls were inoculated with *A. rhizogenes*, leading to prolific hairy root production in culture, eventually generating transformed whole plants (Yazawa et al. 1995). The same group later switched to petioles as leaf explants from several cultivars and obtained up to 31% generation of transformed adventitious buds, again using *A. rhizogenes* protocols (Yamakawa and Chen 1996). Leaf disks and petioles were used to install the stilbene synthase gene (Kobayashi et al. 2000). A smaller, less vigorous, faster flowering species (*Actinidia eriantha*) was transformed and regenerated from leaf strips as a potential system for functional genomics (Wang et al. 2006). Transgenic plants have been regenerated from leaf disks of "Hayward" to test effects of a grape MYB protein on plant pigmentation (Koshita et al. 2008).

2.3.10 Larch

Hypocotyls of developing European larch (*Larix deciduas* Mill.) seedlings were used to introduce transgenes via *A. rhizogenes*-mediated transformation. Wounded hypocotyls from 7-day-old seedlings would produce adventitious shoots 4 weeks after transformation (Shin et al. 1994). An alternative approach introduced genes through "embryogenic masses," cultures formed from long-term maintenance of embryogenic tissue derived from embryos isolated from pollinated cones. Transient expression was observed through microprojectile bombardment (Duchesne 1993) into these masses. Stable transformation of somatic embryos was achieved in *L. laricina* (tamarack). These embryos were both from the precotyledonary stage and those with elongating or developed cotyledons (Klimaszewska et al. 1997). Embryogenic masses were also the preferred starting point for transformation of hybrid larch (Levee et al. 1997). Eventually optimization of particle bombardment protocols would yield stable transformants as zygotic embryos from *L. gmelinii* L. were transformed with this method (Lin et al. 2005). The tissues were cultured to callus that was then induced to form shoots with reasonable frequency.

2.3.11 Peach

Among major tree crops, transformation of peach (*Prunus persica* L. Batsch.) has remained difficult. Despite the efficient systems devised for apricot and plum mentioned elsewhere in this chapter, reports of peach transformation are sparse. Peach has been successfully regenerated from in vitro leaves from plant apices (Gentile et al. 2002), mature and immature cotyledons (Mante et al. 1989; Pooler and Scorza 1995), and zygotic embryos (Hammerschlag et al. 1985). A variety of peach explants, including leaf segments, immature embryos, and embryogenic calli, have been transformed via *Agrobacterium* (Scorza et al. 1990) and biolistics (Ye et al. 1994), but routine regeneration of transformed tissue has remained elusive. Only two reports of successfully reproducing stable transgenic plants were reported and indicate a relatively inefficient transformation and regeneration rate (Smigocki and Hammerschlag 1991; Perez-Clemente et al. 2004). The most successful report demonstrated that regeneration could occur from embryo sections, but was poor or nonexistent from hypoctoyls and cotyledons (Perez-Clemente et al. 2004). Padilla et al. (2006) performed a strategic study using *Agrobacterium*-mediated transformation and GFP markers to assess transformation efficiency of various bacterial strains and explants. The study showed that internodes, cotyledons, and embryonic axes were superior to embryonic hypocotyl slices, the choice material for plum. Still, further optimization will be required to make peach transformation and regeneration routine.

2.3.12 Pear

Pear (*Pyrus communis* L.) has been successfully transformed and regenerated using leaf explants and *Agrobacterium*-mediated gene transfer (Mourgues et al. 1996). Various culture conditions and genotypes were tested, using in vitro-derived leaves from recently transferred plantlets, cut perpendicularly across the midrib (Chevreau et al. 1997; Bell et al. 1999; Yancheva et al. 2006). The same protocols were employed by other studies (Reynoird et al. 1999) demonstrating their utility. Regeneration remained an issue in some genotypes and Matsuda et al. (2005) examined other explant sources to improve efficiency. This report used the same cut leaves from in vitro plants, but then also included 0.5-mm sections of axillary shoot meristems. The meristematic tissues proved superior in otherwise recalcitrant cultivars, and the authors noted that the poor regeneration on selection agents in leaf explants arose from a lack of transformation, not an inherent inability to regenerate. Embryonic cotyledons in mature seeds of the Asian pear (*Pyrus betulaefolia*) have also been amenable to transformation and regeneration (Kaneyoshi et al. 2001).

2.3.13 Pine

Stable transformation of conifers dates back over two decades to efforts of Ron Sederoff and colleagues in loblolly (*Pinus taeda* L.) and sugar pine (Sederoff et al. 1986; Loopstra et al. 1990). Despite these early gains, most reports of pine transformation acknowledged only transient expression, and not the generation of transgenic plants. Efficient generation of transgenic plants is limited by the cell division capacity of the explants in these recalcitrant species. Studies in both pine and spruce (noted later in this chapter) demonstrated increased transformation efficiency when embryogenic tissues are used as explants.

White pine (*Pinus strobus* L.) transgenics were efficiently produced using *Agrobacterium*-mediated transformation against embryogenic tissues (Levee et al. 1999). *Pinus radiata* has also been successfully transformed (Walter et al. 1998).

2.3.14 Plum

Transformation of plum (*Prunus domestica* L) is routine and efficient, at least compared to other closely related stonefruits like apricots, almonds, and peaches. Plum transformation dates back almost two decades to reports of *Agrobacterium*-mediated transformation of hypocotyl segments isolated from the embryonic axes in ungerminated seeds (Mante et al. 1991). Here the surface-sterilized seed was split and the hypocotyl was removed and cut into three sections. The radicle and the epicotyl were discarded, and the central portion was used for transformation by slicing it transversally into several thin (< 1 mm) explants. These protocols have remained generally unchanged, except for a 10× increase that comes from pre-conditioning explants with growth regulators (Petri et al. 2008b). Similar protocols also work with the Japanese plum (*Prunus salicina*) (Urtubia et al. 2008)

2.3.15 Populus

Populus species and related hybrids have become extremely useful model plants in the study of gene function in woody plants, due in part to the full accounting of genes in a sequenced genome and rich expressed sequence tag (EST) resources. Transformation in *Populus* was first achieved over two decades ago (Parsons et al. 1986), and subsequent studies improved on the techniques. Explants used include internode pieces from 6- to 8-week-old in vitro plants (Deblock 1990), and leaf disks from ex vitro plants (Tsai et al. 1994). Transformation of *P. tremula* (Tzfira et al. 1996) and cottonwood varieties (Han et al. 1997) was accelerated using *A. rhizogenes* against stem segments that would develop adventitious roots with great efficiency. Additional protocols were specifically designed for the *P. trichocarpa* genotype Nisqually-1, the line used for genome sequencing. Here internodal stem explants proved superior to midrib or leaf explants in regeneration efficiency (Song et al. 2006). This study showed that explant selection was critical to the transformation process, both in the discrete tissue used and the age of the explant source plant. Specifically, the fifth to eighth stem internode sections from vigorous 5–6 month-old plants performed best in culture. Recent modifications hasten the process in quaking aspen (*P. tremuloides*) inoculating hypocotyl sections leading to the regeneration of transgenic trees in 3–4 months instead of 6–12 months (Cseke et al. 2007).

2.3.16 Spruces

Low transformation rates in spruces were caused by explant materials with limited competence for cell division. A comprehensive assessment of transformation competence during embryo development optimized parameters of white spruce (*Picea glauca*) transformation by particle bombardment (Ellis et al. 1993). Embryogenic callus, embryos themselves, and seedlings were receptive to the treatment, leading to transformed plants. Experiments using cell suspension cultures of Norway spruce further accelerated efficiency. Norway spruce (*Picea abies*) offers the advantage of a well-studied system with prolific cell growth, excellent culture viability, and strong regeneration potential for embryogenic cultures (von Arnold et al. 1996). Studies in *P. abies* demonstrated that biolistics technologies could be used to introduce transgenes, such as GUS (Duchesne and Charest 1991). Biolistics or *Agrobacterium* have been used to generate transgenic plants arising from somatic embryos (Walter et al. 1999).

2.3.17   *Walnut*

The transformation of walnut (*Juglans* spp.) via *Agrobacterium*-associated means was originally tested by Polito et al. (1989) when they were able to produce somatic embryos in culture. At the same time, studies at the University of California-Davis were demonstrating that the somatic embryos of walnut could be transformed and regenerated into plants (McGranahan et al. 1988, 1990; Dandekar et al. 1989). As with many hardwoods, the most efficient transformations have been performed on somatic embryos (Escobar et al. 2000; Tang et al. 2000). Somatic embryos themselves have even been reported to produce secondary embryos (Raemakers et al. 1995), appropriate for transformation. Transformation of agriculturally useful transgenes has been reported for Persian walnut (*Juglans regia*) when somatic embryos derived from a repetitively embryogenic lines were cocultured with *Agrobacterium* bearing the gene encoding Bt toxin (Dandekar et al. 1998). An additional study defined the boundaries of transient and stable transformation of somatic embryos in Persian walnut using GFP (Escobar et al. 2000). This visible marker allowed detection of transformed materials that could be subcultured into plants on appropriate media.

2.4   Concluding Remarks

Explant selection is a critical parameter to consider when performing transformation and regeneration experiments. Different explant types often have varying potential for transformation and certainly for organogenesis or development of somatic embryos. Just as a complete test for transformation and regeneration includes a complete assessment of growth regulators, media constituents, and culture conditions, the choice of explant should be a central consideration in the development of transgenic resources.

References

Ainsley PJ, Collins GG, Sedgley M (2000) Adventitious shoot regeneration from leaf explants of almond (*Prunus dulcis* Mill.). In Vitro Cell Dev Biol Plant 36:470–474

Ainsley PJ, Collins GG, Sedgley M (2001a) Factors affecting *Agrobacterium*-mediated gene transfer and the selection of transgenic calli in paper shell almond (*Prunus dulcis* Mill.). J Hortic Sci Biotechnol 76:522–528

Ainsley PJ, Hammerschlag FA, Bertozzi T, Collins GG, Sedgley M (2001b) Regeneration of almond from immature seed cotyledons. Plant Cell Tissue Organ Cult 67:221–226

Archilletti T, Lauri P, Damiano C (1995) *Agrobacterium*-mediated transformation of almond leaf pieces. Plant Cell Rep 14:267–272

Aragao FJL, Sarokin L, Vianna GR, Rech EL (2000) Selection of transgenic meristematic cells utilizing a herbicidal molecule results in the recovery of fertile transgenic soybean [*Glycine max* (L.) Merril] plants at a high frequency. Theor Appl Genet 101:1–6

Arencibia A, Carmoma E, Téllez P (1998) An efficient method for sugarcane transformation mediated by *Agrobacterium tumefaciens*. Transgen Res 7:213–222

Arencibia A, Carmona E, Cornide MT, Menéndez E, Molina P (2000) Transgenic sugarcane (*Saccarum* sp. L.). In: Bajaj YPS (ed) Biotechnology in Agriculture and Forestry, Vol 46: Transgenic Crops I. Sringer, Berlin, pp 188–206

Arpaia S, Mennella G, Onofaro V, Perri E, Sunseri F, Rotino GL (1997) Production of transgenic eggplant (*Solanum melongena* L.) resistant to Colorado potato beetle (*Leptinotarsa decemlineata* Say) using root explants. Theor Appl Genet 95:329–334

Asaka-Kennedy Y, Tomita K, Ezura H (2004) Efficient plant regeneration and *Agrobacterium*-mediated transformation via somatic embryogenesis in melon (*Cucumis melo* L.). Plant Sci 166:763–769

Bahieldin A, Dyer WE, Qu R (2000) Concentration effects of dicamba on shoot regeneration in wheat. Plant Breed 119:237–439

Bell RL, Scorza R, Srinivasan C, Webb K (1999) Transformation of "Beurre Bosc" pear with the *rol*C gene. J Am Soc Hortic Sci 124:570–574

Bhagwat B, Lane WD (2004) In vitro shoot regeneration from leaves of sweet cherry (*Prunus avium*) "Lapins" and "Sweetheart". Plant Cell Tissue Organ Cult 78:173–181

Bordon Y, Guardiola JL, Garcia-Luis A (2000) Genotype affects the morphogenic response in vitro of epicotyl segments of Citrus rootstocks. Ann Bot 86:159–166

Burgos L, Alburquerque N (2003) Ethylene inhibitors and low kanamycin concentrations improve adventitious regeneration from apricot leaves. Plant Cell Rep 21:1167–1174

Burza W, Zuzga S, Yin ZM, Malepszy S (2006) Cucumber (*Cucumis sativus* L.). In: Wang K (ed) *Agrobacterium* Protocols. USA, Humana Press, Totowa, New Jersey, pp 427–438

Canli FA, Tian L (2008) In vitro shoot regeneration from stored mature cotyledons of sweet cherry (*Prunus avium* L.) cultivars. Sci Hortic 116:34–40

Cao D, Hou W, Song S, Sun H, Wu C, Gao Y, Han T (2008) Assessment of conditions affecting *Agrobacterium rhizogenes*-mediated transformation of soybean. Plant Cell Tissue Organ Cult 96:45–52

Cao X, Liu Q, Rowland LJ, Hammerschlag FA (1998) GUS expression in blueberry (*Vaccinium* spp.): factors influencing *Agrobacterium*-mediated gene transfer efficiency. Plant Cell Rep 18:266–270

Cao XL, Fordham I, Douglass L, Hammerschlag F (2003) Sucrose level influences micropropagation and gene delivery into leaves from in vitro propagated highbush blueberry shoots. Plant Cell Tissue Organ Cult 75:255–259

Cardoza V, Stewart N (2004) Brassica biotechnology: progress in cellular and molecular biology. In Vitro Cell Dev Biol Plant 40:542–551

Chee PP, Slightom JL (1991) Transfer and expression of cucumber mosaic-virus coat protein gene in the genome of *Cucumis sativus*. J Am Soc Hortic Sci 116:1098–1102

Chee PP, Slightom JL (1992) Transformation of cucumber tissues by microprojectile bombardment: identification of plants containing functional genes and non-functional transferred genes. Gene 118:255–260

Cheng M, Hu T, Layton J, Liu C-N, Fry JE (2003) Desiccation of plant tissues post- *Agrobacterium* infection enhances T-DNA delivery and increases stable transformation efficiency in wheat. In Vitro Cell Dev Biol Plant 39(6):595–604

Chevreau E, Mourgues F, Neveu M, Chevalier M (1997) Effect of gelling agents and antibiotics on adventitious bud regeneration from in vitro leaves of pear. In Vitro Cell Dev Biol Plant 33:173–179

Christey MC, Earle ED (1991) Regeneration of *Brassica oleracea* from peduncle explants. HortScience 26:1069–1072

Costa MS, Miguel C, Oliveira MM (2006) An improved selection strategy and the use of acetosyringone in shoot induction medium increase almond transformation efficiency by 100-fold. Plant Cell Tissue Organ Cult 85:205–209

Cseke LJ, Cseke SB, Podila GK (2007) High efficiency poplar transformation. Plant Cell Rep 26:1529–1538

Dandekar A, McGranahan G, Leslie C, Uratsu S (1989) *Agrobacterium*-mediated transformation of somatic embryos as a method for the production of transgenic plants. Methods Cell Sci 12:145–150

Dandekar AM, McGranahan GH, Vail PV, Uratsu SL, Leslie CA, Tebbets JS (1998) High levels of expression of full-length cryIA(c) gene from *Bacillus thuringiensis* in transgenic somatic walnut embryos. Plant Sci 131:181–193

Danilova SA (2007) The technologies for genetic transformation of cereals. Russ J Plant Physiol 54(5):569–581

Deblock M (1990) Factors influencing the tissue-culture and the *Agrobacterium*-mediated transformation of hybrid aspen and poplar clones. Plant Physiol 93:1110–1116

Debondt A, Eggermont K, Druart P, Devil M, Goderis I, Vanderleyden J, Broekaert WE (1994) *Agrobacterium*-mediated transformation of apple (*Malus xdomestica* Borkh.) An assessment of factors affecting gene transfer efficiency during early transformation steps. Plant Cell Rep 13:587–593

Desgagnes R, Laberge S, Allard G, Khoudi H, Castongua YY, Lapointe J, Michaud R, Vezina LP (1995) Genetic transformation of commercial breeding lines of alfalfa (*Medicago sativa*). Plant Cell Tissue Organ Cult 42:129–140

Dong JZ, Yang Z, Jia SR, Chua NH (1991) Transformation of melon (*Cucumis melo* L.) and expression from the cauliflower mosaic virus 35S promoter in transgenic melon plants. Biotechnology 9:858–863

Duchesne LC, Charest PJ (1991) Transient expression of the beta-glucuronidase gene in embryogenic callus of *Picea mariana* following microprojection. Plant Cell Rep 10:191–194

Duchesne LC, Lelu MA, Vonaderkas P, Charest PJ (1993) Microprojectile-mediated DNA delivery in haploid and diploid embryogenic cells of *Larix* spp. Canadian J. Forest Research 23:312–316

El Mansouri I, Mercado JA, Valpuesta V, LopezAranda JM, PliegoAlfaro F, Quesada MA (1996) Shoot regeneration and *Agrobacterium*-mediated transformation of *Fragaria vesca* L. Plant Cell Rep 15:642–646

Ellis DD, McCabe DE, McInnis S, Ramachandran R, Russell DR, Wallace KM, Martinell BJ, Roberts DR, Raffa KF, McCown BH (1993) Stable transformation of Picea glauca by particle acceleration. Biotechnology 11:84–89

Enriquez G, Vázquez R, Prieto D, De la Riva G, Selman D (1998) Herbicide-resistant sugarcane plants (*Saccarum sp.* L.) plants by *Agrobacterium* mediated transformation. Planta 206:20–27

Escalant JV, Teisson C, Cote F (1994) Amplified somatic embryogenesis from male flowers of triploid banana and plantain cultivars (*Musa* spp.). In Vitro Cell Dev Biol Plant 30:181–186

Escobar MA, Park JI, Polito VS, Leslie CA, Uratsu SL, McGranahan GH, Dandekar AM (2000) Using GFP as a scorable marker in walnut somatic embryo transformation. Ann Bot 85:831–835

Eudes F, Acharya S, Laroche A, Selinger LB, Cheng KJ (2003) A novel method to induce direct somatic embryogenesis and regeneration of fertile green cereal plants. Plant Cell Tissue Organ Cult 73:147–157

Feeney M, Bhagwat B, Mitchell JS, Lane WD (2007) Shoot regeneration from organogenic callus of sweet cherry (*Prunus avium* L.). Plant Cell Tissue Organ Cult 90:201–214

Finner JJ, McMullen MD (1991) Transformation of soybean via particle bombardment of embriogenic suspension culture tissue. In Vitro Cell Dev Biol Plant 27:175–182

Firoozabady E, Moy T, Tucker W, Robinson K, Gutterson N (1995) Efficient transformation and regeneration of carnation cultivars using *Agrobacterium*. Mol Breed 1:283–293

Folta K, Dhingra A, Howard L, Stewart P, Chandler C (2006) Characterization of LF9, an octoploid strawberry genotype selected for rapid regeneration and transformation. Planta 224:1058–1067

Franklin G, Sita GL (2003) *Agrobacterium tumefaciens*-mediated transformation of eggplant (*Solanum melongena* L.) using root explants. Plant Cell Rep 21:549–554

Frary A, Van Eck J (2005) Organogenesis from transformed tomato explants. Methods Mol Biol 286:141–150

Galperin M, Patlis L, Ovadia A, Wolf D, Zelcer A, Kenigsbuch D (2003) A melon genotype with superior competence for regeneration and transformation. Plant Breed 122:66–69

Gentile A, Monticelli S, Damiano C (2002) Adventitious shoot regeneration in peach (*Prunus persica* L. Batsch). Plant Cell Rep 20:1011–1016

Gercheva P, Zimmerman RH, Owens LD, Berry C, Hammerschlag FA (1994) Particle bombardment of apple leaf explants influences adventitious shoot formation. HortScience 29:1536–1538

Gonzalez ER, de Andrade A, Bertolo AL, Lacerda GC, Carneiro RT, Defavari VAP, Labate MTV, Labate CA (2002) Production of transgenic Eucalyptus grandis x E-urophylla using the sonication-assisted *Agrobacterium* transformation (SAAT) system. Funct Plant Biol 29:97–102

Graham J, Greig K, McNicol RJ (1996) Transformation of blueberry without antibiotic selection. Ann Appl Biol 128:557–564

Grant NJ, Hammatt N (2000) Adventitious shoot development from wild cherry (*Prunus avium* L.) leaves. New For 20:287–295

Guis M, Roustan JP, Dogimont C, Pitrat M, Pech JC (1998) Melon biotechnology. Biotechnol Genet Eng Rev 15:89–311

Guis M, Ben Amor M, Latché A, Pech JC, Roustan JCA (2000) reliable system for the transformation of cantaloupe Charentais melon (*Cucumis melo* L. var. *cantalupensis*) leading to a majority of diploid regenerants. Sci Hortic 84:91–99

Gómez-Lim MA, Litz R (2004) Genetic transformation of perennial tropical fruits. In Vitro Cell Dev Biol Plant 40:442–449

Grapin A, Ortiz JL, Lescot T, Ferriere N, Cote F (2000) Recovery and regeneration of embryogenic cultures from female flowers of false horn plantain. Plant Cell Tissue Organ Cult 61:237–244

Hammerschlag FA, Bauchan G, Scorza R (1985) Regeneration of peach plants from callus derived from immature embryos. Theor Appl Genet 70:248–251

Han KH, Gordon MP, Strauss SH (1997) High-frequency transformation of cottonwoods (genus *Populus*) by *Agrobacterium* rhizogenes. Can J For Res 27:464–470

Harcourt RL, Kyozuka J, Floyd RB, Bateman KS, Tanaka H, Decroocq V, Llewellyn DJ, Zhu X, Peacock WJ, Dennis ES (2000) Insect- and herbicide-resistant transgenic eucalypts. Mol Breed 6:307–315

Heeres P, Schippers-Rozenboom M, Jacobsen E, Visser RGF (2002) Transformation of a large number of potato varieties: genotype-dependent variation in efficiency and somaclonal variability. Euphytica 124:13–22

Ho CK, Chang SH, Tsay JY, Tsai CJ, Chiang VL, Chen ZZ (1998) *Agrobacterium tumefaciens*-mediated transformation of *Eucalyptus camaldulensis* and production of transgenic plants. Plant Cell Rep 17:675–680

Hsu CY, Liu Y, Luthe DS, Yuceer C (2006) Poplar FT2 shortens the juvenile phase and promotes seasonal flowering. Plant Cell 18:1846–1861

James DJ, Passey AJ, Baker SA (1994) Stable gene expression in transgenic apple tree tissues and segregation of transgenes in the progeny- preliminary evidence. Euphytica 77:119–121

James DJ, Passey AJ, Baker SA, Wilson FM (1996) Transgenes display stable patterns of expression in apple fruit and Mendelian segregation in the progeny. Biotechnology 14:56–60

James DJ, Passey AJ, Barbara DJ, Bevan M (1989) Genetic transformation for apple (*Malus pumila* Mill.) using a disarmed Ti-binary vector. Plant Cell Rep 7:658–661

Kaneyoshi J, Wabiko H, Kobayashi S, Tsuchiya T (2001) *Agrobacterium tumefaciens* AKE10-mediated transformation of an Asian pea pear, *Pyrus betulaefolia* Bunge: host specificity of bacterial strains. Plant Cell Rep 20:622–628

Keulemans J, Dewitte K (1994) Plant regeneration from cotyledons and embryonic axes in apple – sites of reaction and effect of pre-culture in the light. Euphytica 77:135–139

Kinouchi T, Endo R, Yamashita A, Satoh S (2006) Transformation of carnation with genes related to ethylene production and perception: towards generation of potted carnations with a longer display time. Plant Cell Tissue Organ Cult 86:27–35

Khanna H, Becker D, Kleidon J, Dale J (2004) Centrifugation assisted *Agrobacterium tumefaciens*-mediated transformation (CAAT) of embryogenic cell suspensions of banana (*Musa* spp. Cavendish and Lady finger AAB). Mol Breed 14:239–252

Klimaszewska K, Devantier Y, Lachance D, Lelu MA, Charest PJ (1997) *Larix laricina* (tamarack): Somatic embryogenesis and genetic transformation. Can J For Res – Rev Can Rech For 27:538–550

Knittel N, Gruber V, Hahne G, Lenee P (1994) Transformation of sunflower (*Helianthus annuus* L.): a reliable protocol. Plant Cell Rep 14:81–86

Ko TS, Lee S, Krasnyanski S, Korban SS (2003) Two critical factors are required for efficient transformation of multiple soybean cultivars: *Agrobacterium* strain and orientation of immature cotyledonary explant. Theor Appl Genet 107:439–447

Ko TS, Lee S, Farrand SK, Korban SS (2004) A partially disarmed vir helper plasmid, pKYRT1, in conjunction with 2, 4-dichlorophenoxyactic acid promotes emergence of regenerable transgenic somatic embryos from immature cotyledons of soybean. Planta 218:536–541

Kobayashi S, Ding CK, Nakamura Y, Nakajima I, Matsumoto R (2000) Kiwifruits (*Actinidia deliciosa*) transformed with a *Vitis* stilbene synthase gene produce piceid (resveratrol-glucoside). Plant Cell Rep 19:904–910

Koshita Y, Kobayashi S, Ishimaru M, Funamoto Y, Shiraishi M, Azuma A, Yakushiji H, Nakayama M (2008) An anthocyanin regulator from grapes, VlmybA1–2, produces reddish-purple plants. J Jpn Soc Hort Sci 77:33–37

Lambert C, Tepfer D (1992) Use of *Agrobacterium rhizogenes* to create transgenic apple trees having an altered organogenic response to hormones. Theor Appl Genet 85:105–109

Landi L, Mezzetti B (2005) TDZ, auxin and genotype effects on leaf organogenesis in *Fragaria*. Plant Cell Rep 25:281–288

Lemmetyinen J, Keinonen-Mettala K, Lannenpaa M, von Weissenberg K, Sopanen T (1998) Activity of the CaMV 35S promoter in various parts of transgenic early flowering birch clones. Plant Cell Rep 18:243–248

Levee V, Garin E, Klimaszewska K, Seguin A (1999) Stable genetic transformation of white pine (*Pinus strobus* L.) after cocultivation of embryogenic tissues with *Agrobacterium* tumefaciens. Mol Breed 5:429–440

Levee V, Lelu MA, Jouanin L, Cornu D, Pilate G (1997) *Agrobacterium* tumefaciens-mediated transformation of hybrid larch (*Larix kaempferi* x *L. decidua*) and transgenic plant regeneration. Plant Cell Rep 16:680–685

Lin XF, Zhang WB, Takechi K, Takio S, Ono K, Takano H (2005) Stable genetic transformation of *Larix gmelinii* L. by particle bombardment of zygotic embryos. Plant Cell Rep 24:418–425

Liu HK, Yang C, Wei ZM (2004) Efficient *Agrobacterium tumefaciens*-mediated transformation of soybeans using an embryonic tip regeneration system. Planta 219:1042–1049

Liu JR, Sink KC, Dennis FG (1983a) Adventive embryogenesis from leaf explants of apple seedlings. HortScience 18:871–873

Liu JR, Sink KC, Dennis FG (1983b) Plant regeneration from apple seedling explants and callus cultures. Plant Cell Tissue Organ Cult 2:293–304

Liu Q, Salih S, Hammerschlag F (1998) Etiolation of "Royal Gala" apple (*Malus* x *domestica* Borkh.) shoots promotes high-frequency shoot organogenesis and enhanced beta-glucuronidase expression from stem internodes. Plant Cell Rep 18:32–36

Loopstra CA, Stomp AM, Sederoff RR (1990) *Agrobacterium*-mediated DNA transfer in sugar pine. Plant Mol Biol 15:1–9

Lu C-Y, Nugent G, Wardley-Richardson T, Chandler SF, Young R, Dalling MJ (1991) *Agrobacterium-mediated* transformation of carnation *(Dianthus caryophyllus* L.). Biotechnology 9:864–868

Machado AD, Katinger H, Machado MLD (1994) Coat protein-mediated protection against plum pox virus in herbaceous model plants and transformation of apricot and plum. Euphytica 77:129–134

Magioli C, Rocha APM, de Oliveira DE Mansur E (1998) Efficient shoot organogenesis of eggplant (Solanum melongena L.) induced by thidiazuron. Plant Cell Rep 17:661–663

Maheswaran G, Welander M, Hutchinson JF, Graham MW, Richards D (1992) Transformation of apple rootstock M26 with Agrobacterium tumefaciens. J Plant Physiol 139:560–568

Mante S, Morgens PH, Scorza R, Cordts JM, Callahan AM (1991) Agrobacterium-mediated transformation of plum (Prunus domestica L.) hypocotyl slices and regeneration of transgenic plants. Biotechnology 9:853–857

Mante S, Scorza R, Cordts JM (1989) Plant regeneration from cotyledons of Prunus persica, Prunus domestica and Prunus cerasus. Plant Cell Tiss Organ Cult 19:1–11

Matsuda N, Gao M, Isuzugawa K, Takashina T, Nishimura K (2005) Development of an Agrobacterium-mediated transformation method for pear (Pyrus communis L.) with leaf-section and axillary shoot-meristem explants. Plant Cell Rep 24:45–51

May GD, Afza R, Mason HS, Wiecko A, Novak FJ, Arntzen CJ (1995) Generation of transgenic banana (Musa acuminata) plants via Agrobacterium-mediated transformation. Biotechnology 13:486–492

McCabe DE, Swain WF, Martinell BJ, Christou P (1988) Stable transformation of soybean (Glycine max) by particle acceleration. Biotechnology 6:923–926

McGranahan GH, Leslie CA, Uratsu SL, Dandekar AM (1990) Improved efficiency of the walnut somatic embryo gene transfer system. Plant Cell Rep 8:512–516

McGranahan GH, Leslie CA, Uratsu SL, Martin LA, Dandekar AM (1988) Agrobacterium-mediated transformation of walnut somatic embryos and regeneration of transgenic plants. Biotechnology 6:800–804

Metz T, Dixit R, Goldsmith J, Roush R, Earle E (1994) Production of transgenic Brassica oleracea expressing Bacillus thuringiensis insecticidal crystal protein genes. Eucarpia Cruciferae Newsl 16:63–64

Miguel CM, Oliveira MM (1999) Transgenic almond (Prunus dulcis Mill.) plants obtained by Agrobacterium-mediated transformation of leaf explants. Plant Cell Rep 18:387–393

Mohri T, Mukai Y, Shinohara K (1997) Agrobacterium tumefaciens-mediated transformation of Japanese white birch (Betula platyphylla var. japonica). Plant Sci 127:53–60

Molinier J, Thomas C, Brignou M, Hahne G (2002) Transient expression of ipt gene enhances regeneration and transformation rates of sunflower shoot apices (Helianthus annuus L.). Plant Cell Rep 21:251–256

Moralejo M, Rochange F, Boudet AM, Teulieres C (1998) Generation of transgenic Eucalyptus globulus plantlets through Agrobacterium tumefaciens mediated transformation. Aust J Plant Physiol 25:207–212

Mourgues F, Chevreau E, Lambert C, deBondt A (1996) Efficient Agrobacterium-mediated transformation and recovery of transgenic plants from pear (Pyrus communis L). Plant Cell Rep 16:245–249

Msikita W, Sayre RT, White VL, Marks J (2002) Influence of explant source, and light on efficiency of Agrobacterium-mediated transformation of cassava. In: 5th International Meeting of the Cassava Biotecnology Network, Nov 4-9, St. Louis, USA

Nishibayashi S, Hayakawa T, Nakajima T, Suzuki M, Kaneko H (1996) CMV protection in transgenic cucumber plants with an introduced CMV-O cp gene. Theor App Genet 93:672–678

Nishikawa F, Endo T, Shimada T, Fujii H, Shimizu T, Omura M, Ikoma Y (2007) Increased CiFT abundance in the stem correlates with floral induction by low temperature in Satsuma mandarin (Citrus unshiu Marc.). J Exp Bot 58:3915–3927

Nontaswatsri C, Fukai S, Goi M (2004) Revised cocultivation produce effective Agrobacterium-mediated genetic transformation of carnation. Plant Sci 166:59–68

Oridate T, Atsumi H, Ito S, Araki H (1992) Genetic difference in somatic embryogenesis from seeds in melon (Cucumis melo L.). Plant Cell Tissue Organ Cult 29:27–30

Padilla IMG, Golis A, Gentile A, Damiano C, Scorza R (2006) Evaluation of transformation in peach (*Prunus persica*) explants using green fluorescent protein (GFP) and beta-glucuronidase (GUS) reporter genes. Plant Cell Tiss Organ Cult 84:309–314

Parrott WA, Hoffman LM, Hildebrand DF, Williams EG, Collins GB (1989a) Recovery of primary transformants of soybean. Plant Cell Rep 7:615–617

Parrott WA, Williams EG, Hildebrand DF, Collins GB (1989b) Effect of genotype on somatic embryogenesis from immature cotyledons of soybean. Plant Cell Tissue Organ Cult 16:15–21

Parsons TJ, Sinkar VP, Stettler RF, Nester EW, Gordon MP (1986) Transformation of poplar by *Agrobacterium tumefaciens*. Biotechnology 4:533–536

Passey AJ, Barrett KJ, James DJ (2003) Adventitious shoot regeneration from seven commercial strawberry cultivars (*Fragaria* x *ananassa* Duch.) using a range of explant types. Plant Cell Rep 21:397–401

Pena L, Cervera M, Juarez J, Ortega C, Pina JA, Duranvila N, Navarro L (1995) High efficiency *Agrobacterium*-mediated trasnforamtion and regeneration of citrus. Plant Sci 104:183–191

Perez-Clemente RM, Perez-Sanjuan A, Garcia-Ferriz L, Beltran JP, Canas LA (2004) Transgenic peach plants (*Prunus persica* L.) produced by genetic transformation of embryo sections using the green fluorescent protein (GFP) as an in vivo marker. Mol Breed 14:419–427

Perez-Tornero O, Egea J, Vanoostende A, Burgos L (2000) Assessment of factors affecting adventitious shoot regeneration from in vitro cultured leaves of apricot. Plant Sci 158:61–70

Petri C, Alburquerque N, Burgos L (2005a) The effect of aminoglycoside antibiotics on the adventitious regeneration from apricot leaves and selection of nptII-transformed leaf tissues. Plant Cell Tissue Organ Cult 80:271–276

Petri C, Alburquerque N, Garcia-Castillo S, Egea J, Burgos L (2004) Factors affecting gene transfer efficiency to apricot leaves during early *Agrobacterium*-mediated transformation steps. J Hortic Sci Biotechnol 79:704–712

Petri C, Alburquerque N, Perez-Tornero O, Burgos L (2005b) Auxin pulses and a synergistic interaction between polyamines and ethylene inhibitors improve adventitious regeneration from apricot leaves and *Agrobacterium*-mediated transformation of leaf tissues. Plant Cell Tissue Organ Cult 82:105–111

Petri C, Wang H, Alburquerque N, Faize M, Burgos L (2008a) *Agrobacterium*-mediated transformation of apricot (*Prunus armeniaca* L.) leaf explants. Plant Cell Rep 27:1317–1324

Petri C, Webb K, Hily JM, Dardick C, Scorza R (2008b) High transformation efficiency in plum (*Prunus domestica* L.): a new tool for functional genomics studies in *Prunus spp*. Mol Breed 22:581–591

Piqueras A, Debergh PC (1999) Morphogenesis in micropropagation. In: Soh WY, Bhojwani SS (eds) Morphogenesis in Plant Tissue Cultures. Kluwer Acad Publ, Dordrecht, The Netherlands, pp 443–462

Polito VS, McGranahan G, Pinney K, Leslie C (1989) Origin of somatic embryos from repetitively embryogenic cultures of walnut (*Jugulans regia* L) – Implications for Agrobacterium-mediated transformation. Plant Cell Rep 8:219–221

Pooler MR, Scorza R (1995) Regeneration of peach (*Prunus persica* L) rootstock cultivars from cotyledons of mature stored seed. HortScience 30:355–356

Poulsen GB (1996) Genetic transformation of *Brassica*. Plant Breed 115:209–225

Puite KJ, Schaart JG (1996) Genetic modification of the commercial apple cultivars Gala, Golden Delicious and Elstar via an *Agrobacterium* tumefaciens-mediated transformation method. Plant Sci 119:125–133

Puonti-Kaerlas J, Li HQ, Sautter C, Potrykus I (1997) Production of tansgenic cassava (*Manihot esculaenta* Crantz) via organogenesis and *Agrobacterium*-mediated transformation. Afr J Root Tuber Crops 2:181–186

Qiu DL, Diretto G, Tavarza R, Giuliano G (2007) Improved protocol for *Agrobacterium*-mediated transformation of tomato and production of transgenic plants containing carotenoid biosynthetic gene CsZCD. Sci Hortic 112:172–175

Raemakers C, Jacobsen E, Visser RGF (1995) Secondary somatic embryogenesis and applications in plant breeding. Euphytica 81:93–107

Raharjo SHT, Hernandez MO, Zhang YY, Punja ZK (1996) Transformation of pickling cucumber with chitinase-encoding genes using *Agrobacterium tumefaciens*. Plant Cell Rep 15:591–596

Rao KS, Rohini VK (1999) *Agrobacterium*-mediated transformation of sunflower (*Helianthus annus* L.): A simple protocol. Ann Bot 83:347–354

Ramesh SA, Kaiser BN, Franks T, Collins G, Sedgley M (2006) Improved methods in *Agrobacterium*-mediated transformation of almond using positive (mannose/pmi) or negative (kanamycin resistance) selection-based protocols. Plant Cell Rep 25:821–828

Ravelonandro M, Scorza R, Bachelier JC, Labonne G, Levy L, Damsteegt V, Callahan AM, Dunez J (1997) Resistance of transgenic *Prunus domestica* to plum pox virus infection. Plant Dis 81:1231–1235

Rech EL, Vianna GR, Aragao FJL (2008) High-efficiency transformation by biolistics of soybean, common bean and cotton transgenic plants. Nat Protocol 3:410–418

Reynoird JP, Mourgues F, Norelli J, Aldwinckle HS, Brisset MN, Chevreau E (1999) First evidence for improved resistance to fire blight in transgenic pear expressing the attacin E gene from *Hyalophora cecropia*. Plant Sci 149:23–31

Rout JR, Lucas WJ (1996) Characterization and manipulation of embryogenic response from in vitro cultured mature inflorescences of rice (*Oryza sativa* L.) Planta 198:127–138

Ruginl E, Pellegrineschi A, Mencuccini M, Mariotti D (1991) Increase of rooting ability in the woody species kiwi (*Actinidia deliciosa* A. Chev) by transformation with *Agrobacterium rhizogenes* Rol genes. Plant Cell Rep 10:291–295

Samac DA (1995) Strain specificity in transformation of alfalfa by *Agrobacterium tumefaciens*. Plant Cell Tissue Organ Cult 43:271–277

Samac DA, Austin-Phillips S (2006) Alfalfa (*Medicago sativa* L.). In: Wang K (ed) *Agrobacterium* Protocols 343. New Jersey, Humana Press, Totowa, pp 301–311

Samac DA, Temple SJ (2004) Development and utilization of transformation in *Medicago* species. In: Liang GH, Skinner D (eds) Genetically Modified Crops: Their Development. Uses and Risks. Haworth Press, New York, pp 165–2002

Sartoretto LM, Cid LPB, Brasileiro ACM (2002) Biolistic transformation of *Eucalyptus grandis* x *E. urophylla* callus. Funct Plant Biol 29:917–924

Schrammeijer B, Sijmons PC, Vandenelzen PJM, Hoekema A (1990) Meristem transformation of sunflower via *Agrobacterium*. Plant Cell Rep 9:55–60

Scorza R, Morgens PH, Cordts JM, Mante S, Callahan AM (1990) Agrobacterium-mediated transformation of peach (*Prunus persica* Batsch. L) leaf segments, immature embryos and long-term embryogenic callus. In Vitro Cell Dev Biol Plant 26:829–834

Sederoff R, Stomp AM, Chilton WS, Moore LW (1986) Gene transfer into loblolly pine by *Agrobacterium tumefaciens*. Biotechnology 4:647–649

Seong ES, Song KJ (2008) Factors affecting the early gene transfer step in the development of transgenic "Fuji" apple plants. Plant Growth Regul 54:89–95

Shin DI, Podila GK, Huang YH, Karnosky DF (1994) Transgenic larch expressing genes for herbicide and insect resistance. Can J For Res – Rev Can Rech For 24:2059–2067

Shulaev V, Korban SS, Sosinski B, Abbott AG, Aldwinckle HS, Folta KM, Iezzoni A, Main D, Arus P, Dandekar AM, Lewers K, Brown SK, Davis TM, Gardiner SE, Potter D, Veilleux RE (2008) Multiple models for Rosaceae genomics. Plant Physiol 147:985–1003

Sink KC, Song QG (2004) Efficient shoot regeneration, transient expression studies, and *Agrobacterium*-mediated transformation of blueberry (*Vaccinium corymbosum* L.). In Vitro Cell Dev Biol Plant 40:65A–65A

Siritunga D, Arias-Garzón D, White W, Sayre R (2004) Over-expression of hydroxynitrile lyase in transgenic cassava (*Manihot esculenta*, Crantz) roots accelerates embryogenesis. Plant Biotechnol J 2:37–43

Smigocki AC, Hammerschlag FA (1991) Regeneration of plants from peach embryo cells infected with a shooty mutant strain of *Agrobacterium*. J Am Soc Hortic Sci 116:1092–1097

Song GQ, Sink KC (2004) *Agrobacterium tumefaciens*-mediated transformation of blueberry (*Vaccinium corymbosum* L.). Plant Cell Rep 23:475–484

Song GQ, Sink KC (2006) Transformation of Montmorency sour cherry (*Prunus cerasus* L.) and Gisela 6 (*P.cerasus* x *P. canescens*) cherry rootstock mediated by *Agrobacterium tumefaciens*. Plant Cell Rep 25:117–123

Song GQ, Sink KC, Callow PW, Baughan R, Hancock JF (2008) Evaluation of a herbicide-resistant trait conferred by the bar gene driven by four distinct promoters in transgenic blueberry plants. J Am Soc Hortic Sci 133:605–611

Song JY, Lu SF, Chen ZZ, Lourenco R, Chiang VL (2006) Genetic transformation of *Populus trichocarpa* genotype Nisqually-1: A functional genomic tool for woody plants. Plant Cell Physiol 47:1582–1589

Spokevicius AV, Van Beveren K, Leitch MM, Bossinger G (2005) *Agrobacterium*-mediated in vitro transformation of wood-producing stem segments in eucalypts. Plant Cell Rep 23:617–624

Suezawa K, Matsuta N, Omura M, Yamaki S (1988) Plantlet formation from cell suspensions of kiwifruit (*Actinidia chinensis* Planch.). Sci Hortic 37:123–128

Sun HJ, Uchii S, Watanabe S, Ezura H (2006) A highly efficient transformation protocol for Micro-Tom, a model cultivar for tomato functional genomics. Plant Cell Physiol 47:426–431

Tabei Y, Kitade S, Nishizawa Y, Kikuchi N, Kayano T, Hibi T, Akutsu K (1998) Transgenic cucumber plants harboring a rice chitinase gene exhibit enhanced resistance to gray mold (*Botrytis cinerea*). Plant Cell Rep 17:159–164

Tang H, Zen R, Krczal G (2000) An evaluation of antibiotics for the elimination of *Agrobacterium tumefaciens* from walnut somatic embryos and for the effects on the proliferation of somatic embryos and regeneration of transgenic plants. Plant Cell Rep 19:881–887

Tripathi S, Suzuki JY, Ferreira SA, Gonsalves D (2008) Papaya ringspot virus-P: characteristics, pathogenicity, sequence variability and control. Mol Plant Pathol 9:269–280

Trulson AJ, Simpson RB, Shahin EA (1986) Transformation of cucumber (*Cucumis sativus* L.) plants with *Agrobacterium rhizogenes*. Theo Appl Genet 73:11–15

Tsai CJ, Podila GK, Chiang VL (1994) *Agrobacterium* mediated transformation of quaking aspen (Populus tremuloides) and regeneration of transgenic plants. Plant Cell Rep 14:94–97

Tzfira T, BenMeir H, Vainstein A, Altman A (1996) Highly efficient transformation and regeneration of aspen plants through shoot-bud formation in root culture. Plant Cell Rep 15:566–571

Uematsu C, Murase M, Ichikawa H, Imamura J (1991) *Agrobacterium* mediated transformation and regeneration of kiwi fruit. Plant Cell Rep 10:286–290

Urtubia C, Devia J, Castro A, Zamora P, Aguirre C, Tapia E, Barba P, Dell'Orto P, Moynihan MR, Petri C, Scorza R, Prieto H (2008) *Agrobacterium*-mediated genetic transformation of *Prunus salicina*. Plant Cell Rep 27:1333–1340

Valjakka M, Aronen T, Kangasjarvi J, Vapaavuori E, Haggman H (2000) Genetic transformation of silver birch (*Betula pendula*) by particle bombardment. Tree Physiol 20:607–613

Van Altvorst AC, Riksen T, Koehorst H, Dons HJM (1995) Transgenic carnations obtained by *Agrobacterium tumefaciens*-medated transformation of leaf explants. Transgen Res 4:105–113

Van Beveren KS, Spokevicius AV, Tibbits J, Qing W, Bossinger G (2006) Transformation of cambial tissue in vivo provides an efficient means for induced somatic sector analysis and gene testing in stems of woody plant species. Funct Plant Biol 33:629–638

Van Eck J, Kirk DD, Walmsley AM (2006) Tomato (*Lycopersicum esculentum*). In: Wang K (ed) *Agrobacterium* Protocols 343. Humana Press, Totowa, New Jersey, pp 459–473

Van Eck J, Snyder A (2006) Eggplant (*Solanum melongena* L.). In: Wang K (ed) *Agrobacterium* Protocols 343. New Jersey, Humana Press, Totowa, pp 439–447

von Arnold S, Clapham D, Egertsdotter U, Mo LH (1996) Somatic embryogenesis in conifers – a case study of induction and development of somatic embryos in *Picea abies*. Kluwer, Dordrecht, pp 3–9

Visser RGF, Jacobsen E, Hesselingmenders A, Schans ML, Witholt B, Freenstra WJ (1989) transformation of homocigous diploid potato with an *Agrobacterium tumefaciens* binary vector system by adventious shoot regeneration on leaf and stem segments. Plant Mol Biol 121:329–337

Walter C, Grace LJ, Donaldson SS, Moody J, Gemmell JE, van der Maas S, Kvaalen H, Lonneborg A (1999) An efficient biolistic transformation protocol for *Picea abies* embryogenic tissue and regeneration of transgenic plants. Can J For Res – Rev Can Rech For 29:1539–1546

Walter C, Grace LJ, Wagner A, White DWR, Walden AR, Donaldson SS, Hinton H, Gardner RC, Smith DR (1998) Stable transformation and regeneration of transgenic plants of Pinus radiata. Plant Cell Rep 17:460–468

Wang TC, Ran YD, Atkinson RG, Gleave AP, Cohen D (2006) Transformation of *Actinidia eriantha*: A potential species for functional genomics studies in *Actinidia*. Plant Cell Rep 25:425–431

Woodward B, Puonti-Kaerlas J (2001) Somatic embryogenesis from floral tissue of cassava (*Maihot esculenta* Krantz). Euphytica 120:1–6

Yadav JS, Rajam MV (1998) Temporal regulation of somatic embryogenesis by adjusting cellular polyamine content in eggplant. Plant Physiol 116:617–625

Yamakawa Y, Chen LH (1996) *Agrobacterium* rhizogenes-mediated transformation of kiwifruit (*Actinidia deliciosa*) by direct formation of adventitious buds. J Jpn Soc Hort Sci 64:741–747

Yancheva SD, Golubowicz S, Fisher E, Lev-Yadun S, Flaishman MA (2003) Auxin type and timing of application determine the activation of the developmental program during in vitro organogenesis in apple. Plant Sci 165:299–309

Yancheva SD, Shlizerman LA, Golubowicz S, Yabloviz Z, Perl A, Hanania U, Flaishman MA (2006) The use of green fluorescent protein (GFP) improves *Agrobacterium*-mediated transformation of "Spadona" pear (*Pyrus communis* L.). Plant Cell Rep 25:183–189

Yazawa M, Suginuma C, Ichikawa K, Kamada H, Akihama T (1995) Regeneration of transgenic plants from hairy root of kiwi fruit (*Actinidia deliciosa*) induced by Agrobacterium rhizogenes. Breed Sci 45:241–244

Ye XJ, Brown SK, Scorza R, Cordts J, Sanford JC (1994) Genetic transformation of peach tissues by particle bombardment. J Am Soc Hortic Sci 119:367–373

Yepes LM, Aldwinckle HS (1994) Factors that affect leaf regeneration efficiency in apple, and effect of antibiotics in morphogenesis. Plant Cell Tissue Organ Cult 37:257–269

Yin Z, Bartoszewski G, Szwacka M, Malepszy S (2005) Cucumber transformation methods-the review. Biotechnologia 1:95–113

Zhang P, Puonti-kaerlas J (2000) PIG-mediated cassava transformation using positive and negative selection. Plant Cell Rep 19:1041–1048

Chapter 3
Gene Transfer Methods

Seedhabadee Ganeshan and Ravindra N. Chibbar

3.1 Introduction

The ability to alter the genetic composition of a plant is fundamental to crop improvement and development of new cultivars with desirable characters. Plant breeders have utilized the naturally occurring genetic variability in existing germplasm to develop new lines by sexual hybridization. In the absence of natural variation for a trait, chemical and radiation mutagenesis was used to create genetic variability for use in the development of varieties with desirable traits. In another approach, genes for superior traits in close relatives were identified and recombined by wide hybridization, thereby generating interspecific or intergeneric hybrids between the donor and target species. However, all these chromosome-mediated gene transfers need sexual hybridizations. Sexual compatibility and chromosome pairing are key components for the introgression of a desired trait. To overcome limited sexual compatibility, embryo rescue using in vitro culture techniques was used to induce genetic variability for desirable traits (Raghavan 1986). The development of protoplast culture and somatic cell hybridization was one of the first examples to create genetic variability by asexual means. Furthermore, in vitro culture of plant cells in suboptimal conditions was found to induce genetic variations termed somaclonal variation, subsequently exhibiting an altered phenotype (Larkin and Scowcroft 1981). The *Agrobacterium tumefaciens*-mediated integration of foreign DNA into a cell's nuclear genome and production of a transgenic plant in which the inserted gene was inherited following Mendelian genetics was the ultimate method to create genetic variation across species, irrespective of genetic proximity or sexual compatibility (Otten et al. 1981).

R.N. Chibbar (✉)

Department of Plant Sciences, College of Agriculture and Bioresources, University of Saskatchewan, Saskatoon S7N 5A8, Saskatchewan, Canada

e-mail: ravi.chibbar@usask.ca

C. Kole et al. (eds.), *Transgenic Crop Plants*,
DOI 10.1007/978-3-642-04809-8_3, © Springer-Verlag Berlin Heidelberg 2010

3.2 Gene Delivery Methods

The availability and the versatility of different plant DNA delivery methods have become even more pertinent in recent years with the availability of gene sequence data and the need for functional analysis of cloned and sequenced genes. Although the existence of a vast depository of sequences of known functions in the databases can be used for the prediction of gene function of unknown sequences based on homology, limitations are often encountered with respect to precisely determining functions of the genes of interest (Sessions et al. 2002). Therefore, the availability of high-throughput gene transfer systems for economically important crop plants has become highly desirable to expedite gene function analysis. While such trans-formation systems are routine in model systems such as *Arabidopsis*, for many economically important crop plants, extensive effort is still required to achieve routine, high efficiency transformation. It is, therefore, imperative to understand the practicality, usefulness, and versatility of the different gene transfer methods that can contribute to specific transformation projects. This is particularly critical for species which are presently relatively recalcitrant to genetic transformation.

Since the first report of *Agrobacterium*-mediated delivery of genes to produce transgenic plants in the early 1980s, a number of other gene delivery methods have been reported in the literature (Table 3.1). Some of these have been successfully used to produce transgenic plants for commercial applications and/or basic studies to understand plant growth and development. However, there still remains a challeng-ing task ahead to find the best suited transformation method for various plant species. There is also a need to find cost-effective methods for transformation so that laboratories with limited funding resources are capable of conducting such research. Furthermore, due to patenting issues currently covering some of the transformation methods such as *Agrobacterium*, methods need to be developed for availability in the public domain. Notwithstanding these issues, gene transfer to plants overall appears to be simple, but requires careful interphasing of several different systems.

Thus, simply iterated, gene transfer to plants involves the integration of three components, which include a tissue culture system (discussed in Chap. 2), a DNA delivery system, and a vehicle for carrying the DNA to be transferred (Fig. 3.1). An ideal gene delivery method transfers the carrier DNA to a cell with minimum damage to the recipient tissue, allows for stable transgene integration into the recipient genome and sustained cell proliferation of recipient tissue for subsequent regeneration of a transgenic plant. The commonly used gene transfer methods can be classified into several different groups. In this chapter, two broad groups of gene delivery methods will be discussed: (a) biological, and (b) physical. The biological group includes a living organism, such as a bacterium or virus, to deliver a gene to a host cell. The physical methods include direct DNA delivery techniques, which use a chemical alteration or physical force such as pressure or electric discharge to deliver the vector DNA into a host cell. Recently, a third group of techniques, which use a combination of biological and physical techniques to deliver the vector DNA, has been developed. These techniques use the desirable features of both the groups to achieve optimal delivery of vector DNA into host cells. Essentially, the methods

Table 3.1 Summary of important gene transfer methods

Approach	Brief description	Reference
Physical delivery methods		
Bioactive beads	Calcium alginate microbead-immobilized DNA molecules for uptake by protoplasts	Sone et al. (2002), Liu et al. (2004), Murakawa et al. (2008a,b)
Electric discharge	Gold particle acceleration via electric discharge to target cells (ACCELL™ technology)	McCabe and Christou (1993)
Electrofusion	Fusion of protoplasts by electric pulses	Morikawa et al. (1986b)
Electrophoresis	DNA transferred electrophoretically to embryos	Ahokas (1989), Griesbach (1994)
Electroporation	Electric pulses delivered to protoplasts, mesophyll cells, intact tissues	Fromm et al. (1985), Lorz et al. (1985), Morikawa et al. (1986a), Li et al. (1991), Arencibia et al. (1995)
Laser micropuncture	Laser-mediated perforations in cells and tissues allow uptake of exogenous DNA	Guo et al. (1995), Badr et al. (2005), Kajiyama et al. (2008)
Liposomes	Liposome containing DNA taken up by protoplasts	Deshayes et al. (1985)
Macroinjection	Injection of DNA into floral tillers	de la Pena et al. (1987)
Microinjection	Injection of DNA into protoplasts, intact cells such as callus and embryoids	Griesbach (1983), Morikawa and Yamada (1985), Crossway et al. (1986), Griesbach (1987)
Nanoparticles	Honeycomb mesoporous silica nanoparticles containing DNA taken up by protoplasts	Torney et al. (2007)
Particle bombardment	Acceleration of microprojectiles such as tungsten and gold	Klein et al. (1988a), Sanford (1990), Vasil et al. (1991)
PEG-mediated DNA uptake	Protoplasts	Uchimiya et al. (1986)
Pollen	Pollen/plant DNA mixture used for self-fertilization in maize	Ohta (1986)
Pollen tube pathway	DNA applied to cut styles just after pollination and flows along pollen tube to the ovule	Luo and Wu (1989)
Silicon carbide whiskers	Vigorous shaking or vortexing of silicon carbide fibers with DNA and suspension cells, embryogenic callus	Kaeppler et al. (1992), Frame et al. (1994), Petolino et al. (2000)
Somatic cell hybrids	Fusion between protoplasts and generation of somatic hybrids	Carlson et al. (1972), Kao et al. (1974), Wallin et al. (1974)
Biological delivery methods		
Agrobacterium rhizogenes	Carrot disks, tobacco, and morning glory stem segments inoculated	Chilton et al. (1982), Tepfer (1984)
Agrobacterium tumefaciens	Tobacco stem segments, leaf disks	Barton et al. (1983), Fraley and Horsch (1983)

(continued)

Table 3.1 (continued)

Approach	Brief description	Reference
Agroinfection	Turnip leaves inoculated with *Agrobacterium* containing viral DNA engineered into T-DNA	Grimsley et al. (1986)
In planta	By-passes tissue culture, and *Agrobacterium* suspension applied by vacuum infiltration or dip to floral parts, meristems, embryo axis	Bechtold et al. (1993), Clough and Bent (1998)
Other microorganisms	*Rhizobium*, *Sinorhizobium*, and *Mesorhizobium* strains engineered with disarmed Ti plasmid and used to inoculate tobacco and *Arabidopsis*	Broothaerts et al. (2005)
Combination of physical and biological methods		
Agrolistics	Combination of biolistic DNA delivery and *Agrobacterium*, wherein *vir*D1 and *vir*D2 genes are delivered biolistically and cause in planta excision of T-DNA	Hansen and Chilton (1996)
Sonication assisted	Plant tissue subjected to short periods of ultrasound in the presence of *Agrobacterium*	Trick and Finer (1997)

Fig. 3.1 A three-component system required for successful production of transgenic plants. The three components integrate a tissue culture system, a DNA delivery system, and a vector carrying the gene of interest

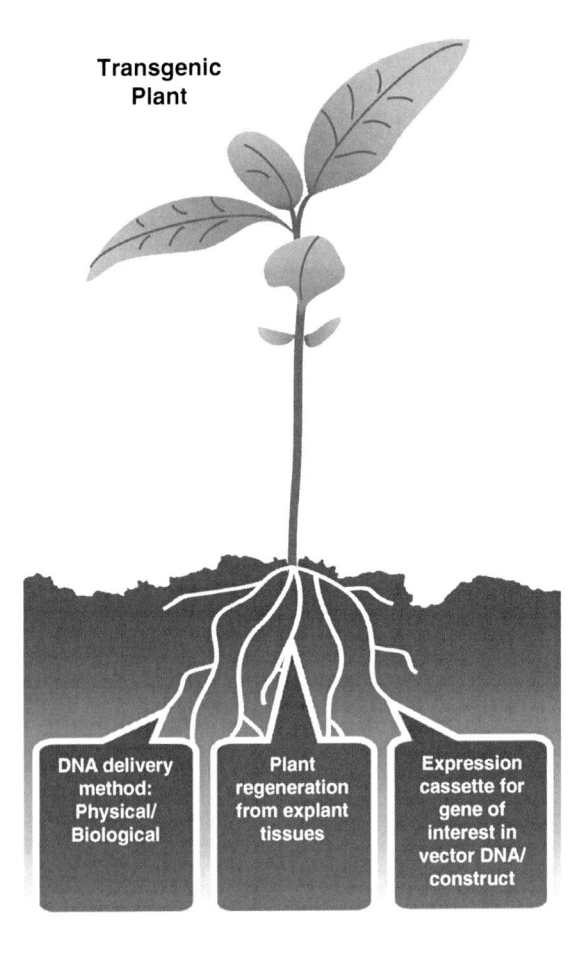

Transgenic Plant

DNA delivery method: Physical/ Biological

Plant regeneration from explant tissues

Expression cassette for gene of interest in vector DNA/ construct

follow a general plan (Fig. 3.2). The initial step is the DNA delivery to cells or tissues, followed by culture and selection to allow only those cells and tissues having a marker gene (e.g., antibiotic or herbicide resistance gene) to survive and proliferate further. Subsequently, plants are regenerated from the surviving cells, rooted and hardened in the soil. Such primary transformants are thereafter used for molecular analyses for determination of integration and copy number of the transgenes of interest.

3.2.1 Biological Methods

3.2.1.1 Agrobacterium-Mediated Gene Transfer

The first report of a transgenic plant was as a result of *A. tumefaciens*-mediated delivery of foreign DNA (Otten et al. 1981; Barton and Chilton 1983; Fraley and Horsch 1983). Since then, *Agrobacterium* has been used to deliver DNA in several

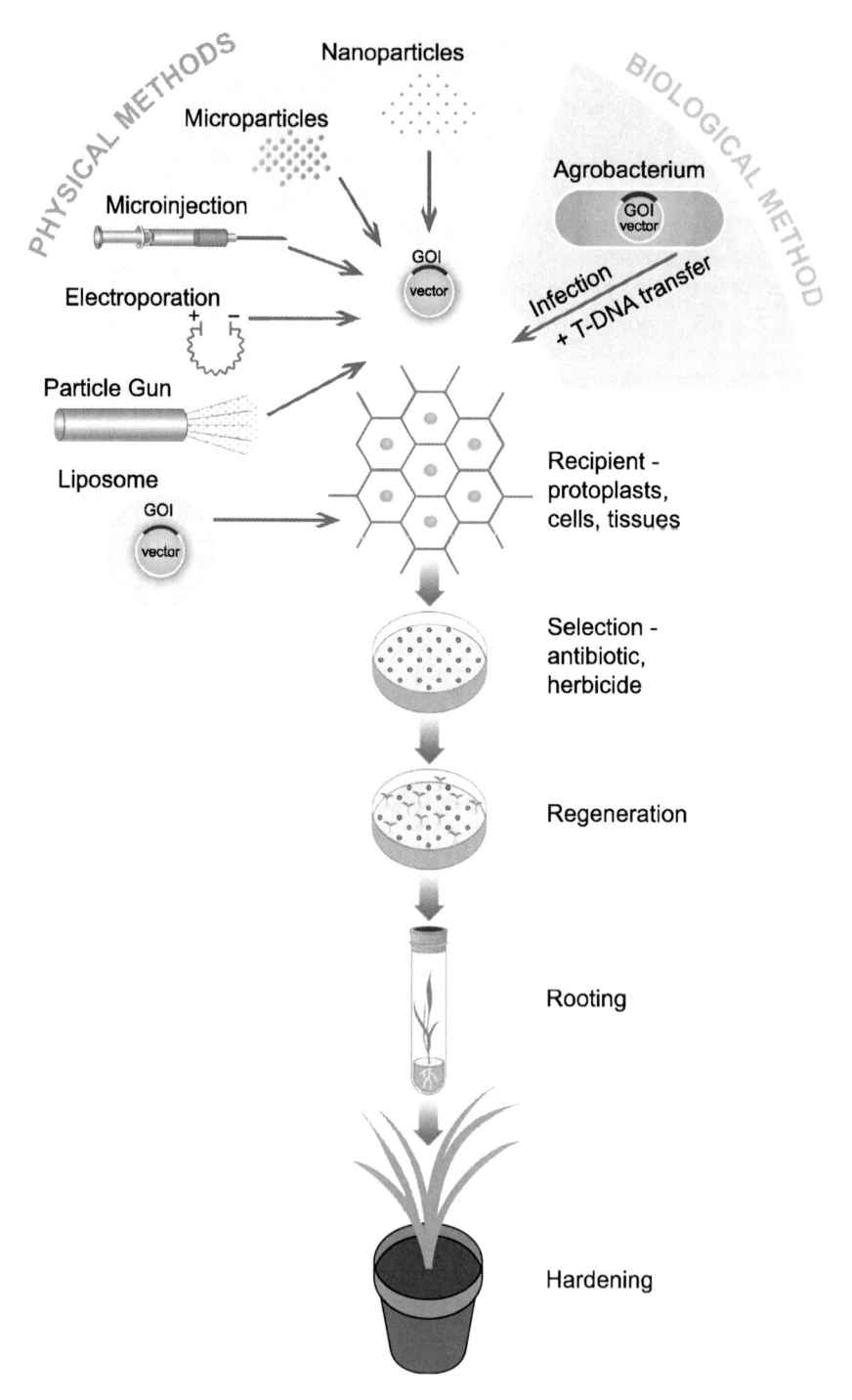

Fig. 3.2 Schematic depiction of physical and biological methods commonly used for gene transfer to plants and the general steps leading to the growth of a putative transgenic plant

plants to understand the basic principles of plant biology and for developing crop cultivars with improved agronomic traits including enhanced crop quality traits. *Agrobacterium* as a genus can transfer DNA to a very broad range of organisms including plants (both monocotyledonous and dicotyledonous angiosperms), gymnosperms, fungi, and recently to human cells (Kunik et al. 2001). Among the five known *Agrobacterium* species, *A. radiobacter* is avirulent, three species are pathogenic and cause crown galls (*A. tumefaciens*, *A. rubi*, and *A. vitis*), while *A. rhizogenes* is responsible for hairy root disease. This chapter will focus on *A. tumefaciens* and *A. rhizogenes*, the two most common species used for DNA delivery to produce transgenic plants.

A. *tumefaciens* causes crown galls on a large number of dicotyledonous and a limited number of monocotyledonous plant species, and gymnosperms (Levee et al. 1997; Levee et al. 1999). The *A. tumefaciens*-induced tumors can be grown in vitro in simple culture media without the bacterium and any added plant growth hormones (Braun 1941; Braun and Laskaris 1942; Braun and Laskaris 1943). The crown gall disease is caused by the transfer into plant cells of a specific DNA fragment (T-DNA), which originates from a tumor-inducing (Ti) plasmid present within the bacterium. The T-DNA becomes integrated into the plant nuclear genome, and expression of the genes present on the T-DNA gives rise to the crown gall phenotype. The T-DNA carries some of the genes responsible for auxin and cytokinin synthesis, which result in the rapid and autonomous growth of crown gall tissue in the absence of added plant hormones. The other T-DNA genes are responsible for the synthesis of specific amino acids or sugars which are normally not present in plant cells. These plant tumor-specific compounds are collectively known as opines, but classified as octopines, nopaline, agropine, succinamopine, or chrysopine produced by specific Ti plasmids. These opines can be metabolized by the respective *Agrobacterium* strains but not by other soil microorganisms thus creating a niche environment and host strain specificity, which results in a very conducive environment for *Agrobacterium*-mediated genetic modification of plant cells. *A. rhizogenes* causes hairy root plant disease, which is characterized by the rapid proliferation of roots at the infection site. *A. rhizogenes* transfers an Ri plasmid in a manner similar to the Ti plasmid of *A. tumefaciens*.

Agrobacterium-mediated T-DNA transfer to plants is governed by three basic genetic elements. A native Ti or Ri plasmid generally varies in size from 200 to 800 kb (Goodner et al. 2001; Wood et al. 2001) and usually contains one T-DNA region, which is usually about one-tenth (10–30 kb in size) of the total plasmid. In some instances, a Ti plasmid may contain multiple T-DNA regions (Merlo et al. 1980). The first major element is defined by border sequences, which are 24–25 bp imperfect direct repeats flanking and defining the T-DNA region (Zambryski et al. 1982; van Haaren et al. 1988). These border sequences are the only DNA sequences required in *cis* for T-DNA transfer (Zambryski et al. 1983a). DNA present in between the border sequences is transferred to the recipient plant cell's nucleus. The second important element is composed of virulence (*vir*) genes also present on the Ti plasmid but outside the T-DNA. *VirA* and *VirG* located on the virulence region make a two-component regulatory system for controlling transcriptional activation of the *vir* operons (Stachel and Zambryski 1986). Some of the *vir* genes are critical in the transfer of T-DNA from the bacterium to host cell, while others help in targeting T-DNA to the nucleus and probably to the precise integration site in the host cell for T-DNA. The third important bacterial genetic element comprises the chromosomal genes critical for attachment of the bacterium to the host plant cell (Sheng and Citovsky 1996). Thus, in the vicinity of wound sites in the plant and the release of signal molecules, bacterial cells are chemotactically attracted to the host cells (Hawes and Smith 1989). It has been found that at the wound site release of monocyclic phenolic compounds such as acetosyringone leads to the induction of the *vir* genes (Stachel et al. 1985). It has generally been proposed that at the initial onset of the infection process, the wound signals are perceived by the VirA protein, which activates the *virG* transcription factor by phosphorylation, leading to the upregulation of other *vir* genes (Citovsky et al. 1992). Among these up-regulated *vir* genes, protein product from the *virD2* gene recognizes the imperfect direct repeats of the T-DNA and in concert with the virD1, virD2, virC1 and virC2 proteins cause a nick in the T-DNA strand (Yanofsky et al. 1986; Yanofsky and Nester 1986; Stachel et al. 1987). The virD2 protein covalently binds the 5′ end of the single-stranded T-DNA (Yanofsky et al. 1986; Vos and Zambryski 1989) and the virE2 protein forms a complex with the T-DNA strand for mobilization into the plant cell nucleus (Citovsky et al. 1988, 1989). It is believed that the type IV secretion system, T4SS, of the *vir* system in *Agrobacterium* is assembled by 11 proteins coded for by the *virB* operon and the virD4 protein leading to the channel bridging the bacterial and plant cell wherein the passage of the T-DNA complex occurs (Christie 1997, 2004; Zupan et al. 1998; Zupan et al. 2000). Once the T-DNA complex is within the plant nucleus, it has been suggested that doubling of the T-DNA occurs and there is integration into plant chromosomes (DeNeve et al. 1997) or transient expression of the genes on the T-DNA.

The utilization of *Agrobacterium* as a gene delivery method was further enhanced by the observations that the disarmed T-DNA lacking functional onco-genes can be transferred and integrated into plant genomes to produce transgenic plants (Barton et al. 1983; Zambryski et al. 1983b). The ability of the *vir* genes to act in *trans* resulted in the development of a small, easy-to-handle binary vector system, which contains two replicons, one containing T-region constituting a binary vector and another replicon containing the *vir* genes termed the *vir* helper (Hoekema et al. 1983). The *vir* helper plasmid contained the disarmed T-DNA and was unable to induce tumors. A number of *Agrobacterium* strains contain-ing nononcogenic *vir* (disarmed) helper plasmids have been developed such as LBA4404 (Ooms et al. 1981), GV301 MP 90 (Koncz and Schell 1986), AGL1 (Lazo et al. 1991), and EHA101, 105 (Hood et al. 1986; Hood et al. 1993). The binary vectors have a variety of restriction sites and carry scorable and selectable markers to estimate transformation events and select transgenic tissues, respec-tively. The host range of *Agrobacterium* defined its utilization as a gene transfer method to produce transgenic plants. Therefore, initial studies focused on host-pathogen interaction to extend *Agrobacterium*'s host range, which was limited to dicotyledonous plants, as monocotyledons were considered outside its host range. In order to find the factors limiting its infectivity, most of the studies were devoted to identifying the factors for *Agrobacterium* infection and expanding its host range.

3.2.1.2 Agroinfection

Introduction of plant infectious agents via *Agrobacterium* has been defined as agroinfection (Grimsley et al. 1986; Grimsley and Bisaro 1987) or agroinoculation (Elmer et al. 1988). This technique is applicable to molecules that can replicate independent of the plant chromosomal DNA and has been used to deliver viral DNA by two different methods. In the first approach, viral DNA is placed in tandem in the bacterial T-DNA, and systemic spread of the virus occurs in a recipient host plant after inoculation. This technique does not require preparation of nucleic acids or insect vectors. In the second approach, *Agrobacterium* carrying viral nucleic acid sequences can be integrated into the nuclear genome of every cell in a transgenic plant. The first technique, which has been used to introduce genomes of cauliflower mosaic virus (CaMV), was agroinfectious on turnips when placed in the T-DNA, but not when on a different plasmid replicon, suggesting that the infection was not as a result of *Agrobacterium* lysis (Grimsley et al. 1986). Agroinfection as a DNA delivery method has been used to study the basic aspects of virology, recombination, and T-DNA transfer.

3.2.1.3 Virus-induced Gene Silencing (VIGS)

Due to their economic importance as some of the most severe crop-disease-causing entities, plant viruses have been extensively studied to develop resistant crop cultivars. As early as the 1920s and 1930s, it was recognized that certain virus-infected plants became resistant to the same virus or closely related strains of the virus (McKinney 1927, 1929, 1937) and eventually the term crossprotection came to be widely used to describe this acquired resistance as a result of prior exposure to viruses (Fulton 1986). During the same period, the concept of pathogen-derived resistance (PDR) was proposed to genetically engineer resistance against pathogens (Sanford and Johnston 1985; Grumet et al. 1987). Pathogen-derived genes coding for coat proteins, replicases, movement proteins, defective interfering RNAs and DNAs, and nontranslated RNAs have been associated with PDR (Beachy 1997). The concept as such gained credibility with the production of transgenic tobacco plants carrying a gene coding for the tobacco mosaic virus (TMV) coat protein, which delayed symptom development (Abel et al. 1986). The molecular basis for the delay or absence of symptoms was ultimately attributed to post-transcriptional gene silencing (PTGS) (Ratcliff et al. 1997). It is now well established that PTGS is a result of occurrence of double-stranded RNA, first reported in *Caenorhabditis elegans* (Fire et al. 1998) and subsequently in plants (Waterhouse et al. 1998).

In principle, virus-induced gene silencing (VIGS), which was the term first used to describe the recovery of a virus-infected transgenic plant from virus infection (van Kammen 1997), is a result of PTGS (Kumagai et al. 1995). VIGS has been recognized as a powerful tool to down-regulate the activity of specific genes and use in high-throughput functional genomics studies in plants (Baulcombe 1999). Initial experiments involved engineering phytoene desaturase (PDS) cDNA into the chimeric sequences derived from the TMV and the tomato mosaic virus and application of the in-vitro-transcribed viral RNA to *Nicotiana benthamiana* leaves by rubbing (Kumagai et al. 1995), which led to down-regulation of the *PDS* gene. The *PDS* gene has also been used in conjunction with the barley stripe mosaic virus (BSMV) for gene silencing studies in monocotyledonous plants such as barley (Holzberg et al. 2002). Furthermore, more pronounced silencing was possible by inserting 40–60 bp direct inverted repeats into viral vectors for both TMV and BSMV (Lacomme et al. 2003).

3.2.1.4   Other Microorganisms for DNA Delivery

Although *Agrobacterium*-mediated transformation of many plant species is now possible, there still remains one major constraint for commercialization aspects of transgenics relating to the numerous patents involving *Agrobacterium*-mediated transformation methodologies (Roa-Rodriguez and Nottenburg 2003). There has, therefore, been an interest in recent years at the possibility of exploring other microorganisms for the transfer of DNA to plants. Even prior to the advent of such patents or before the understanding of the detailed molecular aspects of the T-DNA transfer to the plant genome, it was shown that crown gall induction property could be transferred from virulent *A. tumefaciens* strains to avirulent strains as well as to *A. rubi, A. radiobacter*, and *Rhizobium leguminosarum* (Klein and Klein 1953). It was eventually shown that transfer of the Ti plasmid to avirulent *Agrobacterium* strains or to *Rhizobium* (Hooykaas et al. 1977) and to the bacterium, *Phyllobacterium myrsinacearum* (Veen et al. 1988) caused the tumor formation. In order to explore the feasibility of using such nonagrobacterial microorganisms for transfer of DNA to plants, attempts were made to transfer the disarmed Ti plasmid, pEHA105, to other species of bacteria (Broothaerts et al. 2005). Modified forms of the Ti plasmid were introduced into *Rhizobium* sp. NGR234, *Sinorhizobium meliloti*, and *Mesorhizobium loti* and transformation of plants, such as tobacco, rice, and *Arabidopsis,* indicated GUS activity as well as GUS-positive signals on the Southern blots (Broothaerts et al. 2005). This study has therefore widened the scope of the T-DNA transfer technology to plants. Furthermore, this alternative approach is available under open-source-modeled licenses. Further details regarding this concept of sharing and technology improvement in an open environment are available at Bioforge project (*http://www. bioforge.net*) and Biological Innovation for Open Society (BIOS; *http://www. bios.net*).

3.3 Physical Methods

3.3.1 Liposome-Mediated Delivery

Liposomes are unilamellar phospholipid vesicles ranging in diameter of 0.2–1.6 μm (Olson et al. 1979; Szoka et al. 1980; Jousma et al. 1987) depending on extrusion techniques and measurement approaches. The encapsulation process preserves the structural integrity of small molecules (5–10 kb). The liposomes with the carrier molecules can be fused with the protoplasts using a fusiogenic agent, polyethylene glycol (PEG), or polyvinylalcohol followed by a high calcium ion treatment to promote protoplast fusion (Keller and Melchers 1973). Infection of tobacco mesophyll protoplasts with liposome-mediated delivery of TMV RNA has also been attempted (Nagata et al. 1981). TMV-specific immunofluorescence assay showed the presence of TMV particles in the infected tobacco mesophyll protoplasts. Similar findings were confirmed showing that negatively charged liposomes delivered nucleic acids into protoplasts better than other types of liposomes (Fraley et al. 1982). Liposome-mediated delivery of other viral nucleic acids has also been shown (Caboche 1990). Liposomes were used to transfer a chimeric gene construct encoding chloramphenicol acetyl transferase (CAT) into various plant protoplasts. The marker gene assay showed that this liposome-mediated DNA transfer technique was comparable to PEG-mediated techniques but less efficient than electroporation (Caboche 1990). Liposome-mediated DNA delivery into tobacco mesophyll protoplasts and subsequent regeneration into mature transgenic plants has also been demonstrated. The introduced kanamycin resistance gene was integrated into the host genome and was inherited in a Mendelian fashion (Bellini et al. 1989). The inserted plasmid DNA was integrated into the host tobacco genome in a complex pattern. Liposome-mediated DNA transfer to produce transgenic plants has been tested for only a limited number of plant species, although there are some reports of success, including those encoding cationic liposomes or lipofectin. Cationic liposomes were developed to overcome some of the earlier difficulties associated with DNA delivery into eukaryotic cells and involved synthesis of a cationic lipid N-[1-(2,3-dioleyloxy)propyl]-N, N,N-trimethylammonium chloride (DOTMA), which readily interacted with DNA for delivery to mammalian cell culture lines (Felgner et al. 1987; Felgner and Ringold 1989). Similar cationic liposomes have therefore been used for the transformation of protoplasts from tobacco (Sporlein and Koop 1991) and lentils (Maccarrone et al. 1992), with evidence of transient expression. More recently, DNA encapsulated within a novel cationic vesicle derived from vernonia oil was shown to pass undamaged across isolated plant cuticular membranes (Wiesman et al. 2007). Further studies are required to improve this system and show that indeed physical barriers of the plant cell can be obviated for DNA delivery by this method, possibly not necessitating protoplasts for transformation and regeneration. Generally, the transformation frequencies with liposome-mediated techniques are low compared to other direct DNA delivery techniques. Therefore, liposome-mediated DNA delivery has not gained widespread attention for extensive studies or production of transgenic plants.

3.3.2 Nanoparticles

With the recent explosion in nanotechnology-based research for addressing biological questions and providing solutions, particularly in medicine and therapeutics, there has been increasing interest in pursuing such approaches in resolving plant biology-related research objectives and questions. Thus, for gene transfer to plants, exploration on the use of nanoparticles would be valuable. It can probably be argued that efforts undertaken with the use of liposome-mediated DNA delivery have already heralded nanotechnology-directed gene transfer technology in plants. Nanoparticles have been defined as materials which span the nanometric size range, with one of the dimensional size range being within a few hundred nanometers (Gonzalez-Melendi et al. 2008), or colloidal polymeric systems, either biodegradable or nonbiodegradable, of less than one micrometer in size (Brigger et al. 2002). In animal cells and tissues and for drug therapy, the use of nanoparticles is being viewed as effective carriers for delivery of molecules (Yih and Al-Fandi 2006), circumventing degradation if taken orally or delivered unprotected by other means, and also allowing more targeted delivery and release for drugs (Brigger et al. 2002). With the extensive research to develop delivery systems in drug therapy, there has been similar interest for targeted and controlled DNA delivery to plant cells. However, as mentioned earlier, the presence of the ubiquitous plant cell wall as a barrier often complicates application of nanoparticle delivery systems developed for mammalian cells and tissues. Notwithstanding these obstacles, a recent report on the use of functionalized mesoporous nanoparticles (MSNs) for gene transfer to plant cells appears promising (Torney et al. 2007). In this approach, MSNs were filled with β-estradiol, which is an estrogen-receptor-based transactivator (Zuo et al. 2000), for induction of GFP expression, and capped with gold particles coated with the construct of interest and delivered to tobacco plant cells by particle bombardment (Torney et al. 2007). More recently, the use of a new zeolite-based silicalite nanoparticle system consisting of polyethylene imine-plasmid DNA complex was shown to enhance transfection efficiencies in human embryonic kidney cells (Pearce et al. 2008). Zeolite is a crystalline aluminosilicate of open three-dimensional structure with numerous cavities and channels (Mumpton 1999). The inert nature and holding capacity of zeolite therefore opens up new possibilities for gene delivery approaches to plant cells.

3.3.3 Microinjection

One of the most direct methods for gene transfer is to inject DNA directly into the nucleus. Electromechanical devices are used to control the insertion of fine glass needles into the nucleus of individual cells to deliver the DNA (Spangenberg et al. 1986). Recipient cells or protoplasts are immobilized onto a holding capillary, or in gels, or onto surfaces coated with poly-L-lysine. The nucleus is visualized and the DNA is mechanically injected into it. The recipient cells are cultured and subsequently selected for integration of inserted genes by growth on culture media containing a selective agent. Microinjection has been used to insert genes into individual cells (Nomura and Komamine 1986), microspore-derived embryoids (Neuhaus et al. 1987), or single protoplasts (Steinbiss and Stabel 1983; Morikawa et al. 1986a; Reich et al. 1986). Microinjection method has been used to produce transgenic alfalfa (Reich et al. 1986) by injecting Ti plasmid into alfalfa protoplasts. The microinjection technique is labor intensive and slow, but transformation efficiencies close to 26% were achieved for alfalfa protoplasts (Reich et al. 1986).

3.3.4   Silicon Carbide Whiskers

Silicon carbide is intrinsically very hard and breaks readily to give sharp edges, which are adequate to penetrate plant cell walls. The pores created by the whiskers allow the DNA to be delivered into plant cells. Silicon carbide whiskers are mixed with recipient cells and plasmid DNA by vigorous vortexing and then plated on culture medium. The cultured cells are subsequently assessed for DNA insertion into the cells and integration with nuclear DNA. The silicon carbide whisker-mediated DNA delivery has been shown to produce stably transformed plant cells (Kaeppler et al. 1992, 1994; Kaeppler and Somers 1994) and algae (Dunahay 1993). Silicon carbide whiskers were the most effective in delivering vector DNA as compared to other materials such as silicon nitride or carborundum – a spherical form of silicon carbide and glass beads (Wang et al. 1995). The efficiency of DNA delivery by silicon carbide whiskers could be increased by exposing cells to high molarity of sorbitol or mannitol (Wang et al. 1995). Silicon carbide whiskers have been used to produce transgenic plants of *Lolium*, *Festuca*, and *Agrostis* sp. (Dalton et al. 1998), rice (Matsushita et al. 1999), and maize (Petolino et al. 2000). More recently, a supersonic treatment was combined with the silicon carbide treatment of rice cell suspension cultures, and it was claimed that high efficiency transformation could be obtained (Terakawa et al. 2005).

Although the attributes of silicon carbide whiskers for being simple, cost effective, and less resource demanding for DNA delivery into plants cells have been well recognized, it has also been reported that silicon carbide whiskers exhibit toxic properties and may be harmful to human beings, if not handled with caution (Vaughan et al. 1991; Svensson et al. 1997). There has, therefore, been a search for other possible materials, which could be used for similar DNA transfer purposes. Thus, the use of aluminum borate whiskers (ABW) was suggested as a possible alternative, with previously unreported mutagenic effects and was used to transform scutellar tissues from mature embryos of *Japonica* rice and produced Southern positive transgenic plants (Takahashi et al. 2000). Recent improvements to the ABW method included use of a multidirectional shaker instead of a vortex and the type of ABW for rice callus transformation (Mizuno et al. 2004). Transgenic tobacco plants have also been produced using the ABW method (Mizuno et al. 2005).

3.3.5 Microprojectile Bombardment

The microprojectile method of DNA delivery, called biolistic (biological ballistic), was invented by Sanford et al. in 1987 as a means of bypassing many of the host range limitations of *Agrobacterium* and also overcoming the physical barrier for Z-DNA uptake by the plant cell wall (Weissinger et al. 1987; Klein et al. 1988b, c). The basic principle underlying microprojectile bombardment is to accelerate microparticles to a speed at which they can penetrate the plant cell wall and be incorporated into the interior of a cell (Sanford 1990). This technique can be used to deliver a range of compounds into a cell and has found applications in several disciplines from agriculture to medicine. Microprojectile bombardment has been used to produce transgenic plants in species which were not amenable to *Agrobacterium*-mediated genetic transformation.

The particle bombardment apparatus essentially consists of a mechanism to accelerate the particles to desired velocities and regulate their penetration into the recipient cells. The original apparatus designed by the inventors used a gun powder discharge to accelerate inert metal microprojectiles coated with biologically active compounds (Sanford 1988; Klein et al. 1988b). The gun powder was quickly replaced with the inert gas, helium (Sanford et al. 1991), to provide the force for microprojection. The main component of the most commonly used helium gun (Biolistics® PDS-1000/He, BioRad, Inc.) is a rupture disk assembly that controls the helium pressure at which the microprojectiles carrying the vector DNA are propelled. The rupture disk assembly consists of a gas acceleration tube with a rupture disk placed at the bottom of the tube inside a retaining cap. The microprojectiles coated with the biological compound of interest are placed on a carrier situated below the rupture disk. The chamber is partially evacuated and the helium gas pressure is allowed to build up to the desired level to rupture the rupture disk. The optimized helium gas pressure propels the microprojectiles at the optimized velocity through a metal screen to deliver biologically active compounds into target cells. The main consideration in microprojectile-mediated delivery is to deposit in a cell optimal amounts of biologically active compounds with minimal damage to the cell wall. A major limitation of the microprojectile projection technique has been the uneven penetration and distribution of the microprojectiles.

To regulate microprojectile penetration into cells, several modifications have been made to the power source used to propel the microprojectiles carrying the biologically active compounds. Regulated nitrogen gas pressure (Morikawa et al. 1989), compressed air (Iida et al. 1990), or an air gun (Oard et al. 1990) has been used to propel microprojectiles to deliver biologically active compounds into plant cells. In another approach, a particle in-flow gun (PIG) in which micro-projectiles are accelerated directly into a stream of helium rather than being supported by a macrocarrier has been also developed (Finer et al. 1992). An air gun generating a pressure pulse of approximately 2 ms at a pressure of 60 bars to deliver the microprojectile/DNA suspension into the cells has also been used (Sautter et al. 1991). A major difference in this technique is that the DNA is suspended with the microprojectiles rather than being coated on them. The main function of the microprojectiles is to create holes through which DNA passes into the cells. The inventors consider that the movement of DNA independent of microprojectiles allows them to target small areas of tissues. In another modification, electric-discharge-generated shock waves have been used to propel gold microprojectiles coated with DNA or other biologically active compounds (Christou et al. 1988).

Several factors influence the biolistics-mediated delivery of DNA into plant cells. The main determinant is the delivery of optimal amount of DNA with minimal injury to the recipient plant cell or tissue. Material and size of micro-projectiles, attachment of DNA to microprojectiles, pressure at which micropro-jectiles are propelled and the recipient tissue are all critical factors. Most of these parameters need to be optimized and vary with individual laboratories. By only adjusting the distance that the microparticles travel to the target tissue or changing the target diameter, it can be shown that transformation events may be affected. Using GUS histochemical assays it can be shown that the number of GUS-positive spots was greatly increased when the flight distance of the microparticles was set at six centimeters and the target diameter of one centimeter, consisting in this example of wheat mature embryos tightly packed within this area (Fig. 3.3). Increasing the flight distance and the diameter of the target area led to fewer GUS spots (Fig. 3.3). Thus depending on the type of tissues, parameters for particle bombardment can be

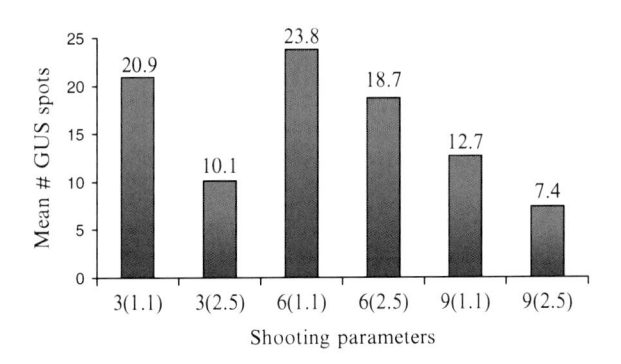

Fig. 3.3 Distance between target tissue and macrocarrier holder containing DNA-coated gold microparticles and their influence on the number of GUS-positive spots on wheat mature embryos. Numbers on x axis indicate target distance in cm and number in parentheses indicate target diameter in cm

accordingly optimized for maximal transient expression, prior to performing actual transformation experiments. Transient expression based on the *Gus* gene or the *GFP* gene is a good indication of adequate coverage of the tissue in terms of possible integration of the transgene across the tissue surface. Biolistics-mediated transformation has been used to genetically transform a large number of plant species including major agricultural crops such as maize, cotton, soybeans, rice, sorghum, and wheat and is likely to be the method of choice for plant species where *Agrobacterium*-mediated transformation is still inefficient.

3.3.6   Electroporation

Short, high-intensity electric pulses reversibly permeabilize the lipid bilayers of cell membranes in all living cells. The electric pulses cause extensive compression and thinning of the plasmalemma, resulting in the transient formation of pores in the plasma membrane (Neumann and Rosenheo 1972). The transiently formed pores allow the diffusion of a range of macromolecules including proteins and nucleic acids. The apparatus used for electroporation is fairly simple. High electric fields are applied to protoplasts, cells, or tissues suspended in a liquid culture medium enclosed in a discharge chamber. The electric field is applied by a capacitor discharge which in some commercial instruments can produce voltage up to 2,500 V. The time and voltage applied depends upon the cell and tissue type used for electroporation.

Electroporation has been used to deliver DNA in a range of plant cells and tissues. However, protoplasts, which lack cell walls, are the most amenable recipient system for electroporation-mediated DNA delivery. Electroporation-mediated gene transfer into plant protoplasts has been shown in several plants such as tobacco (Shillito et al. 1985; Negrutiu et al. 1987), corn (Fromm et al. 1986), rice (Shimamoto et al. 1989), soybean (Dhir et al. 1991), sugarcane (Chowdhury and Vasil 1992), and oilseed rape (Bergman and Glimelius 1993). Even though over the years there have been many other reports of transformation of protoplasts from other plant species by electroporation, the major limitation has been regeneration of fertile transgenic plants from the electroporated protoplasts for a number of species. This was even more problematic for monocotyledonous species. There was, therefore, an interest in bypassing the protoplasts for electroporation. It was reported that tobacco pollen grains subjected to electroporation treatments stayed viable (Mishra et al. 1987) and that electroporation of barley microspores with propidium iodide subsequently produced callus that regenerated plants (Joersbo et al. 1990). Successful DNA delivery into tobacco pollens by electroporation was also reported, including transient GUS expression and positive Southern hybridization signals from blots with DNA extracted from the transformed pollen grains (Abdulbaki et al. 1990; Matthews et al. 1990). Since then there have been numerous reports on optimizing DNA delivery into intact plant cells or tissues, mainly pertaining to the duration of electric pulses and field strength. The first report of stable transformation via electroporation of intact tissues was with rice seeds cultured 2 days prior to being subjected to electroporation with a plasmid containing *NPTII* gene (Li et al. 1991). Subsequently, electroporation of immature corn embryos and callus (D'Halluin et al. 1992) and nodal buds of pea and lentils (Chowrira et al. 1995) led to the production of stable transformants.

3.3.7 Other Potential Physical Methods

Over the years that followed physical methods of transformation of plants by electroporation and biolistic, attempts at developing other versatile and less expensive methods have been made, including improvising existing systems (Table 3.1). Among these developments, some of the approaches based on alginate microbeads in combination with PEG and/or electroporation or DNA-lipofectin complexes (Sone et al. 2002; Liu et al. 2004; Murakawa et al. 2008a, b) are promising. However, only the method using PEG for uptake of the bioactive beads was shown to produce transgenic plants that were Southern positive. Thus, the requirement for protoplasts would still be a major impediment for the bioactive beads-based transformation system due to difficulties still associated with regeneration of fertile plants from protoplasts for a vast majority of plant species.

3.4 Combined Physical and Biological DNA Delivery

3.4.1 Agrolistic

Attempts to maximize the benefits of the different available gene transfer methods in plants have been extensively pursued and are still actively being explored. It has been widely recognized that both the *Agrobacterium*-mediated and the biolistic-mediated gene transfer approaches have their respective advantages and disadvantages, depending on the species and circumstantial requirements. Thus, to utilize the benefits of these two systems, a combination of *Agrobacterium*-mediated transformation with the biolistic delivery was developed and was termed the "agrolistic" transformation system (Hansen and Chilton 1996). In this system, the *virD1/virD2* genes are cotransformed by particle bombardment with a plasmid DNA containing the T-DNA borders flanking the gene of interest. Thus, there is transient expression of the virD1/D2 proteins and excision of the T-DNA occurs in planta similar to that with *Agrobacterium*-mediated transformation. Even though this method has included the underlying benefits of both systems, reports on the successful use of this approach for the generation of stably transformed plants have been lacking.

3.4.2   Sonication-Assisted Delivery

Further improvement to the *Agrobacterium* transformation method included brief treatments of the plant tissues with ultrasound in the presence of *Agrobacterium*, a method called sonication-assisted *Agrobacterium*-mediated transformation (SAAT) (Trick and Finer 1997). Sonication as such has previously been used in conjunction with gene transfer to protoplasts of *Beta vulgaris* and tobacco (Joersbo and Brunstedt 1990). Using the sonication approach, leaf segments of tobacco showed transient expression and regenerated R_1 plants exhibited stable expression of the transgenes (Zhang et al. 1991). Unlike the agrolistic method there have been several reports on the successful production of stably transformed plants using the SAAT method (Christiansen et al. 2000; Pathak and Hamzah 2008; Rashid et al. 2008). It is likely that further refinement using the SAAT method will lead to more reports on the stable transformation of other plant species.

3.5   Concluding Remarks

Development of gene transfer methods in plants has probably been one of the most challenging aspects of plant research. Currently the two methods of choice are undoubtedly the *Agrobacterium*-mediated and the biolistic-mediated DNA delivery systems. In the latter two decades of the last century, development of transformation technology was viewed primarily as an objective to the production of transgenic crops with improved agronomic characteristics for enhanced crop productivity. However, the emergence of functional genomics and the requirement for high-throughput technology for assessment of gene function in plants have generated a whole new focus on generation of transgenic plants. This has become even more evident for economically important crop plants, wherein high efficiency transformation systems are generally lacking, compared to model systems such as *Arabidopsis thaliana*. Nonetheless progress made in rice and corn is encouraging with the use of *Agrobacterium* for transformation in a fairly high-throughput manner. The focus is, therefore, still centered around those other economically important plants, which can be made amenable for high-throughput gene transfer in order to make utmost use of the vast repertoire of data flowing from genome projects.

Thus, gene transfer methods to plants will continue to receive renewed interest in the future. The development of nanoparticles for DNA delivery into plant cells is emerging as the next generation transformation system and is likely to combine the benefits of *Agrobacterium* and biolistic. Meanwhile, the approach of in planta transformation using *Agrobacterium* (Bechtold et al. 1993; Clough and Bent 1998) is likely to continue receiving attention, due to circumvention of a tissue culture step for the regeneration of transformation events. In planta transformation for several species such as *Brassica rapa* ssp. *chinensis* (Qing et al. 2000),

Medicago truncatula (Trieu et al. 2000), *Hibiscus canabinus* (Kojima et al. 2004), rice (Supartana et al. 2005), and wheat (Supartana et al. 2006) has been reported. However the transformation efficiency achieved by this method thus far for all these species is yet to be convincing and requires further refinement and research. Furthermore, several of the other different methods reviewed in this chapter and listed in Table 3.1 are likely to be explored further. The likely successes, however, will come from a combination or refinement of several existing methods, with the most suitable, and most likely popular, gene transfer method possessing the versatility, simplicity, and accessibility for a large number of plant species.

Acknowledgments Canada Research Chairs, Canada Foundation for Innovation, Natural Sciences and Engineering Research Council, Genome Canada, and Genome Prairie are gratefully acknowledged for financial support of research in our laboratory.

References

Abdulbaki AA, Saunders JA, Matthews BF, Pittarelli GW (1990) DNA uptake during electroporation of germinating pollen grains. Plant Sci 70:181–190

Abel PP, Nelson RS, De B, Hoffmann N, Rogers SG, Fraley RT, Beachy RN (1986) Delay of disease development in transgenic plants that express the tobacco mosaic-virus coat protein gene. Science 232:738–743

Ahokas H (1989) Transfection of germinating barley seed electrophoretically with exogenous DNA. Theor Appl Genet 77:469–472

Arencibia A, Molina PR, Delariva G, SelmanHousein G (1995) Production of transgenic sugarcane (*Saccharum officinarum* L.) plants by intact cell electroporation. Plant Cell Rep 14: 305–309

Badr YA, Kereim MA, Yehia MA, Fouad OO, Bahieldin A (2005) Production of fertile transgenic wheat plants by laser micropuncture. Photochem Photobiol Sci 4:803–807

Barton KA, Binns AN, Matzke AJM, Chilton MD (1983) Regeneration of intact tobacco plants containing full length copies of genetically engineered T-DNA and transmission of T-DNA to R1 progeny. Cell 32:1033–1043

Barton KA, Chilton MD (1983) *Agrobacterium* Ti plasmids as vectors for plant genetic engineering. Meth Enzymol 101:527–539

Baulcombe DC (1999) Fast forward genetics based on virus-induced gene silencing. Curr Opin Plant Biol 2:109–113

Beachy RN (1997) Mechanisms and applications of pathogen-derived resistance in transgenic plants. Curr Opin Biotechnol 8:215–220

Bechtold N, Ellis J, Pelletier G (1993) In planta *Agrobacterium*-mediated gene transfer by infiltration of adult *Arabidopsis thaliana* plants. C R Acad Sci III Sci Vie 316:1194–1199

Bellini C, Guerche P, Spielmann A, Goujaud J, Lesaint C, Caboche M (1989) Genetic-analysis of transgenic tobacco plants obtained by liposome-mediated transformation – absence of evidence for the mutagenic effect of inserted sequences in 60 characterized transformants. J Hered 80:361–367

Bergman P, Glimelius K (1993) Electroporation of rapeseed protoplasts – transient and stable transformation. Physiol Plant 88:604–611

Braun A, Laskaris T (1943) Tumor formation by attenuated crown-gall bacteria in the presence of growth-promoting substances. J Bacteriol 45:196

Braun AC (1941) Development of secondary tumors and tumor strands in the crown gall of sunflowers. Phytopathology 31:135–149

Braun AC, Laskaris T (1942) Tumor formation by attenuated crown-gall bacteria in the presence of growth promoting substances. Proc Natl Acad Sci USA 28:468–477

Brigger I, Dubernet C, Couvreur P (2002) Nanoparticles in cancer therapy and diagnosis. Adv Drug Deliv Rev 54:631–651

Broothaerts W, Mitchell HJ, Weir B, Kaines S, Smith LMA, Yang W, Mayer JE, Roa-Rodriguez C, Jefferson RA (2005) Gene transfer to plants by diverse species of bacteria. Nature 433:629–633

Caboche M (1990) Liposome-mediated transfer of nucleic acids in plant protoplasts. Physiol Plant 79:173–176

Carlson PS, Smith HH, Dearing RD (1972) Parasexual interspecific plant hybridization. Proc Natl Acad Sci USA 69:2292

Chilton MD, Tepfer DA, Petit A, David C, Cassedelbart F, Tempe J (1982) *Agrobacterium rhizogenes* inserts T-DNA into the genomes of the host plant root cells. Nature 295:432–434

Chowdhury MKU, Vasil IK (1992) Stably transformed herbicide resistant callus of sugarcane via microprojectile bombardment of cell suspension cultures and electroporation of protoplasts. Plant Cell Rep 11:494–498

Chowrira GM, Akella V, Lurquin PF (1995) Electroporation mediated gene transfer into intact nodal meristems in planta - Generating gransgenic plants without in vitro tissue culture. Mol Biotechnol 3:17–23

Christiansen P, Gibson JM, Moore A, Pedersen C, Tabe L, Larkin PJ (2000) Transgenic *Trifolium repens* with foliage accumulating the high sulphur protein, sunflower seed albumin. Transgenic Res 9:103–113

Christie PJ (1997) *Agrobacterium tumefaciens* T-complex transport apparatus: a paradigm for a new family of multifunctional transporters in eubacteria. J Bacteriol 179:3085–3094

Christie PJ (2004) Type IV secretion: the *Agrobacterium* VirB/D4 and related conjugation systems. Mol Cell Res 1694:219–234

Christou P, McCabe DE, Swain WF (1988) Stable transformation of soybean callus by DNA-coated gold particles. Plant Physiol 87:671–674

Citovsky V, Devos G, Zambryski P (1988) Single-stranded DNA-binding protein encoded by the virE Locus of *Agrobacterium tumefaciens*. Science 240:501–504

Citovsky V, Wong ML, Zambryski P (1989) Cooperative interaction of *Agrobacterium* VirE2 protein with single-stranded DNA – implications for the T-DNA transfer process. Proc Natl Acad Sci USA 86:1193–1197

Citovsky V, Zupan J, Warnick D, Zambryski P (1992) Nuclear localization of *Agrobacterium* VirE2 protein in plant cells. Science 256:1802–1805

Clough SJ, Bent AF (1998) Floral dip: a simplified method for *Agrobacterium*-mediated transformation of *Arabidopsis thaliana*. Plant J 16:735–743

Crossway A, Oakes JV, Irvine JM, Ward B, Knauf VC, Shewmaker CK (1986) Integration of foreign DNA following microinjection of tobacco mesophyll protoplasts. Mol Gen Genet 202:179–185

D'Halluin K, Bonne E, Bossut M, Debeuckeleer M, Leemans J (1992) Transgenic maize plants by tissue electroporation. Plant Cell 4:1495–1505

Dalton SJ, Bettany AJE, Timms E, Morris P (1998) Transgenic plants of *Lolium multiflorum, Lolium perenne, Festuca arundinacea* and *Agrostis stolonifera* by silicon carbide fibre-mediated transformation of cell suspension cultures. Plant Sci 132:31–43

de la pena A, Lorz H, Schell J (1987) Transgenic rye plants obtained by injecting DNA into young floral tillers. Nature 325:274–276

DeNeve M, DeBuck S, Jacobs A, Vanmontagu M, Depicker A (1997) T-DNA integration patterns in co-transformed plant cells suggest that T-DNA repeats originate from co-integration of separate T-DNAs. Plant J 11:15–29

Deshayes A, Herreraestrella L, Caboche M (1985) Liposome-mediated transformation of tobacco mesophyll protoplasts by an *Escherichia coli* plasmid. EMBO J 4:2731–2737

Dhir SK, Dhir S, Sturtevant AP, Widholm JM (1991) Regeneration of transformed shoots from electroporated soybean (*Glycine max* (L) Merr) protoplasts. Plant Cell Rep 10:97–101

Dunahay TG (1993) Transformation of *Chlamydomonas reinhardtii* with silicon carbide whiskers. Biotechniques 15:452

Elmer JS, Sunter G, Gardiner WE, Brand L, Browning CK, Bisaro DM, Rogers SG (1988) *Agrobacterium*-mediated inoculation of plants with tomato golden mosaic virus DNAs. Plant Mol Biol 10:225–234

Felgner PL, Gadek TR, Holm M, Roman R, Chan HW, Wenz M, Northrop JP, Ringold GM, Danielsen M (1987) Lipofection – a highly efficient, lipid-mediated DNA-transfection procedure. Proc Natl Acad Sci USA 84:7413–7417

Felgner PL, Ringold GM (1989) Cationic liposome-mediated transfection. Nature 337:387–388

Finer JJ, Vain P, Jones MW, McMullen MD (1992) Development of the particle inflow gun for DNA delivery to plant cells. Plant Cell Rep 11:323–328

Fire A, Xu SQ, Montgomery MK, Kostas SA, Driver SE, Mello CC (1998) Potent and specific genetic interference by double-stranded RNA in *Caenorhabditis elegans*. Nature 391:806–811

Fraley RT, Dellaporta SL, Papahadjpoulos D (1982) Liposome-mediated delivery of tobacco mosaic virus RNA into tobacco protoplasts – a sensitive assay for monitoring liposome-protoplast interactions. Proc Natl Acad Sci USA 79:1859–1863

Fraley RT, Horsch RB (1983) In vitro transformation of *Petunia* protoplasts by *Agrobacterium tumefaciens*. J Cell Biochem 7(Suppl.):250

Frame BR, Drayton PR, Bagnall SV, Lewnau CJ, Bullock WP, Wilson HM, Dunwell JM, Thompson JA, Wang K (1994) Production of fertile transgenic maize plants by silicon carbide whisker-mediated transformation. Plant J 6:941–948

Fromm M, Taylor LP, Walbot V (1985) Expression of genes transferred into monocot and dicot plant cells by electroporation. Proc Natl Acad Sci USA 82:5824–5828

Fromm ME, Taylor LP, Walbot V (1986) Stable transformation of maize after gene transfer by electroporation. Nature 319:791–793

Fulton RW (1986) Practices and precautions in the use of cross protection for plant virus disease control. Annu Rev Phytopathol 24:67–81

Gonzalez-Melendi P, Fernandez-Pacheco R, Coronado MJ, Corredor E, Testillano PS, Risueno MC, Marquina C, Ibarra MR, Rubiales D, Perez-de-Luque A (2008) Nanoparticles as smart treatment-delivery systems in plants: assessment of different techniques of microscopy for their visualization in plant tissues. Ann Bot 101:187–195

Goodner B, Hinkle G, Gattung S, Miller N, Blanchard M, Qurollo B, Goldman BS, Cao YW, Askenazi M, Halling C, Mullin L, Houmiel K, Gordon J, Vaudin M, Iartchouk O, Epp A, Liu F, Wollam C, Allinger M, Doughty D, Scott C, Lappas C, Markelz B, Flanagan C, Crowell C, Gurson J, Lomo C, Sear C, Strub G, Cielo C, Slater S (2001) Genome sequence of the plant pathogen and biotechnology agent *Agrobacterium tumefaciens* C58. Science 294:2323–2328

Griesbach RJ (1983) Protoplast microinjection. HortScience 18:616

Griesbach RJ (1987) Chromosome-mediated transformation via microinjection. Plant Sci 50:69–77

Griesbach RJ (1994) An improved method for transforming plants through electrophoresis. Plant Sci 102:81–89

Grimsley N, Bisaro D (1987) Agroinfection. In: Hohn T, Schell J (eds) Plant gene research, basic knowledge and application: plant DNA infectious agents. Springer, New York, pp 87–108

Grimsley N, Hohn B, Hohn T, Walden R (1986) Agroinfection, an alternative route for viral infection of plants by using the Ti plasmid. Proc Natl Acad Sci USA 83:3282–3286

Grumet R, Sanford JC, Johnston SA (1987) Pathogen-derived resistance to viral infection using a negative regulatory molecule. Virology 161:561–569

Guo YD, Liang H, Berns MW (1995) Laser-mediated gene transfer in rice. Physiol Plant 93:19–24

Hansen G, Chilton MD (1996) "Agrolistic" transformation of plant cells: integration of T-strands generated in planta. Proc Natl Acad Sci USA 93:14978–14983

Hawes MC, Smith LY (1989) Requirement for chemotaxis in pathogenicity of *Agrobacterium tumefaciens* on roots of soil-grown pea plants. J Bacteriol 171:5668–5671

Hoekema A, Hirsch PR, Hooykaas PJJ, Schilperoort RA (1983) A binary plant vector strategy based on separation of Vir region and T region of the *Agrobacterium tumefaciens* Ti plasmid. Nature 303:179–180

Holzberg S, Brosio P, Gross C, Pogue GP (2002) Barley stripe mosaic virus-induced gene silencing in a monocot plant. Plant J 30:315–327

Hood EE, Gelvin SB, Melchers LS, Hoekema A (1993) New *Agrobacterium* helper plasmids for gene transfer to plants. Transgenic Res 2:208–218

Hood EE, Helmer GL, Fraley RT, Chilton MD (1986) The hypervirulence of *Agrobacterium tumefaciens* A281 is encoded in a region of pTIBO542 outside of transfer DNA. J Bacteriol 168:1291–1301

Hooykaas PJJ, Klapwijk PM, Nuti MP, Schilperoort RA, Rorsch A (1977) Transfer of *Agrobacterium tumefaciens* Ti plasmid to avirulent *Agrobacteria* and to *Rhizobium ex* planta. J Gen Microbiol 98:477–484

Iida A, Morikawa H, Yamada Y (1990) Stable Transformation of cultured tobacco cells by DNA-coated gold particles accelerated by gas pressure-driven particle gun. Appl Microbiol Biotechnol 33:560–563

Joersbo M, Brunstedt J (1990) Direct gene transfer to plant protoplasts by mild sonication. Plant Cell Rep 9:207–210

Joersbo M, Jorgensen RB, Olesen P (1990) Transient electropermeabilization of barley (*Hordeum vulgare* L.) microspores to propidium iodide. Plant Cell Tiss Organ Cult 23:125–129

Jousma H, Talsma H, Spies F, Joosten JGH, Junginger HE, Crommelin DJA (1987) Characterization of liposomes - the influence of extrusion of multilamellar vesicles through polycarbonate membranes on particle size, particle size distribution and number of bilayers. Int J Pharm 35:263–274

Kaeppler H, Somers D (1994) DNA delivery into maize cell cultures using silicon carbide fibers. In: Freeling M, Walbot V (eds) The maize handbook. Springer, New York, pp 610–613

Kaeppler HF, Somers DA, Rines HW, Cockburn AF (1992) Silicon carbide fiber-mediated stable transformation of plant cells. Theor Appl Genet 84:560–566

Kaeppler HF, Pedersen J, Somers DA (1994) Optimization of silicon carbide fiber-mediated DNA delivery into regenerable sorghum and maize tissue cultures. In Vitro Cell Dev Biol Plant 30A:61

Kajiyama S, Joseph B, Inoue F, Shimamura M, Fukusaki E, Tomizawa K, Kobayashi A (2008) Transient gene expression in guard cell chloroplasts of tobacco using ArF excimer laser microablation. J Biosci Bioeng 106:194–198

Kao KN, Constabe F, Michaylu MR, Gamborg OL (1974) Plant protoplast fusion and growth of intergeneric hybrid cells. Planta 120:215–227

Keller WA, Melchers G (1973) Effect of high pH and calcium on tobacco leaf protoplast fusion. Z Naturforsch C 28:737–741

Klein DT, Klein RM (1953) Transmittance of tumor-inducing ability to avirulent crown gall and related bacteria. J Bacteriol 66:220–228

Klein TM, Fromm M, Weissinger A, Tomes D, Schaaf S, Sletten M, Sanford JC (1988a) Transfer of foreign genes into intact maize cells with high velocity microprojectiles. Proc Natl Acad Sci USA 85:4305–4309

Klein TM, Gradziel T, Fromm ME, Sanford JC (1988b) Factors influencing gene delivery into *Zea mays* cells by high velocity microprojectiles. Biotechnology 6:559–563

Klein TM, Harper EC, Svab Z, Sanford JC, Fromm ME, Maliga P (1988c) Stable genetic transformation of intact *Nicotiana* cells by the particle bombardment process. Proc Natl Acad Sci USA 85:8502–8505

Kojima M, Shiojri H, Nogawa M, Nozue M, Matsumoto D, Wada A, Saiki Y, Kiguch K (2004) In planta transformation of kenaf plants (*Hibiscus cannabinus* var. *aokawa* No. 3) by *Agrobacterium tumefaciens*. J Biosci Bioeng 98:136–139

Koncz C, Schell J (1986) The promoter of T_L-DNA gene 5 controls the tissue specific expression of chimeric genes carried by a novel type of *Agrobacterium* binary vector. Mol Gen Genet 204:383–396

Kumagai MH, Donson J, Dellacioppa G, Harvey D, Hanley K, Grill LK (1995) Cytoplasmic inhibition of carotenoid biosynthesis with virus-derived RNA. Proc Natl Acad Sci USA 92:1679–1683

Kunik T, Tzfira T, Kapulnik Y, Gafni Y, Dingwall C, Citovsky V (2001) Genetic transformation of HeLa cells by *Agrobacterium*. Proc Natl Acad Sci USA 98:1871–1876

Lacomme C, Hrubikova K, Hein I (2003) Enhancement of virus-induced gene silencing through viral-based production of inverted repeats. Plant J 34:543–553

Larkin PJ, Scowcroft WR (1981) Somaclonal variation – a novel source of variability from cell cultures for plant improvement. Theor Appl Genet 60:197–214

Lazo GR, Stein PA, Ludwig RA (1991) A DNA transformation-competent *Arabidopsis* genomic library in *Agrobacterium*. Biotechnology 9:963–967

Levee V, Garin E, Klimaszewska K, Seguin A (1999) Stable genetic transformation of white pine (*Pinus strobus* L.) after cocultivation of embryogenic tissues with *Agrobacterium tumefaciens*. Mol Breed 5:429–440

Levee V, Lelu MA, Jouanin L, Cornu D, Pilate G (1997) *Agrobacterium tumefaciens*-mediated transformation of hybrid larch (*Larix kaempferi* x *L. decidua*) and transgenic plant regeneration. Plant Cell Rep 16:680–685

Li BJ, Xu XP, Shi HP, Ke XY (1991) Introduction of foreign genes into the seed embryo cells of rice by electroinjection and the regeneration of transgenic rice plants. Sci China Ser B Chem Life Sci Earth Sci 34:923–931

Liu HB, Kawabe A, Matsunaga S, Murakawa T, Mizukami A, Yanagisawa M, Nagamori E, Harashima S, Kobayashi A, Fukui K (2004) Obtaining transgenic plants using the bio-active beads method. J Plant Res 117:95–99

Lorz H, Baker B, Schell J (1985) Gene transfer to cereal cells mediated by protoplast transformation. Mol Gen Genet 199:178–182

Luo ZX, Wu R (1989) A simple method for the transformation of rice via the pollen tube pathway. Plant Mol Biol Rep 7:69–77

Maccarrone M, Dini L, Dimarzio L, Digiulio A, Rossi A, Mossa G, Finazziagro A (1992) Interaction of DNA with cationic liposomes – ability of transfecting lentil protoplasts. Biochem Biophys Res Commun 186:1417–1422

Matsushita J, Otani M, Wakita Y, Tanaka O, Shimada T (1999) Transgenic plant regeneration through silicon carbide whisker-mediated transformation of rice (*Oryza sativa* L.). Breed Sci 49:21–26

Matthews BF, Abdulbaki AA, Saunders JA (1990) Expression of a foreign gene in electroporated pollen grains of tobacco. Sex Plant Reprod 3:147–151

McCabe D, Christou P (1993) Direct DNA transfer using electric discharge particle-acceleration (ACCELL™) technology). Plant Cell Tissue Organ Cult 33:227–236

McKinney HH (1927) Factors affecting certain properties of a mosaic virus. J Agric Res 35:0001–0012

McKinney HH (1929) Mosaic diseases in the Canary Islands, west Africa, and Gibraltar. J Agric Res 39:0557–0578

McKinney HH (1937) Virus mutation and the gene concept. J Hered 28:51–57

Merlo DJ, Nutter RC, Montoya AL, Garfinkel DJ, Drummond MH, Chilton MD, Gordon MP, Nester EW (1980) The boundaries and copy numbers of Ti plasmid T-DNA vary in crown gall tumors. Mol Gen Genet 177:637–643

Mishra KP, Joshua DC, Bhatia CR (1987) In vitro electroporation of tobacco pollen. Plant Sci 52:135–139

Mizuno K, Takahashi W, Beppu T, Shimada T, Tanaka O (2005) Aluminum borate whisker-mediated production of transgenic tobacco plants. Plant Cell Tissue Organ Cult 80:163–169

Mizuno K, Takahashi W, Ohyama T, Shimada T, Tanaka O (2004) Improvement of the aluminum borate whisker-mediated method of DNA delivery into rice callus. Plant Prod Sci 7:45–49

Morikawa H, Iida A, Matsui C, Ikegami M, Yamada Y (1986a) Gene transfer into intact plant cells by electroinjection through cell walls and membranes. Gene 41:121–124

Morikawa H, Iida A, Yamada Y (1989) Transient expression of foreign genes in plant cells and tissues obtained by a simple biolistic device (particle gun). Appl Microbiol Biotechnol 31:320–322

Morikawa H, Sugino K, Hayashi Y, Takeda J, Senda M, Hirai A, Yamada Y (1986b) Interspecific plant hybridization by electrofusion in *Nicotiana*. Biotechnology 4:57–60

Morikawa H, Yamada Y (1985) Capillary microinjection into protoplasts and intranuclear localization of injected materials. Plant Cell Physiol 26:229–236

Mumpton FA (1999) La roca magica: Uses of natural zeolites in agriculture and industry. Proc Natl Acad Sci USA 96:3463–3470

Murakawa T, Kajiyama S, Ikeuchi T, Kawakami S, Fukui K (2008a) Improvement of transformation efficiency by bioactive-beads-mediated gene transfer using DNA-lipofectin complex as entrapped genetic material. J Biosci Bioeng 105:77–80

Murakawa T, Kajiyama S, Fukui K (2008b) Improvement of bioactive bead-mediated transformation by concomitant application of electroporation. Plant Biotechnol 25:387–390

Nagata T, Okada K, Takebe I, Matsui C (1981) Delivery of tobacco mosaic virus RNA into plant protoplasts mediated by reverse phase evaporation vesicles (liposomes). Mol Gen Genet 184:161 165

Negrutiu I, Shillito R, Potrykus I, Biasini G, Sala F (1987) Hybrid genes in the analysis of transformation conditions. 1. Setting up a simple method for direct gene transfer in plant protoplasts. Plant Mol Biol 8:363–373

Neuhaus G, Spangenberg G, Scheid OM, Schweiger HG (1987) Transgenic rapeseed plants obtained by the microinjection of DNA into microspore-derived embryoids. Theor Appl Genet 75:30–36

Neumann E, Rosenhec K (1972) Permeability changes induced by electric impulses in vesicular membranes. J Membr Biol 10:279–290

Nomura K, Komamine A (1986) Embryogenesis from microinjected single cells in a carrot cell suspension culture. Plant Sci 44:53–58

Oard JH, Paige DF, Simmonds JA, Gradziel TM (1990) Transient gene expression in maize, rice, and wheat cells using an airgun apparatus. Plant Physiol 92:334–339

Ohta Y (1986) High-Efficiency genetic transformation of maize by a mixture of pollen and exogenous DNA. Proc Natl Acad Sci USA 83:715–719

Olson F, Hunt CA, Szoka FC, Vail WJ, Papahadjopoulos D (1979) Preparation of liposomes of defined size distribution by extrusion through polycarbonate membranes. Biochim Biophys Acta 557:9–23

Ooms G, Hooykaas PJJ, Moolenaar G, Schilperoort RA (1981) Crown gall plant tumors of abnormal morphology induced by *Agrobacterium tumefaciens* carrying mutated octopine Ti plasmids - Analysis of T-DNA functions. Gene 14:33–50

Otten L, Degreve H, Hernalsteens JP, Vanmontagu M, Schieder O, Straub J, Schell J (1981) Mendelian transmission of genes introduced into plants by the Ti plasmids of *Agrobacterium tumefaciens*. Mol Gen Genet 183:209–213

Pathak MR, Hamzah RY (2008) An effective method of sonication-assisted *Agrobacterium*-mediated transformation of chickpeas. Plant Cell Tissue Organ Cult 93:65–71

Pearce ME, Mai HQ, Lee N, Larsen SC, Salem AK (2008) Silicalite nanoparticles that promote transgene expression. Nanotechnology 19:175103. doi:doi: 10.1088/0957-4484/19/17/175103

Petolino JF, Hopkins NL, Kosegi BD, Skokut M (2000) Whisker-mediated transformation of embryogenic callus of maize. Plant Cell Rep 19:781–786

Qing CM, Fan L, Lei Y, Bouchez D, Tourneur C, Yan L, Robaglia C (2000) Transformation of pakchoi (*Brassica rapa* L. ssp *chinensis*) by *Agrobacterium* infiltration. Mol Breed 6:67–72

Raghavan V (1986) Embryogenesis in angiosperms. A developmental and experimental study. Cambridge, Cambridge, UK 303 p

Rashid B, Saleem Z, Husnain T, Riazuddin S (2008) Transformation and inheritance of Bt genes in *Gossypium hirsutum*. J Plant Biol 51:248–254

Ratcliff F, Harrison BD, Baulcombe DC (1997) A similarity between viral defense and gene silencing in plants. Science 276:1558–1560

Reich TJ, Iyer VN, Haffner M, Holbrook LA, Miki BL (1986) The use of fluorescent dyes in the microinjection of alfalfa protoplasts. Can J Bot 64:1259–1267

Roa-Rodriguez C, Nottenburg C (2003) *Agrobacterium*-mediated transformation of plants. CAMBIA: http://www.bios.net/Agrobacterium

Sanford JC (1988) The biolistic process. Trends Biotechnol 6:299–302

Sanford JC (1990) Biolistic plant transformation. Physiol Plant 79:206–209

Sanford JC, Devit MJ, Russell JA, Smith FD, Harpending PR, Roy MK, Johnston SA (1991) An improved helium-driven biolistic device. Technique (Philadelphia) 3:3–16

Sanford JC, Johnston SA (1985) The concept of parasite-derived resistance – deriving resistance Genes from the parasites own genome. J Theor Biol 113:395–405

Sautter C, Waldner H, Neuhausurl G, Galli A, Neuhaus G, Potrykus I (1991) Micro-targeting high efficiency gene transfer using a novel approach for the acceleration of micro-projectiles. Biotechnology 9:1080–1085

Sessions A, Burke E, Presting G, Aux G, McElver J, Patton D, Dietrich B, Ho P, Bacwaden J, Ko C, Clarke JD, Cotton D, Bullis D, Snell J, Miguel T, Hutchison D, Kimmerly B, Mitzel T, Katagiri F, Glazebrook J, Law M, Goff SA (2002) A high-throughput *Arabidopsis* reverse genetics system. Plant Cell 14:2985–2994

Sheng JS, Citovsky V (1996) *Agrobacterium* plant cell DNA transport: have virulence proteins, will travel. Plant Cell 8:1699–1710

Shillito RD, Saul MW, Paszkowski J, Muller M, Potrykus I (1985) High efficiency direct gene transfer to plants. Biotechnology 3:1099–1103

Shimamoto K, Terada R, Izawa T, Fujimoto H (1989) Fertile transgenic rice plants regenerated from transformed protoplasts. Nature 338:274–276

Sone T, Nagamori E, Ikeuchi T, Mizukami A, Takakura Y, Kajiyama SI, Fukusaki EI, Harashima S, Kobayashi A, Fukui K (2002) A novel gene delivery system in plants with calcium alginate micro-beads. J Biosci Bioeng 94:87–91

Spangenberg G, Neuhaus G, Schweiger HG (1986) Expression of foreign genes in a higher plant cell after electrofusion-mediated cell reconstitution of a microinjected karyoplast and a cytoplast. Eur J Cell Biol 42:236–238

Sporlein B, Koop HU (1991) Lipofectin – direct gene transfer to higher plants using cationic liposomes. Theor Appl Genet 83:1–5

Stachel SE, Messens E, Vanmontagu M, Zambryski P (1985) Identification of the signal molecules produced by wounded plant cells that activate T-DNA transfer in *Agrobacterium tumefaciens*. Nature 318:624–629

Stachel SE, Timmerman B, Zambryski P (1987) Activation of *Agrobacterium tumefaciens* Vir gene expression generates multiple single-stranded T-strand Mmolecules from the pTia6 T-region - Requirement for 5′ VirD gene products. EMBO J 6:857–863

Stachel SE, Zambryski PC (1986) *Agrobacterium tumefaciens* and the susceptible plant cell - A novel adaptation of extracellular recognition and DNA conjugation. Cell 47:155–157

Steinbiss HH, Stabel P (1983) Protoplast derived tobacco cells can survive capillary microinjection of the fluorescent dye lucifer yellow. Protoplasma 116:223–227

Supartana P, Shimizu T, Nogawa M, Shioiri H, Nakajima T, Haramoto N, Nozue M, Kojima M (2006) Development of simple and efficient in planta transformation method for wheat (*Triticum aestivum* L.) using *Agrobacterium tumefaciens*. J Biosci Bioeng 102:162–170

Supartana P, Shimizu T, Shioiri H, Nogawa M, Nozue M, Kojima M (2005) Development of simple and efficient in Planta transformation method for rice (*Oryza sativa* L.) using *Agrobacterium tumefaciens*. J Biosci Bioeng 100:391–397

Svensson I, Artursson E, Leanderson P, Berglind R, Lindgren F (1997) Toxicity in vitro of some silicon carbides and silicon nitrides: whiskers and powders. Am J Indus Med 31:335–343

Szoka F, Olson F, Heath T, Vail W, Mayhew E, Papahadjopoulos D (1980) Preparation of unilamellar liposomes of intermediate size (0.1–0.2-Mumm) by a combination of reverse phase evaporation and extrusion through polycarbonate membranes. Biochim Biophys Acta 601:559–571

Takahashi W, Shimada T, Matsushita J, Tanaka O (2000) Aluminium borate whisker-mediated DNA delivery into callus of rice and production of transgenic rice plant. Plant Prod Sci 3:219–224

Tepfer D (1984) Transformation of several species of higher plants by *Agrobacterium rhizogenes* – sexual transmission of the transformed genotype and phenotype. Cell 37:959–967

Terakawa T, Hasegawa H, Yamaguchi M (2005) Efficient whisker-mediated gene transformation in a combination with supersonic treatment. Breed Sci 55:465–468

Torney F, Trewyn BG, Lin VSY, Wang K (2007) Mesoporous silica nanoparticles deliver DNA and chemicals into plants. Nat Nanotechnol 2:295–300

Trick HN, Finer JJ (1997) SAAT: sonication-assisted *Agrobacterium*-mediated transformation. Transgenic Res 6:329–336

Trieu AT, Burleigh SH, Kardailsky IV, Maldonado-Mendoza IE, Versaw WK, Blaylock LA, Shin HS, Chiou TJ, Katagi H, Dewbre GR, Weigel D, Harrison MJ (2000) Transformation of *Medicago truncatula* via infiltration of seedlings or flowering plants with *Agrobacterium*. Plant J 22:531–541

Uchimiya H, Hirochika H, Hashimoto H, Hara A, Masuda T, Kasumimoto T, Harada H, Ikeda JE, Yoshioka M (1986) Coexpression and inheritance of foreign genes in transformants obtained by direct DNA transformation of tobacco protoplasts. Mol Gen Genet 205:1–8

van Haaren MJJ, Sedee NJA, Deboer HA, Schilperoort RA, Hooykaas PJJ (1988) Bidirectional transfer from a 24 bp border repeat of *Agrobacterium tumefaciens*. Nucl Acids Res 16: 10225–10236

van Kammen A (1997) Virus-induced gene silencing in infected and transgenic plants. Trends Plant Sci 2:409–411

Vasil V, Brown SM, Re D, Fromm ME, Vasil IK (1991) Stably transformed callus lines from microprojectile bombardment of cell suspension cultures of wheat. Biotechnology 9:743–747

Vaughan GL, Jordan J, Karr S (1991) The toxicity in vitro of silicon carbide whiskers. Environ Res 56:57–67

Veen R, Dulk-Ras H, Bisseling T, Schilperoort R, Hooykaas P (1988) Crown gall tumor and root nodule formation by the bacterium *Phyllobacterium myrsinacearum* after the introduction of an *Agrobacterium* Ti plasmid or a *Rhizobium* Sym plasmid. Mol Plant-Microbe Interact 1:231–234

Vos G, Zambryski P (1989) Expression of *Agrobacterium* nopaline-specific VirD1, VirD2 and VirC1 proteins and their requirement for T-strand production in *E. coli*. Mol Plant-Microbe Interact 2:43–52

Wallin A, Glimeliu K, Eriksson T (1974) Induction of aggregation and fusion of *Daucus carota* protoplasts by polyethylene glycol. Z Pflanzenphysiol 74:64–80

Wang K, Drayton P, Frame B, Dunwell J, Thompson J (1995) Whisker-mediated plant transformation – an alternative technology. In Vitro Cell Dev Biol Plant 31:101–104

Waterhouse PM, Graham HW, Wang MB (1998) Virus resistance and gene silencing in plants can be induced by simultaneous expression of sense and antisense RNA. Proc Natl Acad Sci USA 95:13959–13964

Weissinger A, Tomes D, Sanford J, Kline T, Fromm M (1987) Microprojectile bombardment for maize transformation. In Vitro Cell Dev Biol Plant 23:A75

Wiesman Z, Ben Dom N, Sharvit E, Grinberg S, Linder C, Heldman E, Zaccai M (2007) Novel cationic vesicle platform derived from vernonia oil for efficient delivery of DNA through plant cuticle membranes. J Biotechnol 130:85–94

Wood DW, Setubal JC, Kaul R, Monks DE, Kitajima JP, Okura VK, Zhou Y, Chen L, Wood GE, Almeida NF, Woo L, Chen YC, Paulsen IT, Eisen JA, Karp PD, Bovee D, Chapman P, Clendenning J, Deatherage G, Gillet W, Grant C, Kutyavin T, Levy R, Li MJ, McClelland E, Palmieri A, Raymond C, Rouse G, Saenphimmachak C, Wu ZN, Romero P, Gordon D, Zhang SP, Yoo HY, Tao YM, Biddle P, Jung M, Krespan W, Perry M, Gordon-Kamm B, Liao L, Kim S, Hendrick C, Zhao ZY, Dolan M, Chumley F, Tingey SV, Tomb JF, Gordon MP, Olson MV, Nester EW (2001) The genome of the natural genetic engineer *Agrobacterium tumefaciens* C58. Science 294:2317–2323

Yanofsky MF, Nester EW (1986) Molecular characterization of a host range-determining locus from *Agrobacterium tumefaciens*. J Bacteriol 168:244–250

Yanofsky MF, Porter SG, Young C, Albright LM, Gordon MP, Nester EW (1986) The VirD operon of *Agrobacterium tumefaciens* encodes a site-specific endonuclease. Cell 47:471–477

Yih TC, Al-Fandi M (2006) Engineered nanoparticles as precise drug delivery systems. J Cell Biochem 97:1184–1190

Zambryski P, Depicker A, Kruger K, Goodman HM (1982) Tumor induction by *Agrobacterium tumefaciens* analysis of the boundaries of transferred DNA. J Mol Appl Genet 1:361–370

Zambryski P, Goodman HM, Van MM, Schell J (1983a) *Agrobacterium tumefaciens* tumor induction. In: Shapiro JA (ed) Mobile genetic elements. Academic, New York, USA, pp 505–536

Zambryski P, Joos H, Genetello C, Leemans J, Vanmontagu M, Schell J (1983b) Ti-plasmid vector for the introduction of DNA into plant cells without alteration of their normal regeneration capacity. EMBO J 2:2143–2150

Zhang LJ, Cheng LM, Xu N, Zhao NM, Li CG, Yuan J, Jia SR (1991) Efficient transformation of tobacco by ultrasonication. Biotechnology 9:996–997

Zuo JR, Niu QW, Chua NH (2000) An estrogen receptor-based transactivator XVE mediates highly inducible gene expression in transgenic plants. Plant J 24:265–273

Zupan J, Muth TR, Draper O, Zambryski P (2000) The transfer of DNA from *Agrobacterium tumefaciens* into plants: a feast of fundamental insights. Plant J 23:11–28

Zupan JR, Ward D, Zambryski P (1998) Assembly of the VirB transport complex for DNA transfer from *Agrobacterium tumefaciens* to plant cells. Curr Opin Microbiol 1:649–655

Chapter 4
Selection and Screening Strategies

Haiying Liang, P. Ananda Kumar, Vikrant Nain, William A. Powell, and John E. Carlson

4.1 Introduction

A number of transformation systems have been developed to insert foreign DNA into the appropriate plant genome (nuclear or plastid) (discussed in Chap. 3). However, only a small fraction of the treated cells become transgenic, while the majority of the cells remain untransformed using any of these methods. Thus, effective selection and screening strategies are needed to pick up the rare transgenic lines from a pool of nontransformed cells or plants. To date, more than 50 marker genes and a few molecular techniques have been developed to serve this essential purpose.

In general, a marker gene is cointroduced into a plant genome along with the transgene of interest to help identify the cells that have taken up the foreign DNA, which is especially important when the transformation frequency is low (e.g., 1.0×10^{-3} to 10^{-6}) (Curtis et al. 1995). In some cases, the marker gene is the gene of interest that will express an agronomic characteristic, such as herbicide resistance. Marker genes can be divided into two categories, i.e., selectable markers and screenable (scorable, reporter, visible) markers. Selectable markers are genes that can provide a selective/metabolic advantage to the transformed cells for them to grow under conditions, which inhibit the growth of nontransformed cells. Selectable markers allow selective multiplication of transformed cells by killing or starving the nontransformed ones. Examples of selectable markers include genes that provide either antibiotic or herbicide resistance, or necessary growth regulators. Screenable markers usually encode for specific proteins, such as β-glucuronidase (GUS), green fluorescent protein (GFP), and luciferase (LUX), that can produce distinctive phenotypes, thus enable the identification of transformed cells without adding any toxic compounds. Assays for screenable markers can be destructive or nondestructive. Screenable markers are helpful in monitoring transgenic events and manually

H. Liang (✉)
Department of Genetics and Biochemistry, Clemson University, 108 Biosystem Research Complex, Clemson, SC 29634, USA
e-mail: hliang@clemson.edu

C. Kole et al. (eds.), *Transgenic Crop Plants*,
DOI 10.1007/978-3-642-04809-8_4, © Springer-Verlag Berlin Heidelberg 2010

separating transgenic tissues from nontransformed tissues, but provide no selection pressure on cells or regenerated shoots. It is not a surprise that identifying transgenic cells among many more nontransgenic cells only with screenable markers can be time consuming. Therefore, screenable markers are usually paired with selectable markers in transformation systems. Indeed, studies have shown that the number of transgenic events is reduced almost ten-fold in the absence of selectable markers (Birch 1997; de Vetten et al. 2003; Darbani et al. 2007).

Here we describe the characteristics of individual selectable and screenable marker genes available to date for plant transformation and their applications in production of transgenic plants. The molecular tools that can be utilized for transformant screening and the strategies for marker gene removal after successful transformation events are also described.

4.2 Selection Strategies

Selection strategies can be classified into two categories depending on whether they confer an advantage (positive) or a disadvantage (negative) selection. In negative selection, a more traditional mode of selection, toxic or inhibitory compounds such as antibiotics or herbicides are used to kill or prevent the growth of nontransformed cells. In the case of positive selection, transformed cells are given the ability to grow by using a specific carbon or nitrogen source or a growth regulator as the selection agent. Thus, positive selectable marker genes are defined as those that promote the growth of transformed tissue, whereas negative selectable marker genes result in death or growth inhibition of the nontransformed tissue. As a relatively new mode of selection strategy, positive selection has been demonstrated to be successful in a large variety of monocot and dicot species, and usually provides a higher transformation frequency than negative selection.

4.2.1 Positive Selection

4.2.1.1 Shoot/Root Phenotypic-Based Positive Selection (Table 4.1)

Isopentyl Transferase

The enzyme isopentyl transferase (IPT) contributes to crown gall formation in infected plants and is encoded by the T-DNA of *Agrobacterium tumefaciens* Ti plasmids. IPT catalyzes the first step in cytokinin biosynthesis, the synthesis of isopentyl-adenosine-5'-monophosphate, leading to elevated cytokinin levels in transgenic plants. Since high cytokinin:auxin ratios are required for shoot formation in culture, introduction of the *ipt* gene into plants can enhance regeneration of shoots without the inclusion of exogenous cytokinin in the media. It has been

Table 4.1 Agents and enzymes for shoot/root phenotypic-based positive selection

Agents	Genes	Enzymes	Sources	Genome	References
None	ipt	Isopentyl transferases	Agrobacterium tumefaciens	Nuclear	Endo et al. (2001)
	pga22		Arabidopsis thaliana		Zuo et al. (2002)
None	esr1	Transcription factor (enhancer of shoot regeneration 1)	Arabidopsis thaliana	Nuclear	Banno et al. (2001)
None	cki1	Histidine kinase (cytokinin-independent 1)	Arabidopsis thaliana	Nuclear	Zuo et al. (2002)
None	knotted1	Transcription factor (enhancer of shoot regeneration 1)	Zea mays	Nuclear	Luo et al. (2006)
Benzyladenine-N-3-glucuronide	uidA (gusA)	β-Glucuronidase	Escherichia coli	Nuclear	Joersbo and Okkels (1996)
None	rol	Enzymes involved in rhizogenesis	Agrobacterium rhizogenes	Nuclear	Ebinuma et al. (1997)

demonstrated that the *ipt* gene is effective as a positive selectable marker for transformation in tobacco, tomato, muskmelon, sweet pepper, and citrus (Kunkel et al. 1999; Endo et al. 2001, 2002; Mihálka et al. 2003; Ballester et al. 2007). However, the main issue with this system is that transgenic plants usually show abnormal phenotypes, such as loss of apical dominance and roots, when *ipt* is expressed constitutively. Therefore, the *ipt* gene cannot be constitutively expressed in the same way as traditional markers. Two approaches have been used to avoid this drawback. In one approach, the marker has been excised from the transgenic plant by inducible site-specific recombination (Sugita et al. 2000). The other approach has been to place the *ipt* gene under an inducible promoter, so that cytokinins are produced only when needed. Thus, regeneration is carried out in cytokinin-free medium in the presence of the inducing agent, so that only transformed cells produce the enzyme required to stimulate cytokinin synthesis (Zuo et al. 2002).

Plant *ipt* genes have been cloned from *Arabidopsis* and investigated as possible selectable markers (Kakimoto 2001; Takei et al. 2001). Overexpression of *Arabidopsis* IPT (e.g., *PGA22*) has been found to promote shoot formation from explants in the absence of external cytokinins. Direct selection with plant *ipt* genes needs to be tested in crop species.

Enhancer of Shoot Regeneration 1

The *Arabidopsis* enhancer of shoot regeneration 1 (*ESR1*) gene encodes an AP2/EREBP (APETALA2/ethylene response element binding protein)-domain-containing transcription factor. It can cause high-frequency shoot regeneration in the absence of external cytokinins (Banno et al. 2001). Several *ESR1*-like genes, e.g., *ESR2*, were also found in the *Arabidopsis* genome. The utility of the *ESR1* and *ESR1*-like genes as promising markers in genetic transformation of crop plants has not been demonstrated.

Histidine Kinase

The *Arabidopsis Cytokinin-Independent 1* (*CKI1*) gene appears to be a receptor-like histidine kinase, and was proposed as the putative cytokinin receptor (Kakimoto 1996). Overexpression of *CKI1* was able to promote shoot regeneration independent of exogenous cytokinins and caused typical cytokinin responses. Under the control of the β-estradiol-inducible promoter and in the presence of the inducer β-estradiol, *CKI1* served as a successful selection marker in *Arabidopsis* and tobacco in the absence of external cytokinins. All transformed shoots were found to develop into normal adult plants when transferred onto a noninductive medium and no nontransgenic escapes were found among the regenerated plants (Zuo et al. 2002).

Homeodomain-Containing Knotted1 Protein

Homeobox gene *knotted1* (*kn1*) is normally expressed in shoot meristem and appears to play a critical role in meristem initiation (Hake et al. 2004). Transgenic plants overexpressing *kn1* gene exhibit morphological alterations that are similar to the characteristics of *ipt*-expressing plants, including changes in leaf shape, loss of apical dominance, and production of ectopic meristems on leaves. Although the functional mechanism of the *kn1* gene is not well understood, the maize *kn1* gene produced similar effects in selection of transgenic plants as the *ipt* gene in the absence of cytokinin and auxin (Luo et al. 2006). Under the control of the cauliflower mosaic virus (CaMV) 35S promoter, transformation efficiencies with the maize *kn1* gene as selectable marker were slightly higher than those with the *ipt* gene, while three-fold higher than neomycin phosphotransferase II (*nptII*) gene. Like *ipt* gene, abnormal morphology is a major drawback of using *kn1* as a selectable marker.

β-Glucuronidase

Benzyladenine *N*-3-glucuronide is an inactive, glucuronide derivative of cytokinin. When hydrolyzed by β-glucuronidase (GUS, E.C. 3.2.1.31), benzyladenine is released from benzyladenine *N*-3-glucuronide and stimulates transformed cells to regenerate. Joersbo and Okkels (1996) first applied the *E. coli gusA* gene as a positive selection strategy in tobacco. When paired with 7.5–15 mg L^{-1}

benzyladenine *N*-3-glucuronide, the transformation frequency scored by shoot regeneration was 1.7–2.9-fold higher than that achieved by the *nptII* gene in control experiments. In addition, *gusA* can be used as a scorable marker, which obviates the need for any traditional selectable marker gene.

Root Locus (ROL) Proteins

The *root locus (ROL)* genes from the T-DNA of *Agrobacterium rhizogenes* Ri- (root-inducing) plasmid have profound effects on root development. Vilaine and Casse-Delbart (1987) and Schmulling et al. (1988) reported that a cluster of *rolA*, *rolB*, and *rolC* genes (*rolABC*) is sufficient to induce a typical root proliferation response (hairy root) on *A. rhizogenes*-infected plants in vitro when no exogenous auxins are supplied to the cultivation medium. *Rol* genes have provided a hairy root phenotypic-based selection scheme in transgenic tobacco and *Antirrhinum snapdragon* (Cui et al. 2001; Komarnytsky et al. 2004). Hairy roots are not desirable in standard transformation. Thus, *rol* genes usually are deleted after transformation and selection are done.

4.2.1.2   Carbon-Based Positive Selection (Table 4.2)

Xylose Isomerase

Plant cells from species such as potato, tobacco, tomato, and several coffee species cannot use D-xylose as a sole carbon source, unless D-xylose is isomerized to D-xylulose by xylose isomerase (D-xylose ketol-isomerase, E.C. 5.3.1.5). Xylose isomerase/xylose selection was efficient in these plants (Haldrup et al. 2001; Samson et al. 2004). The xylose isomerase genes (*xyl*A) employed were from *Thermoanaerobacterium thermosulfurogenes* or *Streptomyces rubiginosus*.

Table 4.2 Agents and enzymes for carbon-based positive selection

Agents	Genes	Enzymes	Sources	Genome	References
D-Xylose	*xyl*A	Xylose isomerase	*Streptomyces rubignosus,* *Thermoanaerobacterium thermosulfurogenes*	Nuclear	Haldrup et al. (1998)
D-Mannose	*manA (pmi)*	Phosphomannose isomerase	*Escherichia coli*	Nuclear	Joersbo and Okkels (1996)
D-Arabitol	*atlD*	d-Arabitol dehydrogenase	*Escherichia coli*	Nuclear	LaFayette et al. (2005)

Phosphomannose Isomerase

Mannose is not toxic to plant cells. However, plant cells can take up mannose and use hexokinase to convert it to mannose-6-phosphate, an inhibitor of glycolysis. The production of mannose-6 phosphate also depletes the cell of inorganic phosphate. Phosphomannose isomerase (PMI, E.C. 5.3.1.8) converts mannose-6-phosphate to fructose-6-phosphate, an intermediate of glycolysis, thus allows mannose to become a carbon source. Phosphomannose isomerase is absent in many plants except leguminous plants (Lee and Matheson 1984; Chiang and Kiang 1988). Using mannose as the selective agent (usually in combination with sucrose or glucose), the *E. coli manA* (*pmi*) gene under the control of the CaMV 35S promoter was found to be an effective selectable marker. Since its first demonstration in potato, sugar beet, and corn in 1999 (Bojsen et al. 1999), the *pmi*/mannose selection system has been utilized in many crop species, including rice, sweet orange, wheat, papaya, barley, watermelon, tobacco, sorghum, Chinese cabbage, cucumber, almond, apple, and sugarcane (Reed et al. 2001; Sigareva et al. 2004; Gao et al. 2005; Zhu et al. 2005; Degenhardt et al. 2006; He et al. 2006; Ramesh et al. 2006; Jain et al. 2007; Min et al. 2007; and references therein). In most of the cases, transformation frequencies obtained from the *pmi*/mannose selection scheme were higher than antibiotics selection, and no adverse phenotypes were observed. Besides legumes, which have PMI activity, this *pmi* system is also not suitable for transformation of grapevine and coffee since grapevine and coffee embryos can use mannose or xylose as the sole carbohydrate source (Kieffer et al. 2004; Samson et al. 2004).

D-Arabitol Dehydrogenase

While many plants cannot metabolize most sugar alcohols, including D-arabitol (Stein et al. 1997), some bacteria contain an arabitol gene encoding D-arabitol dehydrogenase (EC 1.1.1.11). This enzyme converts arabitol into xylulose on which plants can grow, since xylulose is an intermediate of the oxidative pentose phosphate pathway (Haldrup et al. 1998). An *E. coli* form of D-arabitol dehydrogenase encoding gene, *atlD*, was plant-codon modified and expressed in rice under the control of a CaMV 35S promoter. Selection with 27.5 g L^{-1} arabitol and 2.5 g L^{-1} sucrose resulted in transformation rate that was comparable to selection with hygromycin and *pmi*, while transgenic rice plants obtained with the arabitol selection scheme appeared morphologically normal during differentiation and regeneration (LaFayette et al. 2005).

4.2.1.3 Auxotrophic Markers

Auxotrophic mutants can be supplemented by transformation with a functional gene, which can serve as a selectable marker gene. Such auxotrophic marker gene enables transformed cells to synthesize an essential component, usually an amino acid, which the cells cannot otherwise produce. The surrounding medium is made to intentionally lack the essential component, which cells require to grow. Cells that have successfully incorporated the selectable marker and the rest of the gene construct will produce the essential components within the cells, and thereby survive. While this technology is commonly used in bacteria and yeast transformation, it is not widely utilized in plants due to the lack of homozygous mutants that require nutritional supplements (Aragão and Brasileiro 2002). Examples include two tobacco mutants that are deficient in threonine dehydratase or nitrate reductase (Vincentz and Caboche 1991).

4.2.1.4   Selection with Biotic and Abiotic Stresses (Table 4.3)

Pathogen Resistance

A sweet pepper ferredoxin-like protein (pflp) has antimicrobial activity in planta (Lin et al. 1997; Tang et al. 2001; Liau et al. 2003). You et al. (2003) and Chan et al. (2005) successfully utilized the *pflp* gene as selection marker and a bacterial pathogen *Erwinia carotovora* as the selection agent for transformation of *Oncidium* orchid. This selection scheme has not been reported in crops.

High Salt Tolerance

Coupled with 200 mM NaCl as the selectable agent, two salt tolerance genes *DREB2A* and *SOS1*, cloned from rice and *Arabidopsis*, respectively, were superior to using an antibiotic or herbicide for selection in producing salt tolerant rice plants (Zhu and Wu 2008).

Heat Shock Tolerance

Overexpression of the plant heat shock protein gene *HSP101* confers basal thermotolerance in crops such tobacco (Chang et al. 2007) and rice (Katiyar-Agarwal et al.

Table 4.3 Agents and enzymes for biotic and abiotic stress-based positive selection

Agents	Genes	Enzymes	Sources	Genome	References
Erwinia carotovora (*Baterial pathogen*)	*pflp*	Ferredoxin-like protein	*Capsicum annuum*	Nuclear	You et al. (2003), Chan et al. (2005)
200 mM NaCl	*dreb2a*	Transcription factor	*Oryza sativa*	Nuclear	Zhu and Wu (2008)
	sos1	Plasma membrane antiporter	*Arabidopsis thaliana*		
47°C, 60 min	*Hsp101*	Heat shock protein	*Oryza sativa*	Nuclear	Chang et al. (2007)

Table 4.4 Agents and enzymes for antibiotics-based positive selection

Agents	Genes	Enzymes	Sources	Genome	References
Spectinomycin,	*aadA*	Aminoglycoside-3‴ adenyl transferase	*Shigella* sp.	Nuclear	Svab et al. (1990)
Streptomycin				Plastid	Svab and Maliga (1993)
Streptomycin	*spt*	Streptomycin phosphotransferase	Tn5	Nuclear	Maliga et al. (1988)

2003). The feasibility of using the rice heat shock protein gene (*osHsp101*) as a selection marker was successfully demonstrated in rice under heat treatment (47°C, 60 min) for selection (Chang et al. 2007).

4.2.1.5 Antibiotics-Based Positive Selection

Under appropriate conditions, some antibiotics like streptomycin and spectinomycin bleach sensitive plant cells instead of killing them, while resistant plants stay green, thus provide a color differentiation between wild-type and transgenic plants (Table 4.4). The streptomycin phosphotransferase (*SPT*) gene from Tn5 provides resistance to streptomycin and has been used to select transgenic tobacco, driven by the T-DNA transcript 2′ promoter (Maliga et al. 1988). The efficiency of transformation using this streptomycin resistance marker was found comparable to the *npt*II gene under the control of the nopaline synthase (*nos*) promoter (Maliga et al. 1988). The *SPT* marker was also successfully applied to monitor transposon excision by providing a cell autonomous resistance phenotype (Ziemienowicz 2001). However, this marker system has not been adopted for general use. The bacterial aminoglycoside-3‴-adenyl-transferase gene *(aadA)* conferring resistance to both streptomycin and spectinomycin (Svab et al. 1990) has been used as a selectable marker in tobacco, white clover, and maize (Miki and McHugh 2004). While this gene has not been broadly utilized as a nuclear selectable marker gene for the production of transgenic plants, it is the most widely used selectable marker for plastid transformation.

4.2.2 Negative Selection

4.2.2.1 Antibiotics (Table 4.5)

Neomycin Phosphotransferase

Neomycin phosphotransferase (also known as aminoglycoside3′-phosphotransferase) confers resistance to various aminoglycoside antibiotics, including kanamycin,

Table 4.5 Toxic antibiotics for negative selection

Antibiotics	Genes	Enzymes	Sources	Genome	References
Neomycin	*neo, nptII*	Neomycin phosphotransferase	*Escherichia coli Tn5*	Nuclear	Fraley et al. (1983)
Kanamycin Paramomycin, G418				Plastid	Carrer et al. (1993)
Hygromycin B	*hph (aphIV)*	Hygromycin phosphotransferase	*Escherichia coli*	Nuclear	Waldron et al. (1985)
Aminoglycosides	*aaC3*	Aminoglycoside-N-acetyl transferases	*Serratia marcesens*	Nuclear	Hayford et al. (1988)
	6′ gat		*Shigella sp.*		Gossele et al. (1994)
Bleomycin	*Ble*	Bleomycin resistance	*Escherichia coli Tn5* *Streptoalloteichus*	Nuclear	Hille et al. (1986) Perez et al. (1989)
Sulfonamides	*sulI*	Dihydropteroate synthase	*Escherichia coli pR46*	Nuclear	Guerineau et al. (1990)
Streptothricin	*sat3*	Acetyl transferase	*Streptomyces sp*	Nuclear	Jelenska et al. (2000)
Chloramphenicol	*cat*	Chloramphenicol acetyl transferase	*Escherichia coli Tn9*	Nuclear	DeBlock et al. (1984)
			Phage p1cm	Plastid	DeBlock et al. (1985)
Kanamycin	*Atwbc19*	ATP binding cassette (ABC) transporter	*Arabidopsis thaliana*	Nuclear	Mentewab and Stewart (2005)

neomycin, geneticin (G418), butirosin, gentamycin B, and paromomycin (Norelli and Aldwinckle 1993), by catalyzing the transfer of the terminal phosphate of ATP to the drug (Jimenez and Davies 1980). In plants, because of this ATP-dependent phosphorylation, binding of the antibiotic to ribosomes in mitochondria and chloroplasts is prevented; thus, protein synthesis is not impaired. Three kinds (I, II, and III) of neomycin phosphotransferase genes (*npt* or *neo*) have been used as selection markers in plants. Among them, *nptII* (neomycin phosphotransferase II, E.C 2.7.1.95) gene from *E. coli* transposon has become the most widely used since it was first established in 1983 (Bevans et al. 1983; Fraley et al. 1983; Herrera-Estrella et al. 1983). Many crop plants, such as maize, cotton, tobacco, soybean, almond, and poplar, have been successfully transformed with the *nptII* gene. Endogenous NPTII activity is very rare in plant tissues. No adverse effects of either NPTII enzyme or the *nptII* gene on humans, animals, or the environment have been reported (Flavell et al. 1992; Nap et al. 1992; US Food and Drug Administration 1998; European Food Safety Authority 2007). In addition, NPTII protein activity can be detected by enzymatic assay.

Hygromycin Phosphotransferase

The hygromycin phosphotransferase (HPT, HPH, E.C. 2.7.1.119) enzyme, also known as aminoglycoside 4′-phosphotransferase (APHIV), gives resistance to hygromycin B antibiotic (Van Den Elzen et al. 1985). Like neomycin phospho-transferase, HPT catalyzes the phosphorylation of the hydroxyl group in the hygromycin antibiotic thus preventing its binding to ribosomes. The *hpt* gene was originally derived from *E. coli* and has been extensively utilized, especially when the use of the *neo* gene is not possible. Hygromycin B is also an aminoglycoside antibiotic, causing the same symptoms as other aminoglycoside antibiotics (Benveniste and Davies 1973). As the second most frequently used antibiotic for selection after kanamycin, hygromycin B has proved very effective in the selection of a wide range of plants, including monocots. Compared to kanamycin, hygromycin B is usually more toxic and kills sensitive cells more quickly. Since hygromycin B exhibits highly toxic effects in mammalian cells, careful working procedures are recommended (McDaniel and Schultz 1993). Working concentrations range from 20 to 200 mg ml^{-1} for plant cells.

Aminoglycoside-3-N-acetyltransferase (ACC3) and Aminoglycoside-6-N-acetyltransferase (ACC6)

The aminoglycoside acetyltransferases comprise four classes of enzymes, designated AAC(1), AAC(2'), AAC(3), and AAC(6'), according to the site of acetylation of the deoxystreptamine core of the aminoglycoside antibiotic (Braeu et al. 1984). These enzymes are common among both gram-negative and gram-positive bacteria. Three genes encoding the AAC(3) enzyme, *acc*(3)-I, *acc*(3)-III, *acc* (3)-IV, have been used successfully as selectable markers for transformation of canola, tobacco, and tomato (Hayford et al. 1988). The enzymes AAC(3)-III and AAC(3)-IV have broad substrate specificity, detoxifying gentamycin, kanamycin, tobramycin, neomycin, and paromomycin by acetylation. AAC(3)-IV also modifies aparmycin and G418. The ACC(3)-I enzyme, on the contrary, modifies only gentamycin and some close derivatives (i.e., fortimicin), and may be useful if one wants to combine it with other selection markers (Shaw et al. 1993). Lastly, an AAC(6) encoding gene *6' gat* from *Shigella* spp., when under the control of the CaMV 35S promoter, was proved as efficient for selection of transformed tobacco protoplasts as *nptII* on high levels of kanamycin (Gossele et al. 1994).

Bleomycin-Binding Protein

Bleomycins are a family of metalloglycopeptide antibiotics. They bind to specific DNA sequences and produce single-stranded and double-stranded DNA breaks. An analog of bleomycin, phleomycin differs in that one of the two thiazole ring moieties is partially saturated (Sugiura et al. 1985). Encoding a bleomycin-bind protein, two bleomycin resistance determinants (ble) from *E. coli* transposon Tn5 (*Tn5Ble*) and chromosome of *Streptoalloteichus hindustanus* (*ShBle*), respectively, have been cloned. The binding of Tn5Ble and ShBle proteins to bleomycin is irreversible, thus rendering them inactive. So far *Tn5Ble* and *ShBle* genes have been introduced into *Nicotiana* (Hille et al. 1986; Perez et al. 1989; El Amrani et al. 2004); however, only selection with phleomycin was able to generate transgenic plants. Several reports have indicated that bleomycin inhibits plant morphogenesis and transgenic plant production and *ShBle* was found to be more effective than Tn5ble under the control of the CaMV 35S promoter (Perez et al. 1989; Singh and Sansavini 1998; Schmidt et al. 2008).

Dihydropteroate Synthase

Dihydropteroate synthase (DHPS, E.C. 2.5.1.15) catalyzes a rate-limiting step for folic acid synthesis in bacteria and plants, and its enzymatic activity can be inhibited by a large number of antimicrobial compounds such as sulfonamides or sulfa drugs. The resistance gene *sul*I from plasmid R46 codes for a mutant form of DHPS that is resistant to inhibition by the sulfonamides. Highly efficient selection systems based on *sul*I and sulfonamides were demonstrated in tobacco (Guerineau et al. 1990) and potato (Wallis et al. 1996). In both cases, the mutant form of dihydropteroate synthase was targeted to the chloroplast.

Streptothricin Acetyltransferase

Streptothricins are antimicrobial agents produced by *Streptomyces* spp. The mechanism of action of streptothricins is similar to that of aminoglycoside antibiotics: inhibition of protein synthesis by binding to the ribosomal small subunit (see for review Jelenska et al. 2000). The *E. coli sat3* gene codes for an acetyl transferase activity that inactivates streptothricins. When controlled by the 35S promoter, the *sat* gene acted as a selectable marker gene in a variety of dicotyledonous plant species including tobacco and carrot (Jelenska et al. 2000).

Chloramphenicol Acetyltransferase

Antibiotic chloramphenicol inhibits protein synthesis and the uptake of cations and anions in higher plants (Jyung et al. 1965). Chloramphenicol acetyltransferase (CAT) (E.C. 2.3.1.2, CAT) catalyzes the transfer of an acetyl group from acetyl-CoA to the $3'$-hydroxy position of chloramphenicol, thus inhibiting chloramphenicol from binding to the ribosome. The *cat* gene from *E. coli* Tn9 or Phage p1cm, driven by the *nos* promoter, has been used for the selection of tobacco transformants by introduction into the nuclear or chloroplast genomes (DeBlock et al. 1984, 1985). However, selection on chloramphenicol was much less efficient than selection on kanamycin conferred by the *nptII* gene. This gene is primarily used as a reporter gene rather than a selectable marker.

ATP-binding Cassette (ABC) Transporter

An *Arabidopsis* ATP-binding cassette (ABC) transporter (*Atwbc19*) gene has been shown recently to confer kanamycin resistance in transgenic tobacco (Mentewab and Stewart 2005). Under the control of the CaMV 35S promoter, the ABC transporter's selection efficiency is comparable to that of *nptII*. This is the first identified plant gene that confers antibiotic resistance. The mechanism of action is not clear. It has been hypothesized that kanamycin is actively sequestered in the vacuole as a substrate of this ABC transporter, where it would not interfere with ribosomes in the cytoplasm, mitochondria, and chloroplasts, thereby mitigating its toxicity (Mentewab and Stewart 2005).

4.2.2.2   Herbicides (Table 4.6)

Phosphinothricin Acetyltransferase

Phosphinothricin (PPT) is an active ingredient in the broad-spectrum herbicide Basta. PPT is an analog of glutamate that inhibits the amino acid biosynthetic enzyme glutamine synthase (GS) of plants and bacteria. In plants, GS is involved in assimilation of ammonia and in regulation of nitrogen metabolism. Inhibition of GS by PPT causes rapid accumulation of intracellular ammonia levels, which leads to disruption of chloroplast structure resulting in inhibition of photosynthesis and plant cell death (Tachibana et al. 1986). PPT resistance genes *bar* and *pat* from *Streptomyces* sp. encode phosphinothricin acetyltransferase (PAT), an enzyme that acetylates the free NH_2 group of PPT, thereby rendering it nontoxic. The PAT enzymes encoded by these two genes are functionally identical and show 85% identity at the amino acid level (Wohlleben et al. 1988; Wehrmann et al. 1996). The *bar* gene is the most widely and successfully used selection marker for all the major cereal species. Basta or PPT can be used to select for PPT-resistant plants by spraying full-grown plants or by adding it to selective medium in earlier stages. In the media, $1–10$ mg L^{-1} PPT is adequate to select for transformed cells in many plant species.

5-Enolpyruvyl-Shikimate-3-Phosphate Synthase and Glyphosate Oxidase

The plastid enzyme 5-enolpyruvylshikimate-3-phosphate synthase (EPSP synthase, E.C. 2.5.1.19) is essential in the shikimate pathway for the biosynthesis of the aromatic amino acids (e.g., tryptophan, tyrosine, and phenylalanine) in plants and bacteria and a primary target of herbicide glyphosate. EPSP synthase uses phosphoenol pyruvate (PEP) and shikimate-3-phosphate as substrates to make EPSP. However, glyphosate competitively interferes with the binding of PEP to the active site of EPSP synthase, hence blocking the pathway (Anderson et al. 1988; Schönbrunn et al. 2001). Both overexpression of wild-type EPSPS and expression of mutant

Table 4.6 Toxic herbicides and selectable marker genes used for the conditional-positive selection of transgenic plants[a]

Herbicides	Genes	Enzyme	Source	Genome	References
Phosphinothricin	*pat, bar*	Phosphinothricin acetyl transferase	*Streptomyces hygroscopicus, Streptomyces viridochromogenes Tu494*	Nuclear	DeBlock et al. (1989)
Glyphosate	*epsps*	5-Enolpyruvylshikimate-3-phosphate synthase	*Petunia hybrida, Zea may, Oryza sativa*	Nuclear	Zhou et al. (1995), Howe et al. (2002), Charng et al. (2008)
	aroa		*Salmonella typhimurium, Escherichia coli*		Comai et al. (1988), Della-Cioppa et al. (1987)
	cp4 epsps		*Agrobacterium tumefaciens*		Barry et al. (1992)
	gox	Glyphosate oxidoreductase	*Ochrobactrum anthropi*		Barry et al. (1992)
Sulfonylureas	*csr1-1*	Acetolactate synthase	*Arabidopsis thaliana*	Nuclear	Olszewski et al. (1988)
Imidazolinones	*csr1-2*		*Arabidopsis thaliana*	Nuclear	Aragão et al. (2000))
Chlorsulfuron, Imazethapyr	*crs 1-4*		*Arabidopsis thaliana*	Nuclear	Ray et al. (2004)
Imidazolinone, sulfonylurea	*mals*		*Gossypium hirsutum*	nuvlear	Rajasekaran et al. (1996)
pyriminobac	*mals*		*Oryza sativa*	Nuclear	Wakasa et al. (2007)
Oxynils	*bxn*	Bromoxynil nitrilase	*Klebsiella pneumoniae subspecies ozanaenae*	Nuclear	Freyssinet et al. (1996)
Amiprophos-methyl, Chlorpropham, Pendimethalin, Norflurazon Acetochlor, Pendimethalin	*cyp1a1 cyp2c19 cyp2b6*	cytochrome P450 monooxygenase	*Homo sapiens*	Nuclear	Inui et al. (2005)
Butafenacil	*ppo*	protoporphyrinogen oxidase	*Arabidopsis thaliana, Myxococcus xanthus*		Li et al. (2003), Lee et al. (2007)
Bensulide	*opd*	organophosphate hydrolase	Pseudomonas diminuta	Nuclear	Pinkerton (2008)
Dinitroaniline	*TUAm*	mutant α-tubulin	Goosegrass (*Eleusine indica*)		Yemets et al. (2008)

[a]Table was adapted from Miki and McHugh (2004)

versions carrying one or more resistance mutations have been shown to be capable of conferring glyphosate tolerance in transformation selection. The utilization of a wild-type petunia EPSP synthase gene has been reported in transgenic petunia (Shah et al. 1986) and tobacco (van Bel et al. 2001) selection. A naturally glyphosate-resistant EPSP synthase gene from the *A. tumefaciens* strain CP4 has been used as a selective marker for transgenic soybean (Clementea et al. 2000), corn (Heck et al. 2005), wheat (Zhou et al. 1995; Hu et al. 2003), and tobacco (Ye et al. 2001). Mutant forms of the EPSP synthase genes (*aroA*) from *E. coli* (Della-Cioppa et al. 1987) or from *Salmonella typhimurium* (Comai et al. 1988) have been proved successful in selection of crop species such as cotton (Zhao et al. 2006), canola (Wang et al. 2006, 2008), and tobacco (Wang et al. 2003). Mutant maize (Howe et al. 2002) and rice (Charng et al. 2008) ESPSP genes also have been employed. The modified EPSPS enzymes encoded by these mutant genes have a decreased affinity for glyphosate while their kinetic efficiency is unaffected. In most of the cases mentioned earlier, the EPSP synthase gene was fused to a transit peptide sequence for chloroplast targeting. Working dosages of 20–200 μM glyphosate have been reported.

Glyphosate oxidase (GOX) is an enzyme that can break down glyphosate into two nontoxic compounds, aminomethylphosphonic acid (AMPA) and glyoxylate. A *GOX* gene cloned from *Ochrobactrum anthropi* strain LBAA (Barry et al. 1992) has been used as a selectable marker in tobacco, *Arabidopsis*, potato, and sugarbeet (Barry and Kishore 1995). A combination of CP4 EPSPS and GOX genes has been successfully used to transform wheat (Zhou et al. 1995) and sugarbeet (Mannerlöf et al. 1997).

Acetolactate Synthase

Acetolactate synthase (ALS, E.C. 4.1.8.13), also known as acetohydroxyacid synthase (AHAS), catalyzes the first reaction in the biosynthesis of the branched-chain amino acids isoleucine, valine, and leucine (Umbarger 1978). Inhibition of AHAS leads to the starvation of these amino acids in plants. The deficiency of these amino acids can also cause secondary effects such as accumulation of a toxic substrate (α-ketobutyrate), disruption of protein synthesis, and disruption of photosynthate transport. Eventually inhibition of AHAS leads to cell death and rapid growth cessation in susceptible species (Chaleff and Mauvais 1984; Ray 1984). Thechemical classes of commercial herbicides that can inhibit ALS include sulfonylureas (SU), imidazolinones (IM), triazolopyrimidines (TP), pyrimidinyl thiobenzoates (Saari et al. 1994 and references therein), and sulfonylamino-carbonyl-triazolinones (Santel et al. 1999). Plant species differ in herbicide susceptibility and can develop resistance to different classes of AHAS inhibitors. In most cases, resistance to AHAS-inhibiting herbicides, in otherwise susceptible species, is caused by point mutations in AHAS genes that reduce the affinity of the enzyme to herbicide inhibition (Kolkman et al. 2004). Consequently, the enzymatic pathway will continue to work, making the plants resistant to the herbicide. The mutant forms of plant AHAS can act as effective selectable marker genes when combined with AHAS-inhibiting herbicide. Mutated-AHAS (mAHAS) genes have been isolated from a number of resistant plant genomes, such as *Arabidopsis* (Haughn and Somerville 1986), lettuce (Eberlein et al. 1999), and rice (Wakasa et al. 2007). The most commonly used mAHAS genes are *crs1* genes isolated from mutants of *Arabidopsis thaliana*, which are resistant to sulfonylurea and imidazolinone herbicides. These *csr1* genes have been used in selection of several transgenic plant species, such as rice (Li et al. 1992), tobacco (Charest et al. 1990), maize (Fromm et al. 1990), canola (Miki et al. 1990), common bean (Bonfim et al. 2007), soybean (Aragão et al. 2000), potato (Andersson et al. 2003), poplar (Brasileiro et al. 1992), and jujube (*Zizyphus jujuba* Mill.) (Gu et al. 2008). It was found that an *Arabidopsis* double mutant (two mutation points) gene (*crs 1-4*) seems to provide more efficient selection than most single mutant genes (Ray et al. 2004). The mAHAS genes from rice (Wakasa et al. 2007) and cotton (Rajasekaran et al. 1996) have also been successfully used as selective marker.

Bromoxynil-Specific Nitrilase

Oxynil herbicides are phenolic molecules that inhibit photosynthesis in plants by binding to electron-transport components of photosystem II in the thylakoid membrane. Two oxynil herbicides are available: bromoxynil and ioxynil. The bromoxynil nitrilase (*bnx*) gene from *Klebsiella ozaenae* codes for a bromoxynil-specific nitrilase 3,5-dibromo-4-hydroxybenzonitrile aminohydrolase (E.C. 3.5.5.6) that hydrolyzes bromoxynil into 3,5-dibromo-4-dihydroxybenzoic acid and ammonia (Stalker et al. 1988), thus confers resistance to bromoxynil. Successful transformation using the *bnx* gene as a selectable marker has been reported in tobacco and canola without using other selectable markers (Freyssinet et al. 1996; Warwick and Miki 2004). However, the *bnx* gene has not been widely used. Cereal plants and several other monocotyledonous crops such as onions are naturally resistant to oxynil herbicides because they are able to metabolize the molecule to the non-phytotoxic benzoic acid (Freyssinet et al. 1996). Thus *bxn* is not a suitable selectable marker for the transformation of monocotyledonous species.

Cytochrome P450 Monooxygenase

P450 monooxygenases are heme proteins that use electrons from NADPH to catalyze the activation of molecular oxygen. Mammalian P450 species show overlapping and broad substrate specificity and confer the ability to metabolize a number of chemicals, including herbicides. Most classes of herbicides are aryl- or alkyl-hydroxylated or *N*-, *S*-, or *O*-dealkylated by P450 species. The phenylurea herbicide chlortoluron is detoxified either via hydroxylation of the ring-methyl or via di-*N*-demethylation (Gonneau et al. 1988). Human P450 species have been used to generate herbicide-tolerant tobacco, potato, and rice plants (Shiota et al. 1994; Inui et al. 2000, 2001). The feasibility of human P450 as a selectable marker has been tested in *Arabidopsis*. A combination of chlorpropham and amiprophos-methyl resulted in a transformation rate that was equal to that of kanamycin selection in the transgenic plants expressing *CYP1A1* and *CYP2C19* cDNAs, respectively (Inui et al. 2005).

Protoporphyrinogen Oxidase

Protoporphyrinogen oxidase (PPO, EC 1.3.3.4), a key enzyme in the chlorophyll/heme biosynthetic pathway, catalyzes the oxidation of protoporphyrinogen IX to protoporphyrin IX (Smith et al. 1993). Inhibition of PPO by the PPO family of herbicides, e.g., butafenacil, causes accumulation of protoporphyrin IX, which then causes light-dependent membrane damage (Lee et al. 1993). Plants overexpressing native PPO genes or naturally tolerant PPO showed resistance to diphenyl ether herbicide (Li et al. 2003 and references therein). A *PPO* double mutant gene was cloned from *Arabidopsis* (Li et al. 2003). In combination with butafenacil, the *Arabidopsis* mutated *PPO* proved to be an effective selectable marker in maize transformation, with transformation frequency comparable to *pat* and *pmi* systems (Li et al. 2003). A *Myxococcus xanthus* native *PPO* (*Mx PPO*, under the control of the constitutive maize ubiquitin promoter) and butafenacil (0.1 μM) selection system was recently demonstrated in rice (Lee et al. 2007).

Organophosphate Hydrolase

Bensulide herbicide is a lipid synthesis inhibitor (not at the acetyl CoA carboxylase site) (Prather et al. 2002). This herbicide can cause precocious vacuolization of meristem cells and inhibit shoot and root development (Cutter et al. 1968). A bacterial organophosphate hydrolase (OPH; EC 3.1.8.1) gene has been recently tested in maize. The encoded enzyme hydrolyzes the toxic organophosphate. It was suggested that this OPH gene may serve as screenable as well as scorable maker (Pinkerton 2008).

Mutant α-Tubulin Genes

Tubulin is the main protein component of microtubules. Antimicrotubule herbicides, such as dinitroanilines and phosphoroamidates, can directly poison microtubule dynamics in plant cells, which results in the cessation of mitosis (Morejohn and Fosket 1984). Thus, antimicrotubule herbicides have been used for chromosome doubling (see for review Khosravi et al. 2007). At higher concentration levels, this type of herbicide inhibits callus growth and plant regeneration and can serve as strong selection agents (see for review Sundar and Sakthivel 2008). The feasibility of using antimicrotubule herbicides as selective reagents in plant transformation has been explored both in monocots and dicots (see for review in Sundar and Sakthivel

2008). According to Yemets et al. (2008), a selection scheme with a combination of a mutant α-tubulin gene from goosegrass (*Eleusine indica*) and dinitroaniline generated an efficiency of transgenic plant selection that was comparable with those using kanamycin or PPT. The effective concentrations of trifluralin range from 3 to10 μM, depending on the species.

4.2.2.3 Other Toxic Compounds (Table 4.7)

2-Deoxyglucose

2-Deoxyglucose (2-DOG) is an analog of glucose. In the cytosol of plant cells, 2-DOG is phosphorylated by hexokinase yielding 2-DOG-6-phosphate (2-DOG-6-P). 2-DOG-6-P acts as a competitor of glucose-6-phosphate. 2-DOG-6-P is known to severely impair plant growth due to multiple effects in metabolism. In addition to inhibiting glycolysis and overall protein synthesis, it interferes with the glycosylation of proteins and the synthesis of cell wall polysaccharides (Kunze et al. 2001 and references therein). Two yeast genes encoding 2-deoxyglucose-6-phosphate phosphatase (EC 3.1.3.68) (DOG^R1 and DOG^R2) have been identified. When overexpressed in yeast, DOG^R1 and DOG^R2 conferred 2-DOG resistance (Randez-Gil et al. 1995). Selection based on yeast DOG^R1 and 2-DOG has been demonstrated successfully in pea (Sonnewald and Ebneth 2004), potato, and tobacco (Kunze et al. 2001). Whereas the use of this selection scheme resulted in lower efficiency for transgenic tobacco plants than the *nptII* gene, comparable efficiency was achieved in the selection of transgenic potato (Kunze et al. 2001). It was also reported that 2-deoxyglucose-6-phosphate phosphatase has narrow substrate specificity and no abnormalities were observed in the transgenic plants.

Betaine Aldehyde Dehydrogenase (BADH)

Betaine aldehyde is phytotoxic to many plant cells and has an adverse effect on growth. According to a study conducted by Rathinasabapathi et al. (1994), shoot regeneration from tobacco leaves, cotyledon expansion, and greening in germinating seedlings were severely inhibited in the presence of 5 mM betaine aldehyde. Genes encoding betaine aldehyde dehydrogenase (BADH, EC 1.2.1.8) have been cloned from several plant species, including spinach, sugarbeet, and amaranth. BADH is highly specific for betaine aldehyde and converts it to glycine betaine, which accumulates in a few crop species as an osmoprotectant. *BADH* is well suited as a chloroplast selectable marker gene. Expression in the chloroplast allowed direct selection and regeneration of transgenic tobacco plants in the presence of betaine aldehyde with an efficiency that was 25-fold higher than spectinomycin resistance conferred by the *aadA* gene (Daniell et al. 2001). But when expressed in the nuclear genome of tomato, *BADH* was not as effective as

Table 4.7 Negative selection based on other toxic compounds[a]

Drugs and analogs	Genes	Enzymes	Sources	Genome	References
2-Deoxyglucose	dogr1	2-Deoxyglucose-6-phosphate phosphatase	Saccharomyces cerevisiae	Nuclear	Kunze et al. (2001)
Betaine aldehyde	badh	Betaine aldehyde dehydrogenase	Spinacia oleracea	Nuclear plastid	Ursin (1996)
S-Aminoethyl l-cysteine (AEC)	dhps	Dihydrodipicolinate synthase	Escherichia coli	Nuclear	Perl et al. (1993)
Lysine and threonine	ak	Aspartate kinase	Escherichia coli	Nuclear	Perl et al. (1993), Tewari-Singh et al. (2004)
4-Methyltryptophan (4-mT)	tdc	Tryptophan decarboxylase	Catharanthus roseus	Nuclear	Goddijn et al. (1993)
Methotrexate	dhfr	Dihydrofolate reductase	Escherichia coli, mouse	Nuclear	Herrera-Estrella et al. (1983), Eichholtz et al. (1987)
5-Methyl-tryptophan (5-mT), CdCl2	tsb1	Tryptophan synthase beta 1	Candida albicans Arabidopsis thaliana	Nuclear Nuclear	Irdani et al (1998) Hsiao et al. (2007)
Actinonin	def2-d	Peptide deformylase	Arabidopsis thaliana	Nuclear	Hou et al. (2007)
L-O-Methylthreonine (OMT)	ilva omr1-5, omr1-7, and omr1-8	Threonine deaminase gene	Escherichia coli, Arabidopsis thaliana	Nuclear	Ebmeier et al. (2004), Garcia and Mourad (2004))
4-Methylindole (4MI), 5-Methyltrypthopan (5MT), and 7-Methyl-DL-tryptophan (7MT)	asa1d asa2	Anthranilate synthase	Oryza sativa Nicotiana tabacum	Nuclear	Yamada et al. (2005) Barone and Widholm (2008)
D-Serine	dsda	D-serine ammonia lyase	Escherichia coli	Nuclear	Erikson et al. (2005)
D-Serine, D-alanine, D-isoleucine, D-valine	daol	D-amino acid oxidase	Rhodotorula gracilis	Nuclear	Erikson et al. (2004)

Gabaculine	*heml*	Glutamate-1-semialdehyde aminotransferase	*Synechococcus PCC6301*	Nuclear	Gough et al. (2001)
Galactose	*galT*	UDP-glucose:galactose-1-phosphate uridyltransferase	*Escherichia coli*	Nuclear	Joersbo et al. (2003)
Cyanamide	*cah*	Cyanamide hydratase	*Myrothecium verrucaria*	Nuclear	Damm (2003), Weeks et al. (2000)
S-aminoethyl l-cysteine (AEC), homo-arginine	*ocs*	Octopine synthase	*Agrobacterium tumefaciens*	Nuclear	Koziel et al. (1984)

[a]Table was adopted from Miki and McHugh (2004)

nptII (Ursin 1996). *BADH* also has been used for salt resistance in several crops, e.g., tomato (Jia et al. 2001) and tobacco (Rathinasabapathi et al. 1994; Liang et al. 1997).

Dihydrodipicolinate Synthase and Aspartate Kinase

Dihydrodipicolinate synthase (DHDPS, EC 4.2.1.52) and aspartate kinase (AK, EC 2.7.2.4) are key enzymes in the aspartate family pathway, which leads to the biosynthesis of lysine, threonine, methionine, and isoleucine. Both enzymes are feedback-regulated: aspartate kinase is feedback-inhibited by lysine and threonine (LT), while dihydrodipicolinate synthase is inhibited by lysine or its toxic analog S-aminoethyl l-cysteine (AEC). Growth in the presence of milli-molar concentration of LT causes methionine starvation due to the complete inhibition of aspartate kinase activity by these two amino acids and results in strong inhibition of growth in a wide range of plant species (Rognes et al. 1983; Arruda et al. 1984; Miao et al. 1988). The DHDPS enzymes from *E. coli* are less sensitive to feedback inhibition. Tobacco plants expressing a bacterial dihydrodipicolinate synthase gene in their chloroplasts had an increased produc-tion of LT (Shaul and Galili 1991). When controlled by the CaMV 35S pro-moter, a bacterial desensitized aspartate kinase gene has been successful as a selectable marker for use in the production of transgenic potato (Perl et al. 1993) and chickpea (Tewari-Singh et al. 2004), coupled with 2 mM of each of LT. A selection system with a bacterial dihydrodipicolinate synthase gene and 0.15 mM AEC was also successful in production of transgenic potato (Perl et al. 1993). One of the potential drawbacks is that the overproduction of lysine or threonine resulting from the modification of metabolism causes abnormalities in some plants (Perl et al. 1993).

Tryptophan Decarboxylase

The enzyme tryptophan decarboxylase (TDC; EC 4.1.1.28) catalyzes the conver-sion of L-tryptophan into tryptamine (Noé et al. 1984), which is an intermediate in the biosynthesis of terpenoid indole alkaloids. Besides L-tryptophan, toxic com-pounds like 4-methyltryptophan (4-mT), 4-fluorotryptophan, and 5-fluorotrypto-phan can be substrates of TDC. When a *Catharanthus roseus* gene coding for TDC was placed under the control of the CaMV 35S promoter and introduced into tobacco, direct selection on 0.1-mM 4-mT yielded transgenic plants with the same efficiency as the *nptII* gene (Goddijn et al. 1993), and the transgenics appeared normal in the greenhouse. To date, this selection system has not been widely employed. A possible disadvantage associated with this *TDC*/4-mT selec-tion scheme is the accumulation of tryptamine in the transformed tissue. The applicability of this selection system in other plant species will depend on their endogenous tryptophan decarboxylase activity as well as their tolerance to elevated TDC-directed tryptamine levels.

Dihydrofolate Reductase

Dihydrofolate reductase (DHFR, E.C. 1.5.1.3) plays an essential role in the metabolic pathways of adenine, histidine, methionine, and thimidilate biosynthesis by reducing dihydrofolate to tetrahydrofolate. Antifolate drugs, such as trimethoprim and methotrexate (Mtx), can bind very tightly to the active site of the enzyme DHFR, therefore impairing protein, RNA, and DNA biosynthesis and, consequently, leading to cell death (Habert et al. 1981). Production of transgenic plants by using a mouse or a fungal dihydrofolate reductase gene (*dhfr*) as a new selectable marker and methotrexate (Mtx) as selection agent was successful in tobacco (Irdani et al. 1998; Aionesei et al. 2006), canola (Pua et al. 1987), and petunia (Eichholtz et al. 1987). A high transformation rate of 10% was reported in canola (Pua et al. 1987). Plant cells are generally very sensitive to low levels of Mtx presumably because of the inherent low activity of the enzyme (Ratnam et al. 1987). Methotrexate dosage ranges of $3-100$ ng mL^{-1} have been used.

Tryptophan Synthase β1

5-methyl-tryptophan, an analog of the essential amino acid tryptophan, is toxic to plants, since it inhibits tryptophan biosynthesis. Hsiao et al. (2007) recently reported that enhanced expression of *Arabidopsis* tryptophan synthase (EC 4.2.1.20) β1 (*AtTSB1*) and the use of 5-methyl-tryptophan and/or $CdCl_2$ as selection agent(s) yielded comparable transformation efficiency in *Arabidopsis* to the conventional hygromycin selection system (Hsiao et al. 2007). Thus, the TSB1 system provides a novel selection system. In addition, overexpression of *AtTSB1* in *Arabidopsis* and tomato confers tolerance to cadmium stress (Hsiao et al. 2008).

Peptide Deformylase

Peptide deformylase (DEF, EC 3.5.1.88) catalyzes the removal of the *N*-formyl group from the initiating methionine in newly translated proteins, thus is essential for all subsequent *N*-terminal protein processing as well as cell survivability. Actinonin, a specific inhibitor of peptide deformylase, has broad-spectrum herbicidal activity against a wide range of plants, including many agriculturally important weed species (Hou et al. 2007). Actinonin has been reported to cause chlorosis and severe inhibition of growth and development, thus having profound herbicidal effects when applied to many plant species both pre- and postemergence (Hou et al. 2006). A direct selection system with an *Arabidopsis Atdef2-D* and actinonin (1.2 mM actinonin) yielded transformation efficiency equal to kanamycin in tobacco (Hou et al. 2007). Cotyledons of plants expressing the *AtDEF2* transgene remained white, but all subsequent growth was normal.

Threonine Deaminase

Threonine deaminase (TD, EC 4.2.1.16) catalyzes the initial step in the synthesis of isoleucine (Ile) by deaminating of threonine to 2-ketobutyrate and ammonia. This enzyme is feedback-regulated by Ile. Overexpression of wild-type or Ile-insensitive mutant threonine deaminase genes in planta increases cellular concentrations of Ile (Slater et al. 1999) and provides resistance to L-O-methylthreonine (OMT) (Mourad et al. 1995; Garcia and Mourad 2004). A structural analog of Ile, OMT is able to compete effectively with Ile during translation and induce cell death. When coupled with OMT as the selection agent, an *E. coli* wild-type threonine deaminase gene, *ilvA,* could be utilized as a selectable marker to identify tobacco transformants (Ebmeier et al. 2004). However, the transformation efficiency was substantially lower than that observed with *nptII* using kanamycin as the selection agent. In addition, a severe off-phenotype was observed under greenhouse conditions, which correlated with increased levels of expression of the *ilvA* transgene in a subset of the transformants. The *Arabidopsis* mutant, feedback-insensitive threonine deaminase alleles (*omr1-5, omr1-7,* and *omr1-8*) may serve as better selectable markers since *Arabidopsis* plants transformed with either of these mutant genes had a normal phenotype, undistinguishable from wild-type (Garcia and Mourad 2004). However, direct selection with these *Arabidopsis* genes and OMT needs to be further demonstrated in crop species.

Anthranilate Synthase

Anthranilate synthase (AS, EC 4.1.3.27) is the first enzyme in the tryptophan biosynthetic pathway and catalyzes the conversion of chorismate to anthranilate. Its catalytic activity is regulated by feedback inhibition of tryptophan (Trp). Feedback-insensitive forms of AS have been found in a number of cell lines of crop species, including tobacco, rice, and potato (Barone and Widholm 2008 and references therein). These cell lines showed resistance to toxic Trp analogs, such as 4-methylindole (4MI), 5-methyltrypthopan (5MT), and 7-methyl-DL-tryptophan (7MT). The feasibility of the selection system with the feedback-insensitive anthranilate synthase α-subunit (*ASA*) gene in combination with the use of Trp or indole analogs as selective agent has been demonstrated in potato, rice, and tobacco (Yamada et al. 2005; Barone and Widholm 2008). Transformed plants grew normally, and a dosage of 300 μM Trp analog was proved effective for selection. Selection with AS system was as effective as hygromycin B selection in rice (monocotyledon) and kanamycin selection in potato (dicotyledon), according to Yamada et al. 2005.

D-Amino Acid Deaminase

Plants have low capacity for D-amino acid metabolism and several D-amino acids are toxic to plants even at relatively low concentrations (Brückner and Westhauser 2003; Forsum et al. 2008). Using D-amino acids and the *dsdA* (D-serine ammonia lyase) gene from *E. coli* and the *dao1* (D-amino acid oxidase) from yeast *Rhodotorula gracilis* as selectable makers has been evaluated in recent years. The *dsdA* gene encodes D-serine ammonia lyase (EC 4.3.1.19), which catalyzes the deamination of D-serine into pyruvate, water, and ammonium. D-amino acid oxidase (DAAO, EC 1.4.3.3) catalyzes the oxidative deamination of a range of D-amino acids (Alonso et al. 1998), including the toxic D-serine and D-alanine. When driven by the CaMV 35S promoter, the bacterial *dsd1* and the yeast *dao1* provided efficient selection in *Arabidopsis* and maize transformation (Erikson et al. 2004, 2005; Lai et al. 2006), coupled with appropriate D-amino acids. Transgenic plants did not exhibit any adverse phenotypes. Both *dsdA* and *dao1* markers allowed flexibility in application of the selective agent in *Arabidopsis*: it can be applied in sterile plates, in foliar sprays, or in liquid culture (Erikson et al. 2004, 2005). The yeast DAAO was able to metabolize nontoxic D-amino acids D-isoleucine and D-valine into toxic compounds and killed transgenic *Arabidopsis* (Erikson et al. 2004). Considering that the natural occurrence of D-amino acids in plants is generally low, especially with no detectable levels of D-valine and D-isoleucine (Brückner and Westhauser 2003; Forsum et al. 2008), *dao1* gene can serve as a substrate-dependent, dual-function, selectable marker in plants.

Glutamate-1-semialdehyde Aminotransferase

Glutamate-1-semialdehyde aminotransferase (GSA-AT, EC. 5.4.3.8) is involved in the C5 pathway and catalyzes the conversion of glutamate-1-semialdehyde into aminolaevulinic acid (ALA). Gabaculine (3-amino-2,3-dihydrobenzoic acid) is a bacterial phototoxin, an irreversible inhibitor of a wide range of pyridoxal-5-phosphate-linked aminotransferases (Rando 1977). A mutant form of GSA-AT, encoded by the *hemL* gene, was discovered in a gabaculine-resistant cyanobacterium, *Synechococcus* PCC6301 strain GR6 (Grimm et al. 1991) and utilized in transformation of tobacco and alfalfa as a selectable marker, where it was driven by the double CaMV 35S promoter and targeted to chloroplasts with the transit peptide of the ribulose bisphosphate carboxylase small subunit (Gough et al. 2001; Rosellini et al. 2007). Gabaculine could be applied in media or by spray. According to Rosellini et al. (2007), the gabaculine-based system is more efficient than the conventional, kanamycin-based system. The inheritance of *hemL* was Mendelian, and no obvious phenotypic effect of its expression was observed.

Galactose-1-phosphate Uridyltransferase

Galactose has long been known to be toxic to a broad range of plant species (Joersbo et al. 2003). The toxicity of galactose is believed to be due to accumulation of galactose-1-phosphate, generated by endogenous galactokinase after uptake. An *E. coli* UDP-glucose:galactose-1-phosphate uridyltransferase (EC 2.7.7.12) (*galT*) gene, driven by a CaMV 35S promoter, was found to allow transgenic shoots of potato and canola to regenerate on galactose-containing selection media, resulting in high transformation frequencies (up to 35% for potato with 1.25 g L^{-1} galactose) (Joersbo et al. 2003). However, use of the *galT*/galactose selection system did not promote regeneration of transgenic apple plants (Degenhardt et al. 2007).

Cyanamide Hydratase

Cyanamide is a nitrogen-rich compound that has been used as a nitrogen fertilizer, defoliant, and herbicide. Due to its toxicity, cyanamide has also been used as a selection agent for plant transformation paired with a gene for the enzyme cyanamide hydratase (E.C. 4.2.1.69) (*Cah*) isolated from the soil fungus *Myrothecium verrucaria* (Maier-Greiner et al. 1991). This enzyme converts cyanamide to the common metabolite urea and has an extremely narrow substrate specificity. The *Cah* gene and cyanamide selection has been used to select transformants of wheat, tobacco, potato, rice, sorghum, soybean, and tomato (Miki and McHugh 2004; Ulanov and Widholm 2007).

Octopine Synthase

Octopine synthase (also called lysopine dehydrogenase) catalyzes the synthesis of opines by the reductive condensation of certain amino acids with pyruvate. This enzyme also metabolizes lysine toxic analog S-aminoethyl l-cysteine (AEC) and toxic arginine analog homo-arginine. The gene encoding octopine synthase is part of the T-DNA component of the *A. tumefaciens* octopine Ti plasmids. Dahl and Tempe (1983) found that callus tissues expressing the enzyme appear to be 20-fold more tolerant to AEC. Selective growth of callus on AEC or homo-arginine was shown in experiments with petunia stem explants (Koziel et al. 1984) or tobacco (Van Slogteren et al. 1982). This selection scheme has not been widely adopted.

4.3 Screening Strategies

Transgenic plants are generally developed by coinsertion of a selectable marker gene with a gene of interest. Putative transgenic plants selected for antibiotic resistance further need to be evaluated for integration of the transgene and its expression because some cells may escape from the effect of antibiotics during tissue culture. Diffusion of selectable marker gene product to neighboring cells may facilitate nontransgenic cells to survive the selection pressure and regenerate. In addition, expression of the gene of interest also varies among different integration events. This necessitates a screening strategy that can confirm integration and expression of the gene of interest. Use of reporter genes (*cat, lacZ, uidA, luc, gfp*) allows discrimination of transformed and nontransformed plants and also monitoring their expression. By employing vital marker (*luc, gfp*) gene expression, protein localization and intracellular protein traffic can be observed in situ, without destroying the plants (Ziemienowicz 2001). Commonly used reporter genes have been summarized in Table 4.8.

4.3.1 Scorable Markers

A fundamental difference between a selectable marker gene and a reporter gene is that selectable markers allow transgenic cells to survive and multiply at lethal concentrations of a selective agent (e.g., antibiotic), while scorable markers produces distinct phenotype that can be easily identified in the background of non-transformed cells (e.g., GUS, GFP, luciferase) (Miki and McHugh 2004). Reporter genes encode proteins that can be detected directly (e.g., GFP), or they catalyze specific reactions the products of which are detectable (e.g., GUS and luciferase).

An ideal scorable marker should have certain desirable features such as, (1) Availability of sensitive detection system with a high signal-to-noise (endogenous background) ratio; (2) Reporter signal should be measurable quantitatively; (3) Reporter gene products should be resistant to chemicals and processes used in histological processing; (4) Histochemical assays should have low diffusion of assay products across the neighboring cells; (5) Reporter gene products should have a short half-life so that it gives a true representation of transcription activity of a cell; and (6) Assays should be nondestructive to plant tissue. Although none of the currently used reporter systems have all these features, a suitable reporter gene can be selected on the basis of experimental requirements.

4.3.1.1 β-Galactosidase

LacZ gene of *E. coli* that encodes β-galactosidase is the most extensively used reporter gene system in microorganisms and animals. β-galactosidase (E.C. 3.2.1.23) a tetramer in its active form has a molecular weight of 116 kDa and optimum pH 7.0–7.5 for its activity. *LacZ* reporter gene has also been used in plant systems (Helmer et al. 1984), but its applications have been hampered by the presence of endogenous galactosidase (David et al. 1998; Stano et al. 2002; Esteban et al. 2005), which is active at pH 7–7.5 suitable for LacZ activity. Consequently

Table 4.8 Genes for the screening of transgenic tissues/plants

Substrates	Genes	Enzymes/protein	Sources	Genome	References
X-Gal	*LacZ*	β-galactosidase	*E. coli*	nuclear	Helmer et al. (1984)
X-Gluc, 4-MUGIuc	*GUS* (*uidA*)	β-glucuronidase	*E. coli*	nuclear	Jefferson et al. (1987)
Luciferin/ coelenterazine	*LUC*/ ruc	Luciferase	*Photinus pyralis/ Renilla reniformis*	nuclear	Ow et al. (1986)
None	*GFP*	Green fluorescent protein	*Aequorea victoria*	nuclear	Chalfie et al. (1994)
oxalic acid+ N,N-dimethylaniline + 4-aminoantipyrine	*OxO*	Oxalate oxidase	*Triticum aestivum*	nuclear	Simmonds et al. (2004)
None	*R, C1* and *B*	R, C1 and B transcription factors	*Zea mays*	nuclear	Ludwig et al. (1990)
None	*Phytoene synthase*	Phytoene synthase	*Lycopersicon esculentum*	nuclear	Trulson et al. (1997)
None	*AsRed, AmCyan, ZsYellow, ZsGreen DsRed eqFP611*	Reef Coral Fluorescent Proteins	*Anemonia sulcata* *Anemonia majano* *Zoanthus* sp. *Zoanthus* sp. *Discosoma* sp. *Entacmaea quadricolour*	nuclear	Wenck et al. (2003)
Kanamycin + (^{32}P) ATP	*NPTII*	neomycin phosphotransferase II	*Escherichia coli* K12	nuclear	Fregien and Davidson (1985)
Ariginine + pyruvate + NADH	*NOS, OCS*	nopaline synthase, octopine synthase	*Agrobacterium tumefaciens*	nuclear	Johnson et al. (1974)
(^{14}C) chloramphenicol + acetyl-coA	*CAT*	Chloramphenicol acetyltransferase	*E. coli* Tn9	nuclear	Herrera-Estrella et al. (1984)

the *LacZ* reporter gene has not been as widely used in plants as in microorganisms and animals.

β-galactosidase can be assayed directly in plant extracts when they contain high levels of LacZ expression. At low levels of expression, endogenous galactosidase and LacZ activities are separated electrophoretically followed by detection of the enzymes with a fluorogenic substrate (4-methyl umbelliferyl-3-D-galactoside). LacZ protein is resistant to various physiochemical factors. Tissue containing LacZ can be fixed with glutaraldehyde without loss of activity. These factors make *LacZ* a good reporter gene for histochemical analysis (Teeri et al. 1989).

4.3.1.2 GUS, β-Glucuronidase (*uidA*)

Among all reporter genes developed so far, *E. coli* gene *uidA*, encoding β-glucuronidase (GUS, E.C. 3.2.1.31) has been the most extensively used reporter gene in transgenic plants. The β-glucuronidase is a homotetramer with a molecular mass of approximately 68 kDa and a pH optimum of 7–8. The *uidA* reporter system possesses most of the features required for use in plants, such as ease of assay, high sensitivity, sufficient specificity of the enzymatic reaction, possibility of histochemical localization, activity of the enzyme in translational fusions, and availability of substrates for spectrometric, fluorometric, and histochemical assays (Jefferson et al. 1987). There is little or no detectable β-glucuronidase activity in almost any higher plant at pH levels used in the assay and endogenous GUS activity is abolished by including methanol in the assay buffer (Hu et al. 1990; Kosugi et al. 1990) (Fig. 4.1a).

For spectrometric, fluorometric, and histochemical assays, p-nitrophenyl-p-D-glucuronide, 4-methylumbelliferyl glucuronide (4-MUGIuc), and 5-bromo-4-chloro-3-indolyl glucuronide (X-Gluc) are commonly used substrates. β-glucuronidase cleaves 4-methylumbelliferyl glucuronide (4-MUGIuc) to fluorescent compound 4-methylumbelliferon, while the colorless, water-soluble product of the enzymatic cleavage of 5-bromo-4-chloro-3-indolyl glucuronide (X-Gluc) undergoes an oxidative dimerization to yield an indigo blue precipitate. The disadvantage of all the GUS assays is that the plant tissue has to be discarded.

One way of performing a nondestructive GUS assay is to test the excreted β-glucuronidase in liquid plant culture media or wound exudates, with florescent substrate 4-MUGIuc. In another viable test, either X-Gluc is applied to solid growth media for root staining or 4-MUGIuc is sprayed on β-glucuronidase expressing leaves (Martin et al. 1992). This GUS assay can be used for screening large transgenic populations, but high toxicity of X-Gluc and weak detection limit its applications.

While the stability of GUS protein is a major advantage to analyze the activities of weak promoters, it can interfere with the correct interpretation of promoter reporter expression data. A rapid change in the transcription of the reporter gene may not be reflected in corresponding rapid changes in GUS activity (Taylor 1997). Secondly, 5-bromo-4-chloro- 3-indoxyl (X-gluc product) diffuses into the

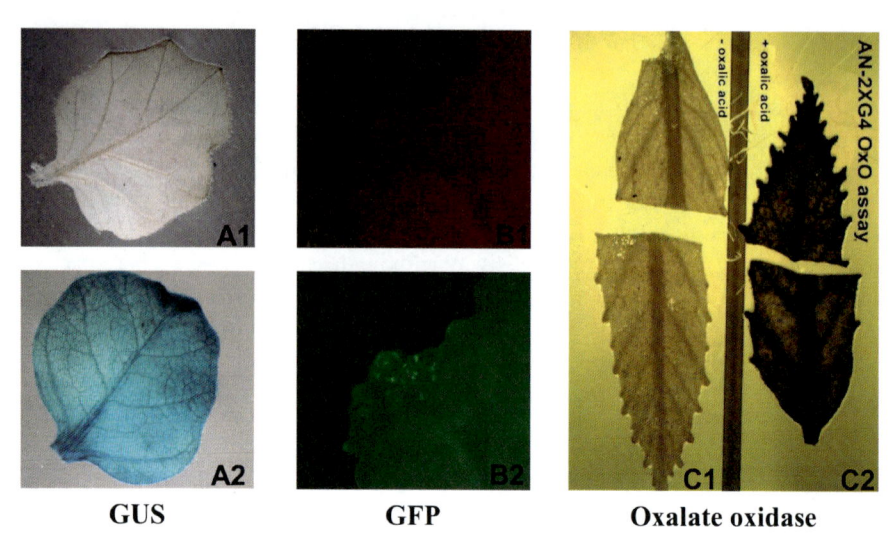

GUS   **GFP**   **Oxalate oxidase**

Fig. 4.1 (**a**) Histochemical analysis of GUS activity driven by the CaMV 35S promoter in nontransgenic (A1) and transgenic eggplant (*Solanum melongena*) (A2) (V Nain, unpublished). (**b**) GFP transformed cells (green spots) on American chestnut (*Castanea dentata*) somatic embryos under different lights and filters (WA Powell & CA Maynard, unpublished). Chlorophyll emits red florescence at 680 nm (B1) while green florescence is emitted at 522 nm (B2) that can be distinguished from background. (**c**) An oxalate oxidase assay: without oxalic acid (negative control) (C1) and with oxalic acid added in assay buffer (C2). This is of tissue culture leaves from transgenic American chestnut transformed with OxO transgene driven by a vascular promoter (LCG Northern, CA Maynard, and WA Powell, unpublished)

neighboring cells, where it precipitates giving a false GUS expression signal in the nonexpressing cells (Mascarenhas and Hamilton 1992). Several reports involving analysis of temporal expression patterns (Hird et al. 1993; Caissard et al. 1994; Treacy et al. 1997; Ariizumi et al. 2002; de Ruijter et al. 2003) have commented on the possible errors in deducing activities using GUS as a reporter. Although technical solutions have been proposed to alleviate these artifacts (De Block and Debrouwer 1993), they have not always been found to be adequate (de Ruijter et al. 2003; Kavita and Burma 2008).

Another artifact in GUS assay arises with residual *Agrobacterium* cells (harboring GUS construct) that have been used for infection of explants, in a genetic transformation experiment. Leaky GUS expression in *Agrobacterium* cells stains blue in a GUS assay, which in turn gives a false GUS signal to the plant tissue as well. This problem is circumvented by inserting an intron (with a stop codon) in the coding region of GUS. As *Agrobacterium* does not have a splicing mechanism and the intron has a stop codon in the reading frame, GUS transcripts formed by leaky expression in *Agrobacterium* do not translate into functional GUS enzyme. Whatever GUS activity is detected comes from the plant tissue only. The pCAMBIA series of vectors such as pCAMBIA1201; 1301; 2201; and 2301 consist of an intron (from the castor bean catalase gene) in the coding sequence to ensure that the activity is derived only from β- Glucuronidase expressed in the plant cell.

4.3.1.3 Luciferase

The search for a reporter gene that could be used in viable plant tissues and whose signal is representative of changes in transcriptional activity of the gene at any given time, culminated with the development of the North American fire fly (*Photinus pyralis*) luciferase (LUC) gene as a reporter system (Ow et al. 1986). The enzyme luciferase (E.C. 1.13.12.7) confers on the organism the ability to glow (exhibits luminescence at 562 nm) in the dark. In the first step, luciferase catalyzes the oxidative carboxylation of luciferin (6-hydroxy-benzothiazole) to excited form of oxyluciferin in the presence of ATP, Mg^{2+}, and O^2. The reaction produces a light flash at a maximum of 562 nm.

Hydrozoan's (*Renilla reniformis*) luciferase is encoded by the *ruc* gene. When used with luciferin and ATP, firefly luciferase/luciferin emits light at 560 nm, while *Renilla* luciferase/coelenterazine emits light at 475 nm. Because these two reporter systems emit light at quite different wave lengths, it is possible to use firefly luciferase/luciferin and *Renilla* luciferase/coelenterazine as a dual reporter system. Luciferases have been isolated from bacteria (*Vibrio harveyi*) also. The bacterial luciferase (LUX, E.C. 1.14.14.3) is a heterodimer, with two peptide subunits encoded by genes *lux* A and *lux* E, while firefly luciferase consists of a single polypeptide encoded by a gene *luc*.

There are two different methods of luciferase substrate application to plant tissues. In in vivo methods plants are grown in substrate supplemented medium, so that the substrate is absorbed through roots and gets distributed in the plant. This method needs extensive physical handling and a long time period for plant growth. It also limits the size of plant that can be analyzed. In the second method luciferin solution containing a mild detergent is sprayed on leaves. It requires only 10-min incubation time before measuring the light emission.

Advantages of the *luc* reporter gene are that it is not destructive to the plant tissue and has a short half-life in vivo that generally reflects real-time gene expression status in the transgenic tissue under investigation. Because of the high reaction efficiency of the firefly luciferase, this reporter gene is excellent for screening purposes.

4.3.1.4 GFPs

Since the first reports of *Aequorea victoria* green fluorescent protein (AvGFP) as a reporter gene in the nematode *Caenorhabditus elegans* (Chalfie et al. 1994), it has found wide applications in transgenic plants. O. Shimomura, M. Chalfie, and R. Y. Tsien won the 2008 Nobel prize in chemistry for discovery and development of GFP. Its high sensitivity, absence of external substrate application, and viability of the tissue under testing make it an ideal reporter gene (Fig. 4.1b). GFPs have several advantages over the previously utilized markers for transformation, gene expression, and protein localization studies (Stewart 2001, 2006).

Because of the presence of a cryptic splice site, native AvGFP does not express up to its detection limit in plants (Haseloff et al. 1997). Removal of the splice site and targeting to endoplasmic reticulum resulted in high expression of AvGFP in plants. Another green fluorescent protein, AcGFP, has been isolated from a nonbioluminescent jellyfish (*Aequorea coerulescens*) (Gurskaya et al. 2006). Moreover, a full range of color variants of fluorescent proteins are now available for transformation, with the opportunity to further customize fluorescence to specific applications (see Sect. 3.1.5 and reviews by Galbraith 2004 and Stewart 2006). Synthetic GFP (*sgfp*) genes that have codon usage and RNA stability optimized for plant expression are even better reporters, especially in comparison to GUS, for assessment of the temporal activities of promoters that have very narrow windows of expression. This is due to the short half-life of the GFP protein (18 h) (de Ruijter et al. 2003) as compared with that of GUS, that has a half-life of 3–4 days in tobacco plants (Weinmann et al. 1994) and about 50 h in tobacco protoplasts (Jefferson 1987; Jefferson et al. 1987).

Another application of GFP is in the determination of zygosity of transgenic plants. Fluorescent proteins can provide instantaneous data on homo- or heterozygosity. Halfhill et al. (2003) have found that heterozygous (hemizygous) transgenic canola plants exhibit half of the green fluorescence of homozygous plants and expression levels are inherited quantitatively at the same heterologous level of florescence in wild relatives also. This finding opened GFP application to the analysis of hybridization and introgression status in transgenic crops.

4.3.1.5 Reef Coral Proteins

Biotechnological advances with AvGFP reporter gene have increased the demand of florescent proteins with different emission colors. Scientists have been successful in cloning *AsRed, AmCyan, ZsYellow, ZsGreen,* and *DsRed* genes from reef corals. The first report of the expression and characterization of one of these reef coral proteins as a marker in plants was DsRed (Jach et al. 2001) isolated from the reef coral *Discosoma* sp. Unlike *A. victoria* GFP, which is a monomer, reef coral florescent proteins are homotetramers that limit their application as fusion tag. Some of the first anthozoan fluorescent proteins were reported by Matz et al. (1999, 2002). Following these reports, Wenck et al. (2003) demonstrated that AsRed, AmCyan, ZsYellow, ZsGreen, and DsRed could be expressed and visualized in several monocotyledonous and dicotyledonous plants in both transient and stable gene integration. The AmCyan1, AsRed, and DsRed transgenic callus appeared to be yellow-green and red under white light. This is significant because, unlike GFP that require low wave length (UV) light source, AmCyan1, AsRed, and DsRed provided passive altering of tissues under room light (Wenck et al. 2003).

Many plants contain phenolic compounds that emit green fluorescence under UV light, used for GFP excitation and mimic GFP results (Stewart 2006). In this context, red fluorescent proteins gain an interest in plant applications because red fluorescence is rarely observed when higher wave lengths' light source (used red fluorescent proteins) is used for analysis of plant samples. As chlorophyll does not autofluoresce at higher wavelengths used for either AsRed or DsRed, these systems may be the best choice for utilization in plant biology.

One limitation of AsRed and DsRed is that in the native form both the proteins are found as tetramers, so they cannot be utilized as fusion tags in localization studies. Recently a monomeric form of DsRed has been produced (Merzlyak et al. 2007). Red fluorescent protein DsReD and its variant DsReD2 from reef corals have been used in transient assay and under stable nuclear genome integration in tobacco and soybean (*Glycine max*), respectively (Jach et al. 2001; Nishizawa et al. 2006). Fertile transgenic plants were regenerated without any negative morphogenic or physiological effect. In confirmation with Wenck et al. (2003) transgenic plants were distinguishable from nontransgenics under white light and worked as a visual marker for transgene expression.

Another red fluorescent protein, eqFP611, has been isolated from the sea anemone (*Entacmaea quadricolor*). Forner and Binder (2007) reported that transient and stable expression of eqFP611 protein had no detrimental effects on cell viability. Targeting of eqFP611 protein to mitochondria inherited mitochondria florescence. Another application of eqFP611 was its compatibility with GFP in dual labeling of plant cell organelles.

4.3.1.6 Oxalate Oxidases

The oxalate oxidases (OxO), which belong to the family of the germin-like proteins, catalyze the oxidation of oxalate, whereby hydrogen peroxide is formed. OxO enzymes are absent in most of the dicotyledonous plants and have a narrow window of expression in monocots (Grzelczak et al. 1985; Caliskan and Cuming 1998). Inexpensive substrates and availability of detection and quantification protocols (Thompson et al. 1995; Zhang et al. 1996) further add to the suitability of *OxO* as a reporter gene for plants (Simmonds et al. 2004). Activity of OxO can be measured histochemically as well as quantitatively (Simmonds et al. 2004) (Fig. 4.1c).

4.3.1.7 Anthocyanin Formation (Maize *R*, *C1*, and *B* Transcription Factors)

Maize *R*, *C1*, *P1*, and *B* transcription factor genes regulate the anthocyanin biosynthesis pathway in a tissue-specific manner in maize. The introduction of *C1*, R, and *B* regulatory genes under the control of constitutive promoters induces cell-autonomous anthocyanin pigmentation and allows for direct visualization of transformed cells and tissues (Ludwig et al. 1990; Radicella et al. 1992). Screening of transgenic tissue on the basis of anthocyanin pigmentation is a reporter system that would not require the application of selection pressure or external substrates for the detection of transgenic cells. Introduction of a plasmid encoding the maize *R* and *C*1 transcriptional factors, each under the control of a separate CaMV 35S promoter with maize *Adh1* intron, into immature wheat embryos resulted in the production of anthocyanin expressing cells. Pigmented cells were observed in the callus derived from these embryos for up to 1 month after bombardment, but these cells failed to proliferate (McKinnon et al. 1996). In a similar study, in which suspension cells of wheat were cotransformed with an anthocyanin marker and a selectable marker, anthocyanin expressing callus was isolated (Dhir et al. 1994). Toxicity of these genes toward transformed cells and the requirement of environmental factors for expression (Chawla et al. 1999) limit the use of these genes as a reporter system.

4.3.1.8 Phytoene Synthase

Genes involved in carotenoid biosynthesis can be used as visual markers for identification of transgenic cells. The phytoene synthase enzyme catalyzes a reaction to produce phytoene from geranylgeranyl pyrophosphate. In the carotenoid biosynthesis pathway, phytoene is a precursor of the red carotenoid lycopene, a carotenoid that gives tomato fruit red color, followed by β-carotene in the next reaction. Trulson et al. (1997) reported that expression of phytoene synthase (from *Erwinia herbicola*) under transcription regulation of tomato callus-specific E8 promoter resulted in orange pigmentation in the callus. E8-phytoene synthase transgenic tomatoes were phenotypically similar to nontransgenic tomato plants, with the only difference being that fruits of transgenic plants developed color earlier than nontransgenic plants. Although phytoene synthase seems to be a good visual marker, its general application in other transgenic systems has been hampered because the other genes required for carotenoid synthesis are not present in all other plant species.

4.3.1.9 NPTII

NPTII, the most commonly used selectable gene, from transposon 5 (Tn5) of *E. coli* K12, encodes aminoglycoside 3-phosphotransferase II (APHII), commonly known as neomycin phosphotransferase II (NPTII, E.C 2.7.1.95), which inactivates the sugar-containing antibiotics, neomycin, kanamycin, geneticin (G418), and paromomycin by phosphorylation. Endogenous NPTII activity is rarely observed in plant tissues, making it a suitable reporter gene for plant applications. For an enzyme assay of NPTII, protein sample is first fractionated using nondenaturing polyacrylamide gel electrophoresis (PAGE), followed by phosphorylation of kanamycin with radioactively labeled ATP (^{32}P), by layering kanamycin containing agar over the enzyme containing polyacrylamide gel. The whole set is incubated at 35°C and the phosphorylation leading to incorporation of ^{32}P in kanamycin is detected by autoradiography (Fregien and Davidson 1985). Alternatively, filters with dot blots of the protein sample can be incubated with the substrates and then subjected to autoradiography. Presence of *NPTII* gene product can also be easily quantified by using NPTII-specific antibodies in an ELISA test (Nagel et al. 1992.)

4.3.1.10   Opines

The presence of opines in tumors in plants was discovered long before the identification of pathogenic Ti (tumor inducing) and Ri (root inducing) plasmids of

Agrobacterium or the demonstration of T-DNA transfer (Johnson et al. 1974). An advantage of opines as screenable markers for plant transformation is that they are natural markers of genetic transformation (crown gall or hairy root cells). The presence of opines in any plant material clearly indicates the transformed status of the plant cells. As a result, opine synthesis and the related genes have been widely used to construct numerous *Agrobacterium*-based vectors designed to engineer plant cells. Most of these vectors carry a wild-type or a modified *nos* (nopaline synthase) or *ocs* (octopine synthase) genes.

Nopaline and octopine are generally detected in plant extracts by high-voltage paper electrophoresis followed by reaction with phenanthrenequinone. Presence of UV-fluorescent products indicates the presence of opines. Yang et al. (1987) introduced a heat treatment step, compatible with paper electrophoresis that results in rapid production of a red-purple pigment. This colorimetric assay is sensitive to 1.25-µg quantities of opine and eliminates problems of background fluorescence encountered with crude plant extract.

4.3.1.11 Chloramphenicol Acetyl Transferase

Chloramphenicol acetyltransferase (CAT, E.C. 2.3.1.2), from the *E. coli Tn9* gene, neutralizes the antibiotic chloramphenicol by transfers of acetyl groups and thus changes its structure and prevents the antibiotic from inhibiting protein synthesis. CAT was the first bacterial gene to be introduced in plant cells (Herrera-Estrella et al. 1984), and it is still widely used as a reporter gene today, because of the stability of the enzyme and high sensitivity and ease of the enzymatic assay. The *CAT* gene is absent in mammals and higher plants, so its activity can be measured in the plant extract without any electrophoretic separation.

To measure the CAT activity, extracts from CAT transgenic plants are incubated with radiolabeled chloramphenicol. The acetylated products generated by the action of CAT are separated from the unmodified chloramphenicol by thin-layer chromatography and quantified by scraping the spots from the thin-layer plates and counting them by scintillation spectroscopy. Another assay substitutes the standard acetyl donor with butyl-CoA; the higher hydrophobicity of the butyl-chloramphenicol allowing for good separation of the butyl-compound and easy quantification of a large number of samples. Some plants have a nonspecific acetylase that can acetylate chloramphenicol as well, that necessitate use of controls. Extracts of some plant species may contain CAT enzyme inhibitors, although this problem can be eliminated in most cases with a heat treatment at 65°C.

4.3.2 PCR-Based Screening

The polymerase chain reaction (PCR) method is most sensitive among all molecular biology techniques used for testing the presence of the specific DNA sequence of a gene. Because of its sensitivity, the presence of a small quantity of contaminating

DNA can show false positives. For the screening of transgenic plants, PCR is generally performed with primers specific to the selectable marker and gene of interest, used for developing the transgenic plants. When large numbers of transgenic plants are screened, isolation of DNA becomes labor- and time intensive. As the PCR reaction does not require highly purified DNA, protocols have been standardized for rapid DNA isolation without freezing the plant samples in liquid nitrogen. A duplex PCR reaction is performed using two set of primers – one for the gene of interest (transgene or marker gene) and another for an enodgenous plant gene to confirm the DNA quality for PCR amplification (Mannerlof and Tenning 1997; Xu et al. 2005).

The advent of real-time PCR has made multiplexing and relative quantification easy. As compared with traditional transgene copy number detection technologies such as Southern blot analysis, real-time PCR provides a fast, inexpensive, and high-throughput alternative. Real-time PCR can be used to determine copy number and zygosity in transgenic plants (as reviewed by Bubner and Baldwin 2004; Prior et al. 2006; Yuan et al. 2007). The availability of different chemistries for fluorescence detection in real-time PCR created some confusion about their relative merits. In a comparative study, molecular beacon, SYBR Green, TaqMan, and MGB assays were designed for the event-specific detection and quantification of the 3' integration junction of GTS 40-3-2 (Roundup Ready) soybean. Sensitivity as well as robustness in the presence of background DNA was tested. None of the PCR-based approaches appeared to be significantly better than any of the other (Andersen et al. 2006). In another study, five different chemistries employing TaqMan, Lux, Plexor, Cycling Probe Technology, and LNA were tested, and it was concluded that none of chemistries outperformed the others (Gasparic et al. 2008).

Adaptor ligation PCR (AL-PCR) of genomic DNA flanking T-DNA borders is another attractive alternative to genomic DNA blot hybridization. An adapter is ligated to restriction enzyme-digested genomic DNA and PCR is carried out using primers specific for a T-DNA border and the adapter sequence. Each independent integration event shows a PCR amplicon of different size (Spertini et al. 1999). Alternatively PCR amplicons can be sequenced to find the locus of integration. The AL-PCR patterns obtained in *Allium cepa* were specific and reproducible for a given transgenic line and gave insight in the number of T-DNA copies (Zheng et al. 2001).

The presence of latent *Agrobacterium* in the plant can give false-positive results in a PCR analysis of putative transgenic plants. Nain et al. (2005) have reported a simple protocol for PCR analysis of *Agrobacterium*-contaminated transgenic plants that is based on denaturation and renaturation of DNA in a time-dependant manner. The contaminating plasmid vector becomes double stranded most quickly during renaturation and is cut by a restriction enzyme having site(s) within the PCR amplicon. Once this plasmid DNA is digested, it will eliminate PCR amplification from contaminating plasmid DNA. The genomic DNA with a few copies of the transgene remains single stranded and unaffected by the restriction enzyme, leading to amplification by PCR. Hence, only the transgene present in the genomic DNA is amplified by PCR (Fig. 4.2).

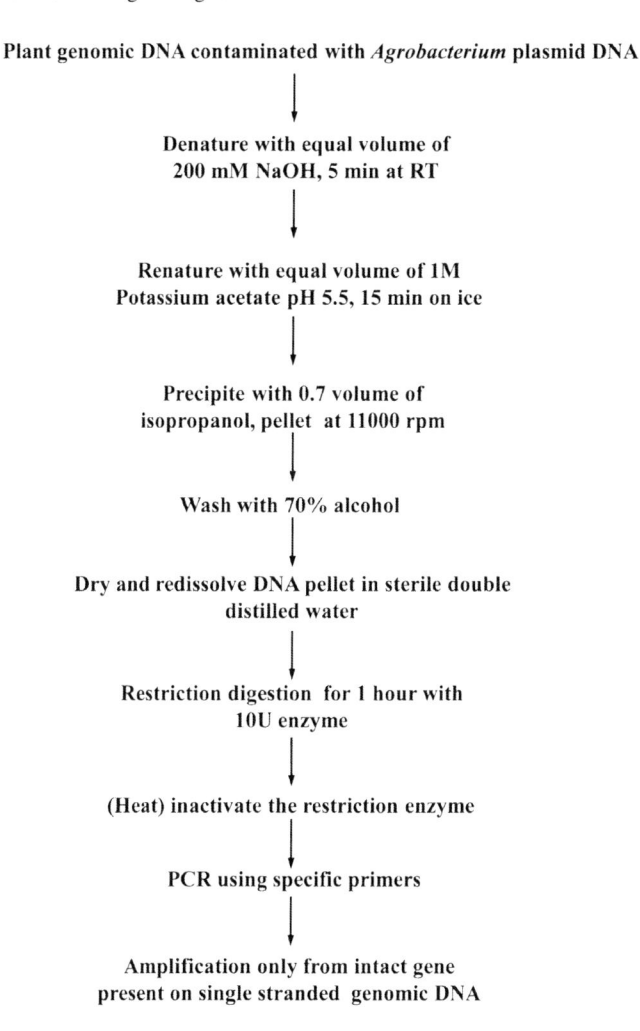

Plant genomic DNA contaminated with *Agrobacterium* plasmid DNA

↓

**Denature with equal volume of
200 mM NaOH, 5 min at RT**

↓

**Renature with equal volume of 1M
Potassium acetate pH 5.5, 15 min on ice**

↓

**Precipite with 0.7 volume of
isopropanol, pellet at 11000 rpm**

↓

Wash with 70% alcohol

↓

**Dry and redissolve DNA pellet in sterile double
distilled water**

↓

**Restriction digestion for 1 hour with
10U enzyme**

↓

(Heat) inactivate the restriction enzyme

↓

PCR using specific primers

↓

**Amplification only from intact gene
present on single stranded genomic DNA**

Fig. 4.2 Confirmation of transgene integration in plant genome by PCR. After processing the genomic DNA from putative transgenic plants, PCR amplification comes only from transgene integrated in plant genome Nain et al. (2005)

4.3.3   *Southern Hybridization Analysis*

Southern blotting (Southern 1975) is a technique for transfer of single-stranded (denatured) DNA molecules from an electrophoresis gel to a nitrocellulose or nylon membrane. The nitrocellulose membrane is incubated with a specific radiolabeled probe and the location of the DNA fragment that hybridizes with the probe is detected by autoradiography (Sambrook et al. 1989) or with a Phosphor Imager (for Quantitative Filmless Autoradiography). High sensitivity, low background, and

1.   Restriction digestion of DNA samples (5-10 µg)
     Resolve on agarose gel electrophoresis

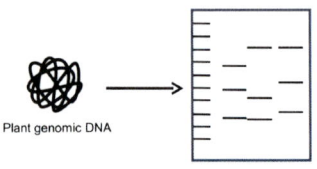

Plant genomic DNA

Agarose gel electrophoresis

2.   Set up transfer assembly
     Allow capillary transfers for 16hrs
     Rinse membrane in 2x SSC buffer
     UV-crosslink Nylon membrane

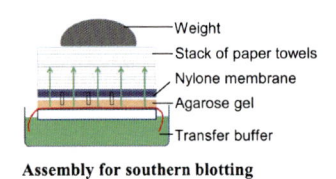

Assembly for southern blotting

3.   Add prehybridization solution 4-6 hrs to block the membrane
     Synthesis of labeled double-stranded DNA probes
     Denaturation to single-stranded probes

4.   Remove prehybridization solution and add hybridization solution
     Hybridize overnight at appropriate temperature.

5.   Autoradioragraphy
     Non-isotope detection system
     •   Chemiluminescent detection
     •   Flurerscence detection

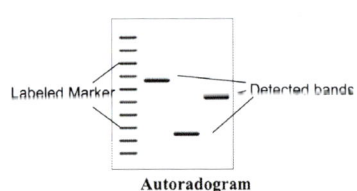

Autoradogram

Fig. 4.3 Methodology for confirmation of transgene integration in plant genome by Southern hybridization

determination of transgene copy number integrated to plant genome make this technique indispensable for molecular analysis of transgenic plants (Fig. 4.3).

The standard radioactively labeled DNA detection system used in such experiments requires licensing to handle radioactive materials. The development of a nonradioactive biotin, digoxigenin (DIG), and fluorescein isothiocyanate (FITC) labeling-based nucleic acid detection system has made such DNA hybridization experiments safe and more generally feasible (Leary et al. 1983; Peterhaensel et al. 2007). Bound biotin labeled probe is detected by an alkaline phosphatase conjugate of streptavidin, while for digoxigenin detection an antidigoxigenin antibody conjugate is used. Alkaline phosphatase conjugate is visualized with chromogenic alkaline phosphatase substrates. Chromogenic substrates produce a colored signal directly on the membrane. More sensitive chemiluminescence substrates for alkaline phosphatase produce light that can be conveniently recorded with X-ray film (as with ^{32}P probes). Fluorescence signal emitted by FITC labeled probe is recorded with chemiluminescence detection system. Although various modifications of Southern hybridization have been reported, the use of radiolabeled probes is still the most common and the most reliable method.

When determining the integration of transgenes into the plant genome, Southern hybridization is the most reliable technique. Sometimes reporter gene signal may be absent even after the integration of transgene in the plant genome. This may be because of silencing of the transgene after integration, low RNA stability, poor translation efficiency or integration of a truncated gene. Under these conditions only Southern hybridization can give a true picture of an integration event. Furthermore, Southern hybridization can reveal the presence of genomic DNA flanking the transgene, which is the ultimate confirmation of insertion, and which is not always possible using PCR. However, Southern hybridization requires a large quantity (5 μg) of DNA from each transgenic plant that can limit its application when plant tissue is not available in adequate quantity.

To determine the transgene copy number and distinguish different integration events, the plant genomic DNA is digested with a restriction enzyme that does not have a restriction site in the T-DNA and when the T-DNA or transgene is used as a probe, each independent integration event will appear as a unique band. Presence of more than one band will indicate a multicopy transgenic event. If the restriction enzyme used for genomic DNA digestion has a unique site between the marker gene and gene of interest (transgene) and the membrane is probed with the marker gene and the transgene separately, then each integration event will show up as a single band. The size of the band identified with each probe will be different, but the number of bands in each plant sample will remain the same. Any mismatch will represent an integration of a truncated copy. To identify a multiple-copy tandem integration at one locus, genomic DNA is digested with restriction enzymes flanking the gene of interest in the T-DNA. Probing the blot with the transgene will give a single band of size of the transgene in all the samples, but intensity of the band will be proportionate to the number of gene copies integrated (Bhat and Srinivasan 2002).

4.3.4 ELISA

Enzyme-linked immunosorbent assay (ELISA) tests for the presence of the specific protein that the transgene produces in the plant. ELISA procedures use antibodies that react specific with the new protein(s) produced in the transgenic plants. There are different versions of the ELISA method used for detection of heterologous protein expressed in transgenic plants. One method uses lateral flow strips that deliver results in 2–5 min. This "strip test" technology is also marketed as the "dipstick" procedure. Advantages of the ELISA strip tests are speed, relative ease of use, and low cost. On the other hand, a major disadvantage is that it cannot quantify the protein of interest in a transgenic plant sample. Another version of the ELISA test, the "plate test," allows quantification of protein of interest. A standard ELISA plate can test 96 samples at a time, including positive and negative controls. Intensity of color indicates the amount of the protein present (Fig. 4.4). The plate test can take 2–4 h and is more laborious and costly than the strip test.

4.4 Marker-Removal Strategies

Marker genes play a crucial role during plant transformation for identifying rare transgenic cells/plants. However, the presence of marker DNA sequences in the final transgenic plants is often problematic for commercial biotechnology products

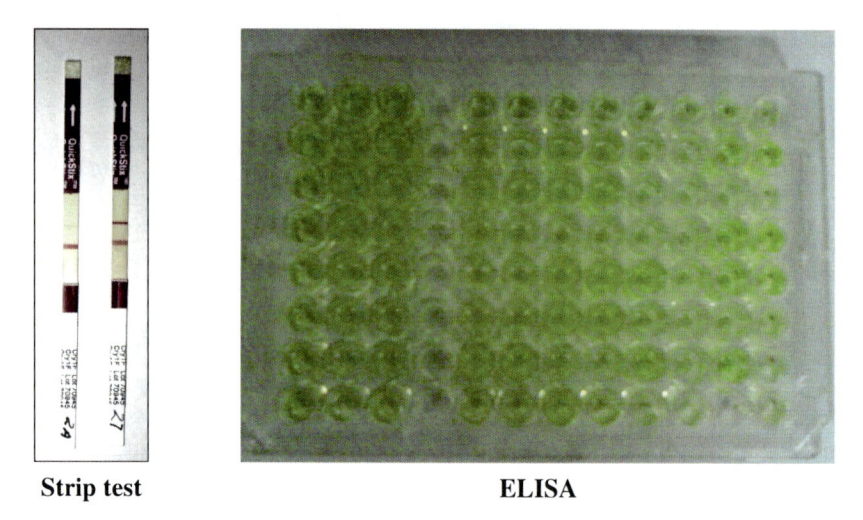

Strip test   **ELISA**

Fig. 4.4 Confirmation of transgene expression by strip test and ELISA plate method (V Nain, unpublished)

because of biosafety regulatory requirements and public concerns, especially when marker genes of nonplant origin are employed. Additionally, elimination of marker genes, whose functions are no longer in need, can save the cell from the burden of maintaining the unwanted transgene, allow gene stacking through reuse of the same marker gene in subsequent transformations, and avoid negative pleiotropic effects that may be associated with the marker gene in plants. These concerns have prompted the development of several approaches to generate marker-free transgenic crops. Technologies in this area have advanced greatly in recent years and new marker-removal strategies are expected to be continually explored. Currently, the use of site-specific-recombinases under the control of inducible promoter presents the greatest promise in terms of efficiency, preciseness, and time period required.

4.4.1 Cotransformation and Subsequent Segregation

The simplest strategy to eliminate marker gene is the cotransformation of genes of interest with marker genes followed by segregation through sexual crosses. Both desirable transgene and marker gene can be delivered into plant genomes by separate plasmids in one or two *Agrobacterium* strains or with single plasmids carrying multiple T-DNA regions (transposon-based), based on *Agrobacterium*- or biolistics-mediated transformation (Miki and McHugh 2004; Zhao et al. 2007). The unlinked marker gene can subsequently be removed from the plant genome during segregation and recombination that occurs during sexual reproduction by selecting on the transgene of interest, and not the marker gene, in progeny. Screening for the

Table 4.9 Chemicals and enzymes for the conditional-negative selection of transgenic tissues[a]

Substrates	Genes	Enzymes	Sources	Genome	References
5-Fluorocytosine	*codA*	Cytosine deaminase	*Escherichia coli*	Nuclear,	Stougaard (1993)
				plastid	Serino and Maliga (1997)
Naphthalene acetamide	*aux2*	Amido hydrolase	*Agrobacterium rhizogenes*	Nuclear	Béclin et al. (1993)
Indole-3-acetamide	*tms 2*	Indoleacetic acid hydrolase	*Agrobacterium tumefaciens*		Depicker et al. (1988)
Dihaloalkanes	*dhlA*	Dehalogenase	*Xanthobacter autotrophicus*	Nuclear	Naested et al. (1999), Moore and Srivastava (2008)
Sulfonylurea R7402	*cyp105a*	Cytochrome P450 mono-oxygenase	*Streptomyces griseolus*	Nuclear	O'Keefe et al. (1994)
Allyl alcohol	*cue*	Alcohol dehydrogenase	*Arabidopsis thaliana*	Nuclear	Lopez-Juez et al. (1998)
ganciclovir	*HSVtk*	thymidine kinase type 1 gene	*Homo sapiens* virus	nuclear	Czakó et al. (1995)
glyceryl glyphosate	*pehA*	phosphonate monoesterase	*Burkholderia caryophilli* PG2982	Nuclear	Dotson et al. (1996)

[a]Table was adapted from Miki and McHugh (2004)

progeny without the marker gene can be assisted by PCR or Southern hybridization analysis, as well as by incorporating an extra conditionally negative or scorable marker gene next to the original selective marker gene in the same construct (Darbani et al. 2007). Some selective marker genes have been developed for this purpose (Table 4.9). Each of these selective markers encodes an enzyme that converts nontoxic agents to toxic agents resulting in the death of the transformed cells containing the marker genes used for transformation selection or screening.

Recently, a novel marker gene, *dao1* from yeast, has been established for either positive or negative marker in *Arabidopsis*, depending on the substrate (Erikson et al. 2004). D-amino acid oxidase (DAAO, EC 1.4.3.3) catalyzes the oxidative deamination of a range of D-amino acids. This enzyme can metabolize toxic D-alanine and D-serine into nontoxic products, whereas it converts D-isoleucine and D-valine, which have low toxicity, into the toxic keto acids 3-methyl-2-oxopentanoate and 3-methyl-2-oxobutanoate, respectively. Hence, both positive and negative selection is possible with the same marker gene through changing D-alanine or D-serine to D-isoleucine or D-valine for the substrates. The bifunctional *dao1* marker gene yielded unambiguous results and allowed selection immediately after germination in *Arabidopsis* (Erikson et al. 2004).

The removal of selectable markers by cotransformation can be time consuming and labor intensive. The requirement of sexual crosses limits its application in woody, vegetatively propagated, or sterile plant species or cultivars. The efficiency of this strategy also depends on the loose linkage between the cointegrated DNAs.

4.4.2 Transposon-based Marker Gene Removal

The maize *Ac/Ds* transposable element system has been used to transpose marker genes and transposases from the T-DNA, thus leaving only the gene of interest in the inserted copy of the T-DNA (Yoder and Goldsbrough 1994). In one such system, the maize *Ac* transposable element was engineered to contain the *ipt* gene, which conferred an extreme shooty phenotype in the first positive selective step and also served as a negative selectable marker in the second selection step (due to the elimination of the *ipt* and transposase genes by transposition, normal shoots appeared after several weeks or months in culture) (Ebinuma et al. 1997). A marker-free frequency of about 5% was obtained in tobacco and hybrid aspen (Ebinuma et al. 1997). Since this system does not require a sexual reproduction step, it is an alternative for vegetatively propagated cultivars and plants with a long reproductive cycle. However, several drawbacks of this system, such as imprecise excision, low excision frequency, and genomic instability of transgenic plants, have undermined its efficacy (Scutt et al. 2002; Darbani et al. 2007). The maize *Ac/Ds* transposable element system has also been used to transpose (separate) the gene of interest that is previously linked together with marker gene and transposase in the T-DNA. However, this technology relies on sexual segregation to remove the marker gene and the transposase (Miki and McHugh 2004).

4.4.3 Site-specific Recombination-mediated Marker Deletion

The site-specific recombination systems mediate control of a variety of biological functions by carrying out precise excision, inversion, or integration of defined DNA units in their natural prokaryote and lower eukaryote hosts. Due to their accuracy and relative simplicity, the Cre/*lox* genes of bacteriophage P1 of *E. coli*, R/*RS* from the SR1 plasmid of *Zygosaccharomyces rouxii*, and FLP/*FRT* from the 2-μm plasmid of *Saccharomyces cerevisiae* have been the focus of the most intense studies in plants and other organisms. These systems function through the interactions of a single recombinase (e.g., Cre, R, FLP) with a pair identical recognition target sites (34 bp *lox* and *FRT*; 31 bp *RS*) in a "cut and paste" recombination process. The required recombinase can be introduced to target sites in transformants by retransformation to activate the marker gene excision. The recombinase locus is then removed by sexual segregation (see for review Miki and McHugh 2004; Darbani et al. 2007; Gidoni et al. 2008). Since constitutive expression of recombinase may cause aberrant developmental phenotypes in plants, the use of site-specific recombinases under the control of inducible promoters provides a more promising avenue. Table 4.10 lists the inducible promoters that have been paired with recombinase genes for marker gene excision. Depending on the type of inducible promoters being utilized, this inducible recombination strategy can provide further control

Table 4.10 Induced recombination-mediated auto-excision of markers and recombinase genes

Induction signal/ promoter system	Recombination system/plant tested	Stage of induction/ treatment	Excision rate	References
Chemical/ herbicide antidote-inducible maize GST	R-RS/tobacco	Shoot explants/ inducer in the medium	14–58%[a]	Sugita et al. (2000); Ebinuma et al. (2001)
Chemical/β-estradiol inducible transactivator XVE	Cre-lox/*Arabidopsis*, rice, tomato	Shoot explants, callus-germinating seeds/inducer in the medium	15–66%[a,b,c]	Zuo et al. (2001); Sreekala et al. (2005); Zhang et al. (2006)
Chemical/ dexamethasone-activated R-LBD	R-RS/strawberry	Regenerating leaf explants/ inducer in the medium	N.D.	Schaart et al. (2004)
Chemical/Salicylic acid-induced activated PR-1a	Cre–loxP/tomato	Shoot explants/ inducer in the medium	38.7%[a]	Ma et al. (2009)
Heat-shock/ *Arabidopsis* HSP81–1	Cre-lox/*Arabidopsis*	Whole seedlings/ ×2 alternate 37°C-16 h and recovery	N.D.	Hoff et al. (2001)
Heat-shock/ *Arabidopsis* HSP81–1	Cre-lox/tobacco	Whole seedlings/ ×2 alternate 37°C-16 h and recovery	< 100%[b]	Liu et al. (2005)
Heat-shock/ soybean HSP17.5E	Cre-lox/maize	Callus and immature embryos/42°C for 3–5 h	< 100%[b,c]	Zhang et al. (2003)
Heat-shock/ soybean HSP17.5E	Cre-lox/tobacco	Seed, leaf/×3 alternate 42°C-2 h and recovery	40–80%[a,b,c]	Wang et al. (2005)
Heat-shock/ *Drosophila* hsp70	Cre-lox/potato	Shoot internodes and mini-tubers/ 42°C for 2–3 h	5–14%[a]	Cuellar et al. (2006)
Heat-shock/ *Arabidopsis* HSP18.2	FLP-loxP, FRT/ tobacco	Regenerating leaf explants/65°C for at least 1 h	60%[a]	Luo et al. (2008)
Embryo-specific/ *Arabidopsis* app1	Cre-lox/soybean	Somatic embryogenesis	5–59%[a,b,c]	Li et al. (2007)
Microspore-specific/tobacco NTM19	Cre-lox/*Arabidopsis*, tobacco	Early pollen development	< 99.98%[a,c]	Mlynárová et al. (2006)
Pollen-specific/*B. campestris*	Cre-or FLP-lox-FRT fusion/tobacco	Mature pollen development	< 100%[a,c]	Luo et al. (2007)

(continued)

Table 4.10 (continued)

Induction signal/ promoter system	Recombination system/plant tested	Stage of induction/ treatment	Excision rate	References
BGP1 and tomato LAT5				
Pollen and seed specific/ *Arabidopsis* PAB5		Pollen and seed development		
Male germline-specific/ *Arabidopsis* SDS	Cre-lox/*Arabidopsis*	Male gamete development	83–100%[a,c]	Verweire et al. (2007)
Floral meristem-specific/ *Arabidopsis* AP1		Flower meristem-male and female germ lines		
Floral-specific/rice OsMADS45	Cre-lox/rice	Floral organs	13–100%[a,c]	Bai et al. (2008)

GST Glutathione-S-transferase (GST-II-27) promoter-MAT vector, *R-LBD* glucocorticoid receptor ligand binding domain fused to the C-terminus of R recombinase gene, *AP1* Apetala1, *SDS* solo dancers, *N.D.* not determined

Table was adapted from Gidoni et al. (2008)

[a]Complete plant gene excision events

[b]Incomplete plant (chimeric) gene excision events

[c]Provided indications for germline excision events

of the excision process on timing, flexibility and tissue specificity. In addition, this inducible system allows a single recombination construct where a marker gene and a recombinase gene coreside between the recombination target sites; therefore, genetic segregation is not required to remove the recombinase gene since excised circular DNA containing both marker and recombinase genes are expected to be lost via cellular degradation. Transient expression of recombinase through cocultivated or infiltration of *Agrobacterium* T-DNA vectors or systemic infection with plant virus vector has also been developed for tobacco (Gleave et al. 1999; Kopertekh and Schiemann 2004; Jia et al. 2006) and maize (Kerbach et al. 2005). Employment of two site-specific recombination systems has been recently explored in tobacco (Luo et al. 2007) and maize (Djukanovic et al. 2008). It was found that the two site-specific recombination systems could provide more efficient and complete excision. In addition to deletion of a marker gene from the transgenic plant, site-specific recombination systems can also be useful in site-specific integration of a gene into a predetermined genomic location in a precise, single copy pattern.

Besides the Cre/*lox*, R/*RS*, and FLP/*FRT* mentioned earlier, new recombination systems are being discovered and explored, including the β/six from *Streptococcus pyogenes* (Grønlund et al. 2007) and a small serine resolvase ParA from bacterial plasmids RK2 and RP4 catalyses (Thomson et al. 2009), which have proven efficient in marker gene deletion in tobacco and *Arabidopsis*. Site-specific recombination holds great promise for marker deletion and has advanced quickly in recent years. However, the commercial potential of this technology has not been yet demonstrated.

4.4.4 *Intrachromosomal Homologous Recombination System*

The intrachromosomal homologous recombination system is very similar to the site-specific recombination systems described above, except that there is no requirement for recombinase, whose action on cryptic excision sites in the plant genomes may cause pleiotropic effects. When placed between two 352 bp attachment P (attP) regions in bacteriophage, three marker genes (*nptII*, *gfp*, and *tms2*) were deleted from transgenic tobacco with a deletion frequency of 44% (Zubko et al. 2000). Interestingly, most plantlets that have lost the marker gene region also have lost transgene regions outside the attP cassette, suggesting that intrachromosomal homologous recombination is not always associated with precise homologous recombination between the two attP regions but that it can generate larger deletions probably as a result of illegitimate recombination (Zubko et al. 2000). The attP intrachromosomal excision system has also been utilized in transformation of tobacco plastids (Kittiwongwattana et al. 2007).

4.4.5 *Cytokinin-Based Backbone-Free Approach*

Recently, Richael et al. (2008) have described a new method that was based on transient expression of the bacterial isopentenyltransferase (*ipt*) gene that was positioned within the backbone (outside the T-DNA region) of binary vectors. It was found that the resulting temporary production of the natural cytokinin isopentenyl adenosine induced explants to produce shoots on media containing neither a selection agent nor synthetic hormones. This approach has been tested in various Solanaceous plant species including potato, tomato, tomatillo, and tobacco, as well as canola (Richael et al. 2008; Rommens et al. 2008). Transformation frequencies achieved were similar to conventional backbone-free transformation with marker-containing T-DNAs and higher than conventional methods that simply omit a selection step (de Vetten et al. 2003). Since shoots displaying a cytokinin overexpression phenotype were ignored and only shoots with a wild-type appearance were selected, the abnormal shoot morphology usually associated with the *ipt* gene in transformants is not an issue in this approach. This approach provides both marker-free and backbone-free transformation.

4.4.6 *Radiation Method*

The use of γ-radiation to physically remove a marker gene previously introduced into the soybean genome was evaluated by Tinoco et al. (2006). Preliminary data indicated a very low success rate of marker gene removal. In most cases, the gene of interest was deleted along with the marker gene. In addition, abnormal phenotypes were observed

4.5   Conclusions

Selectable and scorable marker genes play a vital role in identifying transformed plant cells. A wide range of marker genes have been employed for successful plant genetic transformations. Due to the substantial public concern about the potential spread of marker genes of nonplant origin, there is a momentum for research toward environment-friendly selection system involving natural plant materials and precise marker gene removal.

References

Aionesei T, Hosp J, Voronin V, Heberle-Bors E, Touraev A (2006) Methotrexate is a new selectable marker for tobacco immature pollen transformation. Plant Cell Rep 25:410–416

Alonso J, Barredo JL, Díez B, Mellado E, Salto F, García JL, Cortés E (1998) D-amino acid oxidase gene from *Rhodotorula gracilis* (*Rhodosporidium toruloides*) ATCC 26217. Microbiology 144:1095–1101

Andersen CB, Holst-Jensen A, Berdal KG, Thorstensen T, Tengs T (2006) Equal performance of TaqMan, MGB, molecular beacon, and SYBR green-based detection assays in detection and quantification of roundup ready soybean. J Agric Food Chem 54:9658–9663

Anderson KS, Sikorski JA, Johnson KA (1988) Evaluation of 5- enolpyruvylshikimate-3-phosphate synthase substrate and inhibitor binding by stopped-flow and equilibrium fluorescence measurements. Biochem J 27:1604–1610

Andersson M, Trifonova A, Andersson AB, Johansson M, Bülow L, Hofvander P (2003) A novel selection system for potato transformation using a mutated *AHAS* gene. Plant Cell Rep 22:261–267

Aragão FJL, Sarokin L, Vianna GR, Rech EL (2000) Selection of transgenic meristematic cells utilizing a herbicidal molecule results in the recovery of fertile transgenic soybean (*Glycine max* L. Merril) plants at a high frequency. Theor Appl Genet 101:1–6

Aragão FJL, Brasileiro ACM (2002) Positive, negative and marker-free strategies for transgenic plant selection. Braz J Plant Physiol 14:1–10

Ariizumi T, Amagai M, Shibata D, Hatakeyama K, Watanabe M, Toriyama K (2002) Comparative study of promoter activity of three anther-specific genes encoding lipid transfer protein, xyloglucan endotransglucosylase/hydrolase and polygalacturonase in transgenic *Arabidopsis thaliana*. Plant Cell Rep 21:90–96

Arruda P, Bright SW, Kuch JSH, Lea P, Rognes S (1984) Regulation of aspartate kinase isozymes in barley mutants resistant to lysine plus threonine. Plant Physiol 76:442–446

Bai X, Wang Q, Chu C (2008) Excision of a selective marker in transgenic rice using a novel Cre/loxP system controlled by a floral specific promoter. Transgenic Res 17:1035–1043

Ballester A, Cervera M, Peña L (2007) Efficient production of transgenic citrus plants using isopentenyl transferase positive selection and removal of the marker gene by site-specific recombination. Plant Cell Rep 26:39–45

Banno H, Ikeda Y, Niu QW, Chua NH (2001) Overexpression of *Arabidopsis esr1* induces initiation of shoot regeneration. Plant Cell 13:2609–2618

Barone P, Widholm JM (2008) Use of 4-methylindole or 7-methyl-dl -tryptophan in a transformant selection system based on the feedback-insensitive anthranilate synthase α-subunit of tobacco (*asa2*). Plant Cell Rep 27:509–517

Barry G, Kishore G, Padgette S, Talor M, Kolacz K, Weldon M, Re D, Eichholtz D, Fincher K, Hallas L (1992) Inhibitors of amino acid biosynthesis: strategies for imparting glyphosate tolerance to plants. In: Singh B, Flores HE, Shannon JC (eds) Biosynthesis and Molecular Regulation of Amino Acids in Plants. Am Soc Plant Physiol, pp 139–145

Barry GF, Kishore GM (1995) Glyphosate tolerant plants. US Patent No 5,463,175: http://www.patentstorm.us/patents/5463175.html

Béclin C, Charlot F, Botton E, Jouanin L, Dore C (1993) Potential use of the *aux2* gene from *Agrobacterium rhizogenes* as a conditional negative marker in transgenic cabbage. Transgenic Res 2:48–55

Beneviste R, Davies J (1973) Mechanisms of antibiotic resistance in bacteria. Annu Rev Biochem 42:471–506

Bevans MW, Flavell RB, Chilton M-D (1983) A chimaeric antibiotic resistance gene as a selectable marker for plant cell transformation. Nature 304:184–187

Bhat SR, Srinivasan S (2002) Molecular and genetic analyses of transgenic plants: considerations and approaches. Plant Sci 163:673–681

Birch RG (1997) Plant transformation: problems and strategies for practical application. Annu Rev Plant Physiol Plant Mol Biol 48:297–326

Bojsen K, Donaldson I, Haldrup A, Joersboe M, Kreierg JD, Nielsen J, Okkels FT, Peterson SG, Whenham RJ (1999) Positive selection. US Patent No 5,994,629: http://www.patentstorm.us/patents/5994629.html

Bonfim K, Faria JC, Nogueira EOPL, Mendes VA, Aragão FJL (2007) RNAi-mediated resistance to bean golden mosaic virus in genetically engineered common bean (*Phaseolus vulgaris*). Mol Plant Microbe Interact 20:717–726

Braeu B, Pilz U, Piepersberg W (1984) Genes for gentamicin-(3)-N-acetyltransferases III and IV. Mol Gen Genet 193:179–187

Brasileiro ACM, Tourneur C, Leple J-C, Combes V, Jouanin L (1992) Expression of the mutant *Arabidopsis thaliana* acetolactate synthase gene confers chlorsulfuron resistance to transgenic poplar plants. Transgenic Res 1:133–141

Brückner H, Westhauser H (2003) Chromatographic determination of L-and D-amino acids in plants. Amino Acids 24:43–55

Bubner B, Baldwin IT (2004) Use of real-time PCR for determining copy number and zygosity in transgenic plants. Plant Cell Rep 23:263–271

Caissard J-C, Guivarc'h A, Rembur J, Azmi A, Chriqui D (1994) Spurious localizations of diX-indigo microcrystals generated by the histochemical GUS assay. Transgenic Res 3:176–181

Caliskan M, Cuming AC (1998) Spatial specificity of H_2O_2-generating oxalate oxidase gene expression during wheat embryo germination. Plant J 15:165–171

Carrer H, Hockenberry TN, Svab Z, Maliga P (1993) Kanamycin resistance as a selectable marker for plastid transformation in tobacco. Mol Gen Genet 241:49–56

Chaleff RS, Mauvais CJ (1984) Acetolactate synthase is the site of action of two sulfonylurea herbicides in higher plants. Science 224:1443–1445

Chalfie M, Tu Y, Euskirchen G, Ward WW, Prasher DC (1994) Green fluorescent protein as a marker for gene expression. Science 263:802–805

Chan YL, Lin KH, Sanjaya LLJ, Chen WH, Chan MT (2005) Gene stacking in *Phalaenopsis* orchid enhances dual tolerance to pathogen attack. Transgenic Res 14:279–288

Chang CC, Huang PS, Lin HR, Lu CH (2007) Transactivation of protein expression by rice *hsp101* *in planta* and using *hsp101* as a selection marker for transformation. Plant Cell Physiol 48:1098–1107

Charest PJ, Hattori J, DeMoor J, Iyer VN, Miki NL (1990) *In vitro* study of transgenic tobacco expressing *Arabidopsis* wild type and mutant acetohydroxyacid synthase genes. Plant Cell Rep 8:643–646

Charng Y-C, Li K-T, Tai H-K, Lin N-S, Tu J (2008) An inducible transposon system to terminate the function of a selectable marker in transgenic plants. Mol Breed 21:359–368

Chawla HS, Leslie AC, Simmonds JA (1999) Developmental and environmental regulations of anthocyanin pigmentation in wheat tissues transformed with anthocynin regulatory genes. In Vitro Cell Dev Biol Plant 35:403–408

Chiang YC, Kiang YT (1988) Genetic analysis of mannose-6-phosphate isomerase in soybeans. Genome 30:808–811

Clementea TE, LaValleeb BJ, Howeb AR, Conner-Wardb D, Rozmanb RJ, Hunterb PE, Broylesb DL, Kastenb DS, Hinchee MA (2000) Progeny analysis of glyphosate selected transgenic soybeans derived from *Agrobacterium*-mediated transformation. Crop Sci 40: 797–803

Comai L, Larson-Kelly N, Kiser J, Mau CJD, Pokalsky AR, Shewmaker CK, McBride K, Jones A, Stalker DM (1988) Chloroplast transport of a ribulose bisphosphate carboxylase small subunit-5-enolpyruvyl 3- phosphoshikimate synthase chimeric protein requires part of the mature small subunit in addition to the transit peptide. J Biol Chem 263:15104–15109

Cuellar W, Gaudin A, Solórzano D, Casas A, Nopo L, Chudalayandi P, Medrano G, Kreuze J, Ghislain M (2006) Self-excision of the antibiotic resistance gene *nptII* using a heat inducible Cre-loxP system from transgenic potato. Plant Mol Biol 62:71–82

Cui M, Takayanagi K, Kamada H, Nishimura S, Handa T (2001) Efficient shoot regeneration from hairy roots of *Antirrhinum majus* L. transformed by the *rol*-type mat vector system. Plant Cell Rep 20:55–59

Curtis IS, Power JB, Davey MR (1995) NPTII assays for measuring gene expression and enzyme activity in transgenic plants. Methods Mol Biol 49:149–159

Cutter EG, Ashton FM, Huffstutter D (1968) The effect of bensulide on the growth, morphology and anatomy of oat roots. Weed Res 8:346–352

Czakó M, Marathe RP, Xiang C, Guerra DJ, Bishop GJ, JonesJDG ML (1995) Variable expression of the herpes simplex virus thymidine kinase gene in *Nicotiana tabacum* affects negative selection. Theor Appl Genet 91:1242–1247

Dahl GA, Tempe J (1983) Studies on the use of toxic precursor analogs of opines to select transformed plant cells. Theor Appl Genet 66:233–239

Damm B (2003) Selection marker. US Patent No 6,660,910: http://www.patentstorm.us/patents/6660910/description.html

Daniell H, Muthukumar B, Lee SB (2001) Marker free transgenic plants: engineering the chloroplast genome without the use of antibiotic selection. Curr Genet 39:109–116

Darbani B, Eimanifar A, Stewart CNJ, Camargo WN (2007) Methods to produce marker-free transgenic plants. Biotechnol J 2:83–90

David LS, David AS, Kenneth C, Gross A (1998) Gene coding for tomato fruit β-galactosidase II is expressed during fruit ripening. Plant Physiol 117:417–423

de Vetten N, Wolters AM, Raemakers K, van der Meer I, der Stege R, Heeres E, Heeres P, Visser R (2003) A transformation method for obtaining marker-free plants of a cross-pollinating and vegetatively propagated crop. Nat Biotechnol 21:439–442

DeBlock M, Herrera-Estrella L, Van Montagu M, Schell J, Zambryski P (1984) Expression of foreign genes in regenerated plants and in their progeny. EMBO J 3:1681–1689

DeBlock M, Schell J, Van Montagu M (1985) Chloroplast transformation by *Agrobacterium tumefaciens*. EMBO J 4:1367–1372

DeBlock M, De Brower D, Tenning P (1989) Transformation of *Brassica napus* and *Brassica oleracea* using *Agrobacterium tumefaciens* and the expression of the *bar* and *neo* genes in the transgenic plants. Plant Physiol 91:694–701

De Block M, Debrouwer D (1993) Engineered fertility control in transgenic *Brassica napus* L.: histochemical analysis of anther development. Planta 189:218–225

Degenhardt J, Poppe A, Montag J, Szankowski I (2006) The use of the phosphomannose-isomerase/mannose selection system to recover transgenic apple plants. Plant Cell Rep 25:1149–1156

Degenhardt J, Poppe A, Rösner L, Sza I (2007) Alternative selection systems in apple transformation. Acta Hortic 738:287–292

Della-Cioppa G, Bauer SC, Taylor ML, Rochester DE, Klein BK, Shah DM, Fraley RT, Kishore GM (1987) Targeting a herbicide-resistant enzyme from *Escherichia coli* to chloroplasts of higher plants. Biotechnology 5:579–584

Depicker A, Jacobs AM, Van Montagu MC (1988) A negative selection scheme for tobacco protoplast-derived cells expressing the T-DNA gene 2. Plant Cell Rep 7:63–66

Dhir SK, Pajeau ME, Fromm ME, Fry JE (1994) In: Henry JR, Ronalds JA (eds) Improvement of Cereal Quality by Genetic Engineering. Plenum Press, New York, USA, pp 71–75

Djukanovic V, Lenderts B, Bidney D, Lyznik LA (2008) A Cre::Flp fusion protein recombines FRT or LOXP sites in transgenic maize plants. Plant Biotechnol J 6:770–781

Dotson S, Lanahan MB, Smith AG, Kishore GM (1996) A phosphonate monoester hydrolase from *Burkholderia caryophilli* PG2982 is useful as a conditional lethal gene in plants. Plant J 10:383–392

Eberlein CV, Guttieri MJ, Berger PH, Fellman JK, Mallory-Smith CA, Thill DC, Baerg RJ, Belknap WR (1999) Physiological consequences of mutation for ALS-inhibitor resistance. Weed Sci 47:383–392

Ebinuma H, Sugita K, Matsunaga E, Yamakado M (1997) Selection of marker-free transgenic plants using the isopentenyl transferase gene. Proc Natl Acad Sci USA 94:2117–2121

Ebinuma H, Sugita K, Matsunaga E, Endo S, Yamada K, Komamine A (2001) Systems for the removal of a selection marker and their combination with a positive marker. Plant Cell Rep 20:383–392

Ebmeier A, Allison L, Cerutti H, Clemente T (2004) Evaluation of the *Escherichia coli* threonine deaminase gene as a selectable marker for plant transformation. Planta 218:751–758

Eichholtz DA, Rogers SG, Horsch RB, Klee HJ, Hayford M, Hoffman NL, Braford SB, Fink CF, Flick J, O'Connell KM, Fraley RT (1987) Expression of mouse dihydrofolate reductase gene confers methotrexate resistance in transgenic petunia plants. Somat Cell Mol Genet 13:67–76

El Amrani A, Barakate A, Askari BM, Li X, Roberts AG, Ryan MD, Halpin C (2004) Coordinate expression and independent subcellular targeting of multiple proteins from a single transgene. Plant Physiol 135:16–24

Endo S, Kasahara T, Sugita K, Matsunaga E, Ebinuma H (2001) The isopentenyl transferase gene is effective as a selectable marker gene for plant transformation in tobacco (*Nicotiana tabacum* cv. Petite Havana SRI). Plant Cell Rep 20:60–66

Endo S, Sugita K, Sakai M, Tanaka H, Ebinuma H (2002) Single-step transformation for generating marker-free transgenic rice using the *ipt*-type mat vector system. Plant J 30: 115–122

Erikson O, Hertzberg M, Näsholm T (2004) A conditional marker gene allowing both positive and negative selection in plants. Nat Biotechnol 22:445–458

Erikson O, Hertzberg M, Näsholm T (2005) The *dsdA* gene from *Escherichia coli* provides a novel selectable marker for plant transformation. Plant Mol Biol Rep 57:425–433

Esteban R, Emilia L, Berta DA (2005) Family of β-galactosidase cDNAs related to development of vegetative tissue in *Cicer arietinum*. Plant Sci 168:457–466

European Food Safety Authority (2007) Statement on the safe use of the *NPTII* antibiotic resistance marker gene in genetically modified plants by the scientific panel on genetically modified organisms (GMO): http://www.efsa.europa.eu/EFSA/efsa_locale-1178620753812_1178620775641.htm

Flavell RB, Dart E, Fuchs RL, Fraley RT (1992) Selectable marker genes: safe for plants. Biotechnology 10:141–144

Forner J, Binder S (2007) The red fluorescent protein eqFP611: application in subcellular localization studies in higher plants. BMC Plant Biol 7:28

Forsum O, Svennerstam H, Ganeteg U, Näsholm T (2008) Capacities and constraints of amino acid utilization in *Arabidopsis*. New Phytol 179:1058–1069

Fraley RT, Rogers SG, Horsch RB, Sanders PR, Flick JS, Adams SP, Bittner ML, Brand LA, Fink CL, Fry JS, Galluppi GR, Goldberg SB, Hoffmann NL, Woo SC (1983) Expression of bacterial genes in plant cells. Proc Natl Acad Sci USA 80:4803–4807

Fregien N, Davidson N (1985) Quantitative in situ gel electrophoretic assay for neomycin phosphotransferase activity in mammalian cell lysates. Ann Biochem 148:101–104

Freyssinet G, Pelissier B, Freyssinet M, Delon R (1996) Crops resistant to oxynils: from the laboratory to the market. Field Crops Res 45:125–133

Fromm ME, Morrish F, Armstrong C, Williams R, Thomas J, Klein TM (1990) Inheritance and expression of chimeric genes in the progeny of transgenic maize plants. Biotechnology 8:833–839

Galbraith DW (2004) The rainbow of fluorescent proteins. Methods Cell Biol 75:153–169

Gao Z, Xie X, Ling Y, Muthukrishnan S, Liang GH (2005) *Agrobacterium tumefaciens*-mediated sorghum transformation using a mannose selection system. Plant Biotechnol J 3:591–599

Garcia EL, Mourad GS (2004) A site-directed mutagenesis interrogation of the carboxy-terminal end of *Arabidopsis thaliana* threonine dehydratase/deaminase reveals a synergistic interaction between two effector-binding sites and contributes to the development of a novel selectable marker. Plant Mol Biol 55:121–134

Gasparic MB, Katarina C, Jana Z, Kristina G (2008) Comparison of different real-time PCR chemistries and their suitability for detection and quantification of genetically modified organisms. BMC Biotechnol 8:26

Gidoni D, Srivastava V, Carmi N (2008) Site-specific excisional recombination strategies for elimination of undesirable transgenes from crop plants. In Vitro Cell Dev Biol Plant 44:457–467

Gleave AP, Mitra DS, Mudge SR, Morris BA (1999) Selectable marker-free transgenic plants without sexual crossing: transient expression of *cre* recombinase and use of a conditional lethal dominant gene. Plant Mol Biol 40:223–235

Goddijn OJM, van der Duyn Schouten PM, Schilperoort RA, Hoge JHC (1993) A chimeric tryptophan decarboxylase gene as a novel selectable marker in plant cells. Plant Mol Biol 22:907–912

Gonneau M, Pasquette B, Cabanne F, Scalla R (1988) Metabolism of chlortoluron in tolerant species: possible role of cytochrome p-450 mono-oxygenases. Weed Res 28:19–25

Gossele V, van Aarssen R, Cornelissen M (1994) A 6' gentamicin acetyltransferase gene allows effective selection of tobacco transformants using kanamycin as a substrate. Plant Mol Biol 26:2009–2012

Gough KC, Hawes WS, Kilpatrick J, Whitelam GC (2001) Cyanobacterial *gr6* glutamate-1-semialdehyde aminotransferase: a novel enzyme-based selectable marker for plant transformation. Plant Cell Rep 20:296–300

Grimm B, Smith AJ, Kannangara CG, Smith M (1991) Gabaculine-resistant glutamate 1-semialdehyde aminotransferase of *Synechococcus*: deletion of a tripeptide close to the NH2 terminus and internal amino acid substitution. J Biol Chem 266:12495–12501

Grønlund JT, Stemmer C, Lichota J, Merkle T, Grasser KD (2007) Functionality of the β/six site-specific recombination system in tobacco and arabidopsis: a novel tool for genetic engineering of plant genomes. Plant Mol Biol 63:545–556

Grzelczak ZF, Rahman S, Kennedy TD, Lane BG (1985) Germin. Compartmentation of the protein, its translatable mRNA, and its biosynthesis among roots, stems, and leaves of wheat seedlings. Can J Biochem Cell Biol 63:1003–1013

Gu X-F, Meng H, Qi Q, Zhang JR (2008) *Agrobacterium* -mediated transformation of the winter jujube (*Zizyphus jujuba* mill.). Plant Cell Tissue Organ Cult 94:23–32

Guerineau F, Brooks L, Meadows J, Lucy A, Robinson C, Mullineaux P (1990) Sulfonamide resistance gene for plant transformation. Plant Mol Biol 15:127–136

Gurskaya NG, Verkhusha VV, Shcheglov AS, Staroverov DB, Chepurnykh TV, Fradkov AF, Lukyanov S, Lukyanov KA (2006) Engineering of a monomeric green-to-red photoactivatable fluorescent protein induced by blue light. Nat Biotechnol 24:461–465

Habert DA, Beverley SM, Kiely ML, Schimke RT (1981) Properties of an altered dihydrofolate reductase encoded by amplified genes in cultures mouse fibroblasts. J Biol Chem 256:9501–9510

Hake S, Smith HMS, Magnani E, Holtan H, Mele G, Ramirez J (2004) The role of *knox* genes in plant development. Annu Rev Cell Dev Biol 20:125–151

Haldrup A, Petersen SG, Okkels FT (1998) The xylose isomerase gene from thermoanaerobacterium thermosulfurogenes allows effective selection of transgenic plant cells using d-xylose as the selection agent. Plant Mol Biol 37:287–296

Haldrup A, Noerremark M, Okkels FT (2001) Plant selection principle based on xylose isomerase. In Vitro Cell Dev Biol Plant 37:114–119

Halfhill MD, Millwood RJ, Weissinger AK, Warwick SI, Stewart CN (2003) Additive transgene expression and genetic introgression in multiple GFP transgenic crop x weed hybrid generations. Theor Appl Genet 107:1533–1540

Haseloff J, Siemering KR, Prasher DC, Hodge S (1997) Removal of a cryptic intron and subcellular localization of green fluorescent protein are required to mark transgenic Arabidopsis plants brightly. Proc Natl Acad Sci USA 94:2122–2127

Haughn GW, Somerville C (1986) Sulfonylurea-resistant mutants of *Arabidopsis thaliana*. Mol Gen Genet 204:430–434

Hayford MB, Medford JI, Hoffman NI, Rogers SG, Klee HJ (1988) Development of a plant transformation selection system based on expression of genes encoding gentamicin acetyltransferases. Plant Physiol 86:1216–1222

He Z, Duan Z, Liang W, Chen F, Yao W, Liang H, Yue C, Sun Z, Chen F, Dai J (2006) Mannose selection system used for cucumber transformation. Plant Cell Rep 25:953–958

Heck GR, Armstrong CL, Astwood JD, Behr CF, Bookout JT, Brown SM, Cavato TA, DeBoer DL, Deng MY, George C, Hillyard JR, Hironaka CM, Howe AR, Jakse EH, Ledesma BE, Lee TC, Lirette RP, Mangano ML, Mutz JN, Qi Y, Rodriguez RE, Sidhu SR, Silvanovich A, Stoecker MA, Yingling RA, You J (2005) Development and characterization of a CP4 EPSPS-based, glyphosate-tolerant corn event. Crop Sci 45:329–339

Helmer G, Casadaban M, Bevan M, Kayes L, Chilton ML (1984) A new chimeric gene as a marker for plant transformation:the expression of *Escherichia coli* β-galactosidase in sunflower and tobacco cells. Biotechnology 2:520–527

Herrera-Estrella L, De Block M, Messens E, Hernalsteens P-P, Van Montagu M, Schell J (1983) Chimeric genes as dominant selectable markers in plant cells. EMBO J 2:987–995

Herrera-Estrella L, Broeck GVD, Maenhaut R, Montagu MV, Schell J, Timko M, Cashmore A (1984) Light inducible and chloroplast associated expression of a chimeric gene introduced into *Nicotiana tabacum* using a Ti plasmid vector. Nature 310:115–120

Hille J, Verheggen F, Roelvink P, Franssen H, van Kammen A, Zabel P (1986) Bleomycin resistance: a new dominant selectable marker for plant cell transformation. Plant Mol Biol Rep 7:171–176

Hird DL, Worrall D, Hodge R, Smartt S, Paul W, Scott R (1993) The anther-specific protein encoded by the *Brassica napus* and *Arabidopsis thaliana* A6 gene displays similarity to β-1, 3-glucanases. Plant J 4:1023–1033

Hoff T, Schnorr KM, Mundy JA (2001) Recombinase-mediated transcriptional induction system in transgenic plants. Plant Mol Biol 45:41–49

Hou CX, Dirk LMA, Goodman JP, Williams MA (2006) Metabolism of the peptide deformylase inhibitor actinonin in tobacco. Weed Sci 54:246–254

Hou CX, Lynnette MA, Dirk SP, Das NC, Maiti IB, Houtz RL, Williams MA (2007) Plant peptide deformylase: a novel selectable marker and herbicide target based on essential cotranslational chloroplast protein processing. Plant Biotechnol J 5:275–281

Howe AR, Gasser CS, Brown SM, Padgette SR, Hart J, Parker GB, Fromm ME, Armstrong CL (2002) Glyphosate as a selective agent for the production of fertile transgenic maize (*Zea mays* L.). Plant Mol Breed 10:153–164

Hsiao S, Su R-C, da Silva JAT, Chan M-T (2007) Plant native tryptophan synthase beta 1 gene is a non-antibiotic selection marker for plant transformation. Planta 225:897–906

Hsiao P, Su R-C, Ko S-S, Tong C-G, Yang RY, Chan M-T (2008) Overexpression of *Arabidopsis thaliana* tryptophan synthase beta 1 (AtTSB1) in *Arabidopsis* and tomato confers tolerance to cadmium stress. Plant Cell Environ 31:1074–1085

Hu C, Chee PP, Chee KPP, Chesney RH, Zhou JH, Miller PD, O'brien WT (1990) Intrinsic GUS-like activities in seed plants. Plant Cell Rep 9:1–5

Hu T, Metz S, Chay C, Zhou HP, Biest N, Chen G, Cheng M, Feng X, Radionenko M, Lu F, Fry J (2003) *Agrobacterium*-mediated large-scale transformation of wheat (*Triticum aestivum* L.) using glyphosate selection. Plant Cell Rep 21:1010–1019

Inui H, Kodama T, Ohkawa Y, Ohkawa H (2000) Herbicide metabolism and cross-tolerance in transgenic potato plants coexpressing human *CYP1A1*, *CYP2B6*, and *CYP2C19*. Pestic Biochem Physiol 66:116–129

Inui H, Shiota N, Ido Y, Inoue T, Hirose S, Kawahigashi H, Ohkawa Y, Ohkawa H (2001) Herbicide metabolism and tolerance in the transgenic rice plants expressing human *CYP2C9* and *CYP2C19*. Pestic Biochem Physiol 71:156–169

Inui H, Yamada R, Yamada T, Ohkawa Y, Ohkawa H (2005) A selectable marker using cytochrome p450 monooxygenases for *Arabidopsis* transformation. Plant Biotechnol J 22:281–286

Irdani T, Bogani P, Mengoni A, Mastromei G, Buiatti M (1998) Construction of a new vector conferring methotrexate resistance in nicotiana tabacum plants. Plant Mol Biol 37:1079–1084

Jach G, Binot E, Frings S, Luxa K, Jeff S (2001) Use of red fluorescent protein from *Discosoma sp.* (dsReD) as a reporter for plant gene expression. Plant J 28:483–491

Jain M, Chengalrayan K, Abouzid A, Gallo M (2007) Prospecting the utility of a PMI/mannose selection system for the recovery of transgenic sugarcane (*Saccharum* spp. Hybrid) plants. Plant Cell Rep 26:581–590

Jefferson RA (1987) Assaying chimeric genes in plants: the GUS gene fusion system. Plant Mol Biol Rep 5:387–405

Jefferson RA, Kavanagh TA, Bevan MW (1987) GUS fusions: β-glucuronidase as a sensitive and versatile gene fusion marker in higher plants. EMBO J 6:3901–3907

Jelenska J, Tietze E, Tempe J, Brevet J (2000) Streptothricin resistance as a novel selectable marker for transgenic plant cells. Plant Cell Rep 19:298–303

Jia G-X, Zhu Z-Q, Chang F-Q, Li Y-X (2001) Transformation of tomato with the *BADH* gene from *Atriplex* improves salt tolerance. Plant Cell Rep 21:141–146

Jia H, Pang Y, Chen X, Fang R (2006) Removal of the selectable marker gene from transgenic tobacco plants by expression of Cre recombinase from a tobacco mosaic virus vector through Agroinfection. Transgenic Res 15:375–384

Jimenez A, Davies J (1980) Expression of a transposable antibiotic resistance element in *Saccharomyces*. Nature 287:869–871

Joersbo M, Okkels FT (1996) A novel principle for selection of transgenic plant cells: positive selection. Plant Cell Rep 16:219–221

Joersbo M, Jørgensen K, Brunstedt J (2003) A selection system for transgenic plants based on galactose as selective agent and a udp-glucose: Galactose-1-phosphate uridyltransferase gene as selective gene. Mol Breed 11:315–323

Johnson R, Guderian RH, Eden F, Chilton MD, Gordon MP, Nester EW (1974) Detectionand quantitation of octopine in normal plant tissue and in crown gall tumors. Proc Nat Acad Sci USA 71:536–539

Jyung WH, Wittwer SH, Bukovac MJ (1965) Ion uptake and protein synthesis in enzymatically isolated plant cells. Nature 205:921–922

Kakimoto T (1996) Cki1, a histidine kinase homolog implicated in cytokinin signal transduction. Science 274:982–985

Kakimoto T (2001) Identification of plant cytokinin biosynthetic enzymes as dimethylallyl diphosphate:ATP/ADP isopentenyltransferases. Plant Cell Physiol 42:677–685

Katiyar-Agarwal S, Agarwal M, Grover A (2003) Heat-tolerant basmati rice engineered by overexpression of *hsp101*. Plant Mol Biol 51:677–686

Kavita P, Burma PK (2008) A comparative analysis of green fluorescent protein and β-glucuroni-dase protein-encoding genes as a reporter system for studying the temporal expression profiles of promoters. J Biosci 33:337–343

Kerbach S, Lörz H, Becker D (2005) Site-specific recombination in *Zea mays*. Theor Appl Genet 111:1608–1616

Khosravi P, Kermani MJ, Nematzadeh GA, Bihamta MR, Yokoya K (2007) Role of mitotic inhibitors and genotype on chromosome doubling of *Rosa*. Euphytica 160:267–275

Kieffer F, Triouleyre C, Bertsch C, Farine S, Leva Y, Walter B (2004) Mannose and xylose cannot be used as selectable agents for *Vitis vinifera* L. Vitis 43:35–39

Kittiwongwattana C, Lutz KA, Clark M, Maliga P (2007) Plastid marker gene excision by the phiC31 phage site-specific recombinase. Plant Mol Biol 64:137–143

Kolkman JM, Slabaugh MB, Bruniard JM, Berry S, Bushman BS, Olungu C, Maes N, Abratt G, Zambelli A, Miller JF, Leon A, Knapp SJ (2004) Acetohydroxyacid synthase mutations conferring resistance to imidazolinone or sulfonylurea herbicides in sunflower. Theor Appl Genet 109:1147–1159

Komarnytsky S, Gaume A, Garvey A, Borisjuk N, Raskin A (2004) A quick and efficient system for antibiotic-free expression of heterologous genes in tobacco roots. Plant Cell Rep 22:765–773

Kopertekh LJG, Schiemann J (2004) PVX-Cre-mediated marker gene elimination from transgenic plants. Plant Mol Biol 55:491–500

Kosugi S, Ohashi Y, Nakajima K, Arai Y (1990) An improved assay for β-glucuronidase in transformed cells: methanol almost completely suppresses a putative endogenous β-glucuroni-dase activity. Plant Sci 70:133–140

Koziel MG, Adams TL, Hazlet MA, Damm D, Miller J, Dahlbeck D, Jayne S, Staskawics BJ (1984) A cauliflower mosaic virus promoter directs expression of kanamycin resistance in morphogenic transformed plant cells. J Mol Appl Genet 2:549–562

Kunkel T, Niu Q-W, Chan Y-S, Chua N-H (1999) Inducible isopentenyl transferase as a high-efficiency marker for plant transformation. Nat Biotechnol 17:916–919

Kunze I, Ebneth M, Heim U, Geiger M, Sonnewald U, Herbers K (2001) 2-deoxyglucose resistance: a novel selection marker for plant transformation. Mol Breed 7:221–227

LaFayette PR, Kane PM, Phan BH, Parrott WA (2005) Arabitol dehydrogenase as a selectable marker for rice. Plant Cell Rep 24:596–602

Lai FM, Mei K, Mankin L, Jones T (2006) Application of two new selectable marker genes, *dsdA* and *dao1* in maize transformation. In: Xu Z, Li J, Xue Y, Yang W (eds) Biotechnology and Sustainable Agriculture 2006 and Beyond. Springer, Berlin, pp 141–142

Leary JL, Brigati DJ, Ward DC (1983) Rapid and sensitive colorimetric method for visualizing biotin-labeled DNA probes hybridized to DNA or RNA immobilized on nitrocellulose: Bio-blots. Proc Natl Acad Sci USA 80:4045–4049

Lee BT, Matheson NK (1984) Phosphomannoisomerase and phosphoglucoisomerase in seeds of *Cassia coluteoides* and some other legumes that synthesize galactomannan. Phytochemistry 23:983–987

Lee HJ, Duke MV, Duke SO (1993) Cellular localization of protoporphyrinogen-oxidizing activities of etiolated barley (*Hordeum vulgare* L.) leaves (relationship to mechanism of action of protoporphyrinogen oxidase-inhibiting herbicides). Plant Physiol 102:881–889

Lee K, Yang K, Kang K, Kang S, Lee N, Back K (2007) Use of *Myxococcus xanthus* proto-porphyrinogen oxidase as a selectable marker for transformation of rice. Pest Biochem Physiol 88:31–35

Li X, Volrath SL, Nicholl DBG, Chilcott CE, Johnson MA, Ward ER, Law MD (2003) Develop-ment of protoporphyrinogen oxidase as an efficient selection marker for *Agrobacterium tumefaciens*-mediated transformation of maize. Plant Physiol 133:736–747

Li Z, Hayashimoto A, Murai N (1992) A sulfonylurea herbicide resistance gene from *Arabidopsis thaliana* as a new selectable marker for production of fertile transgenic rice plants. Plant Physiol 100:662–668

Li Z, Xing A, Moon BP, Burgoyne SA, Guida AD, Liang H, Lee C, Caster CS, Barton JE, Klein TM, Falco SC (2007) A Cre/loxP-mediated self-activating gene excision system to produce marker gene free transgenic soybean plants. Plant Mol Biol 65:329–341

Liang Z, Ma D, Tang L, Hong Y, Luo A, Zhou J, Dai X (1997) Expression of the spinach betaine aldehyde dehydrogenase (*BADH*) gene in transgenic tobacco plants. China J Biotechnol 13:153–159

Liau C-H, Lu J-C, Prasad V, Hsiao H-H, You S-J, Lee J-T, Yang N-S, Huang H-E, Feng T-Y, Chen W-H, Chan M-T (2003) The sweet pepper ferredoxin-like protein (pflp) conferred resistance against soft rot disease in *Oncidium* orchid. Transgenic Res 12:329–336

Lin HJ, Cheng HY, Chen CH, Huang HC, Feng TY (1997) Plant amphipathic proteins delay the hypersensitive response caused by harpin$_{Pss}$ and *Pseudomonas syringae* pv. *syringae*. Physiol Mol Plant Pathol 51:367–376

Liu H-K, Yang C, Wei Z-H (2005) Heat shock-regulated site-specific excision of extraneous DNA in transgenic plants. Plant Sci 168:997–1003

Lopez-Juez E, Jarvis RP, Takeuchi A, Page AM, Choury J (1998) New *Arabidopsis cue* mutants suggest a close connection between plastid- and phytochrome-regulation of nuclear gene expression. Plant Physiol 118:803–815

Ludwig SR, Ben B, Larry B, Susan RW (1990) A regulatory gene as a novel visible marker for maize transformation. Science 247:449–450

Luo K, Zheng X, Chen Y, Xiao Y, Zhao D, Richard M, Pei Y, Li Y (2006) The maize *knotted1* gene is an effective positive selectable marker gene for *Agrobacterium*-mediated tobacco transformation. Plant Cell Rep 25:403–409

Luo K, Duan H, Zhao D, Zheng X, Deng W, Chen Y, Stewart CNJ, Richard M, Jiang X, Wu Y, He A, Pei Y, Li Y (2007) 'GM-gene-deletor': Fused *loxP*-FRT recognition sequences dramatically improve the efficiency of FLP or CRE recombinase on transgene excision from pollen and seed of tobacco plants. Plant Biotechnol J 5:263–374

Luo K, Sun M, Deng W, Xu S (2008) Excision of selectable marker gene from transgenic tobacco using the gm-gene-deletor system regulated by a heat-inducible promoter. Biotechnol Lett 30:1295–1302

Ma BG, Duan XY, Niu JX, Ma C, Hao QN, Zhang LX, Zhang HP (2009) Expression of stilbene synthase gene in transgenic tomato using salicylic acid-inducible Cre/loxP recombination system with self-excision of selectable marker. Biotechnol Lett 31:163–169

Maier-Greiner UH, Obermaier-Skrobranek BM, Estermaier LM, Kammerloher W, Freund C, Wülfing C, Burkert UI, Matern DH, Breuer M, Eulitz M (1991) Isolation and properties of a nitrile hydratase from the soil fungus *Myrothecium verrucaria* that is highly specific for the fertilizer cyanamide and cloning of its gene. Proc Natl Acad Sci USA 88:4260–4264

Maliga P, Svab Z, Harper EC, Jones JDG (1988) Improved expression of streptomycin resistance in plants due to a deletion in the streptomycin phosphotransferase coding sequence. Mol Gen Genet 214:456–459

Mannerlof M, Tenning P (1997) Screening of transgenic plants by multiplex PCR. Plant Mol Biol Rep 15:38–45

Mannerlöf M, Tuvesson S, Steen P, Tenning P (1997) Transgenic sugar beet tolerant to glyphosate. Euphytica 94:83–91

Martin T, Schmidt R, Altmann T, Frommer WB (1992) Non-destructive assay systems for detection of β-glucuronidase activity in higher plants. Plant Mol Biol Rep 10:37–46

Mascarenhas JP, Hamilton DA (1992) Artifacts in the localization of GUS activity in anthers of petunia transformed with a CaMV 35S-GUS construct. Plant J 2:405–408

Matz MV, Fradkov AF, Labas YA, Savitsky AP, Zaraisky AG, Markelov ML, Lukyanov SA (1999) Fluorescent proteins from nonbioluminescent Anthozoa species. Nat Biotechnol 17:969–973

Matz MV, Konstantin A, Lukyanov SA (2002) Family of the green fluorescent protein: journey to the end of the rainbow. BioEssays 24:953–959

McDaniel LD, Schultz RA (1993) Elevation of sister chromatid exchange frequency in transformed human fibroblasts following exposure to widely used aminoglycosides. Environ Mol Mutagen 21:67–72

McKinnon GE, Abedinia M, Henry RJ (1996) Expression of non-selectable markers in wheat and rice tissues. Plant Tiss Cult Biotechnol 2:24–32

Mentewab A, Stewart CNJ (2005) Overexpression of an *Arabidopsis thaliana* ABC transporter confers kanamycin resistance to transgenic plants. Nat Biotechnol 23:1177–1180

Merzlyak EM, Goedhart J, Shcherbo D, Bulina ME, Shcheglov AS, Fradkov AF, Gaintzeva A, Lukyanov KA, Lukyanov S, Gadella TW, Chudakov DM (2007) Bright monomeric red fluorescent protein with an extended fluorescence lifetime. Nat Meth 4:555–557

Miao S, Duncan DR, Widholm J (1988) Selection of regenerable maize callus cultures resistant to 5-methyl-DL-tryptophan, S-2-aminoethyl-L-cysteine and high levels of L-lysine plus L-threonine. Plant Cell Tiss Org Cult 14:3–14

Mihálka V, Balázs E, Nagy I (2003) Binary transformation systems based on 'shooter' mutants of *Agrobacterium tumefaciens*: A simple, efficient, and universal gene transfer technology that permits marker gene elimination. Plant Cell Rep 21:778–784

Miki BL, Labbé H, Hattori J, Ouellet T, Sunohara G, Charest PJ, Iyer VN (1990) Transformation of *Brassica napus* canola cultivars with *Arabidopsis thaliana* acetohydroxyacid synthase genes and analysis of herbicide resistance. Theor Appl Genet 80:449–458

Miki B, McHugh S (2004) Selectable marker genes in transgenic plants: applications, alternatives and biosafety. J Biotechnol 107:193–232

Min B-W, Cho Y-N, Song M-J, Noh T-K, Kim B-K, Chae W-K, Park Y-S, Choi Y-D, Harn C-H (2007) Successful genetic transformation of Chinese cabbage using phosphomannose isomerase as a selection marker. Plant Cell Rep 26:337–344

Mlynárová L, Conner AJ, Nap JP (2006) Directed microspore-specific recombination of transgenic alleles to prevent pollen-mediated transmission of transgenes. Plant Biotechnol J 4:445–452

Moore S, Srivastava V (2008) A bacterial haloalkane dehalogenase (dhlA) gene as conditional negative selection marker for rice callus cells. In Vitro Cell Dev Biol Plant 44:468–473

Morejohn LC, Fosket DE (1984) Inhibition of plant microtubule polymerization *in vitro* by the phosphoric amide herbicide amiprophos-methyl. Science 224:874–876

Mourad G, Williams D, King J (1995) A double mutant allele, crs1–4, of Arabidopsis thaliana encodes an acetolactase synthase with altered kinetics. Planta 196:64–68

Naested H, Fennema M, Hao L, Andersen M, Janssen DB, Mundy J (1999) A bacterial haloalkane dehalogenase gene as a negative selectable marker in *Arabidopsis*. Plant J 18:571–576

Nagel RJ, Manners JM, Birch RG (1992) Evaluation of an ELISA assay for rapid detection and quantification of neomycin phosphotransferase II in transgenic plants. Plant Mol Biol Rep 10:263–272

Nain V, Jaiswal R, Dalal M, Ramesh B, Kumar PA (2005) Polymerase chain reaction analysis of transgenic plants contaminated by *Agrobacterium*. Plant Mol Biol Rep 23:59–65

Nap JP, Bijvoet J, Stiekema WJ (1992) Biosafety of kanamycin-resistant transgenic plants. Transgenic Res 1:239–249

Nishizawa K, Kita Y, Kitayama M, Ishimoto MA (2006) Red fluorescent protein, DsRed2, as a visual reporter for transient expression and stable transformation in soybean. Plant Cell Rep 25:1355–1361

Noé W, Mollenschott C, Berlin J (1984) Tryptophan decarboxylase from *Catharanthus roseus* cell suspension cultures purification molecular and kinetic data of the homogenous protein. Plant Mol Biol 3:281–288

Norelli JL, Aldwinckle HS (1993) The role of aminoglycoside antibiotics in the regeneration and selection of neomycin phosphotransferase-transgenic apple tissue. J Am Soc Hortic Sci 118:311–316

O'Keefe DP, Tepperman JM, Dean C, Leto KJ, Erbes DL, Odell JT (1994) Plant expression of a bacterial cytochrome P450 that catalyzes activation of a sulfonylurea pro-herbicide. Plant Physiol 105:473–482

Olszewski NE, Martin FB, Ausubel FM (1988) Specialized binary vector for plant transformation: expression of the *Arabidopsis thaliana AHAS* gene in *Nicotiana tabacum*. Nucl Acids Res 16:10765–10781

Ow DW, Wood KV, DeLuca M, Wet JRD, Helinski DR, Howell SH (1986) Transient and stable expression of the firefly luciferase gene in plant cells and transgenic plants. Science 234:856–859

Perez P, Tiraby G, Kallerhoff J, Perret J (1989) Phleomycin resistance as a dominant selectable marker for plant cell transformation. Plant Mol Biol Rep 13:365–373

Perl A, Galili S, Shaul O, Ben-Tzvi I, Galili G (1993) Bacterial dihydrodipicolinate synthase and desensitized aspartic kinase: two novel selectable markers for plant transformation. Biotechnology 11:715–718

Peterhaensel C, Weier D, Lahaye T (2007) Nonradioactive northern and southern analyses from plant samples. Methods Mol Biol 353:69–78

Pinkerton TS (2008) Genetically engineered resistance to organophosphate herbicides provides a new scorable and selectable marker system for transgenic plants. Mol Breed 21:27–36

Prather TS, DiTomaso JM, Holt JS (2002) History, mechanisms, and strategies for prevention and management of herbicide resistant weeds. Proc Calif Weed Sci Soc 52:155–163

Prior FA, Tackaberry ES, Aubin RA, Casley WL (2006) Accurate determination of zygosity in transgenic rice by real-time PCR does not require standard curves or efficiency correction. Transgenic Res 15:261–265

Pua EC, Mehra-Palta A, Nagy F, Chua NH (1987) Transgenic plants of *brassica napus* l. Biotechnology 5:815–817

Radicella JP, Brown D, Tolar LA, Chandler VL (1992) Allelic diversity of the maize B regulatory gene: different leader and promoter sequences of two B alleles determine distinct tissue specificities of anthocyanin production. Gene Dev 6:2152–2164

Rajasekaran K, Grula JW, Hudspeth RL, Pofelis S, Anderson DM (1996) Herbicide-resistant acala and coker cottons transformed with a native gene encoding mutant forms of acetohydroxyacid synthase. Mol Breed 2:307–319

Ramesh SA, Kaiser BN, Franks T, Collins G, Sedgley M (2006) Improved methods in *Agrobacterium*-mediated transformation of almond using positive (mannose/pmi) or negative (kanamycin resistance) selection-based protocols. Plant Cell Rep 25:821–828

Randez-Gil F, Blasco A, Prieto JA, Sanz P (1995) *DOGR1* and *DOGR2*: two genes from *Saccharomyces cerevisiae* that confer 2-deoxyglucose resistance when overexpressed. Yeast 11:1233–1240

Rando RR (1977) Mechanism of the irreversible inhibition of γ- aminobutyric acid α-ketoglutaric acid transaminase by the neurotoxin gabaculine. Biochemistry 16:4604–4610

Rathinasabapathi B, McCue KF, Gage DA, Hanson AD (1994) Metabolic engineering of glycine betaine synthesis: plant betaine aldehyde dehydrogenases lacking typical transit peptides are targeted to tobacco chloroplasts where they confer betaine aldehyde resistance. Planta 193:155–162

Ratnam S, Delcamp TJ, Hynes JB, Freisheim JH (1987) Purification and characterization of dihydrofolate reductase from soybean seedlings. Arch Biochem Biophys 255:279–289

Ray K, Jagannath A, Gangwani SA, Burma PK, Pental D (2004) Mutant acetolactate synthase gene is an efficient in vitro selectable marker for the genetic transformation of *Brassica juncea* (oilseed mustard). J Plant Physiol 161:1079–1083

Ray TB (1984) Site of action of chlorsulfuron: inhibition of valine andi Biosynthesis in plants. Plant Physiol 75:827–831

Reed J, Privalle L, Powell ML, Meghji M, Dawson J, Dunder E, Suttie J, Wenck A, Launis K, Kramer C, Chang Y-F, Hansen G, Wright M (2001) Phosphomannose isomerase: an efficient selectable marker for plant transformation. In Vitro Cell Dev Biol Plant 37:127–132

Richael CM, Kalyaeva M, Chretien RC, Yan H, Adimulam S, Stivison A, Weeks JT, Rommens CM (2008) Cytokinin vectors mediate marker-free and backbone-free plant transformation. Transgenic Res 17:905–917

Rognes SE, Bright SWJ, Miflin BJ (1983) Feedback-insensitive aspartate kinase isoenzymes in barley mutants resistant to lysine plus threonine. Planta 157:32–38

Rommens CM, Yan H, Swords K, Richael C, Ye J (2008) Low-acrylamide French fries and potato chips. Plant Biotechnol J 6:843–853

Rosellini D, Capomaccio S, Ferradini N, Sardaro MLS, Nicolia A, Veronesi F (2007) Non-antibiotic, efficient selection for alfalfa genetic engineering. Plant Cell Rep 26:1035–1044

de Ruijter NCA, Verhees J, Leeuwen Tenning VW, van der Krol AR (2003) Evaluation and comparison of the GUS, LUC and GFP reporter system for gene expression studies in plants. Plant Biol 5:103–511

Saari LL, Cotterman JC, Thill DC (1994) Resistance to acetolactate synthase inhibiting herbicide. In: Powles S, Holtum JAM (eds) Herbicide Resistance in Plants: Biology and Biochemistry. Lewis, Boca Raton, FL, pp 83–139

Sambrook J, Fritsch EF, Maniatis T (1989) Molecular cloning. Cold Spring Harbor Lab Press, New York vol 2, pp 9.1-9.60

Samson NP, Campa C, Noirot M, Kochko A (2004) Potential use of d-xylose for coffee plant transformation. In: Proc 19th Assoc Sci Infor Coffee Colloq, Bangalore, India, pp 707-713

Santel HJ, Bowden BA, Sorensen VM, Mueller KH (1999) Flucarbazone-sodium-a new herbicide for the control of wild oat and green foxtail in wheat. In: Proc Brighton Crop Protec Conf Weeds. British Crop Protec Coun, Farnham, UK, p 23

Schaart JG, Krens FA, Pelgrom KT, Mendes O, Rouwendal GJ (2004) Effective production of marker-free transgenic strawberry plants using inducible site-specific recombination and a bifunctional selectable marker gene. Plant Biotechnol J 2:233–240

Schmidt MA, LaFayette PR, Artelt BA, Parrott WA (2008) A comparison of strategies for transformation with multiple genes via microprojectile-mediated bombardment. In Vitro Cell Dev Biol-Plant 44:162–168

Schmulling T, Schell J, Spena A (1988) Single genes from *Agrobacterium rhizogenes* influence plant development. EMBO J 7:2621–2629

Schönbrunn E, Eschenburg S, Schuttleworth WA, Schloss JV, Amrhein N, Evans JNS, Kabsch W (2001) Interaction of the herbicide glyphosate with its target enzyme 5-enolpyruvylshikimate 3-phosphate synthase in atomic detail. Proc Natl Acad Sci USA 98:1376–1380

Scutt CP, Zubko E, Meyer P (2002) Techniques for the removal of marker genes from transgenic plants. Biochimie 84:1119–1126

Serino G, Maliga P (1997) A negative selection scheme based on the expression of cytosine deaminase in plastids. Plant J 12:697–701

Shah DM, Horsch RB, Klee HJ, Kishore GM, Winter JA, Tumer NE, Hironaka CM, Sanders PR, Gasser CS, Aykent S, Siegel NR, Rogers SG, Fraley RT (1986) Engineering herbicide tolerance in transgenic plants. Science 233:478–481

Shaul O, Galili G (1991) Increased lysine synthesis in transgenic tobacco plants expressing bacterial dihydrodipicolinate synthas in their chloroplasts. Plant J 2:203–209

Shaw KJ, Rather PN, Hare RS, Miller G (1993) Molecular genetics of aminoglycoside resistance genes and familial relationships of the aminoglycoside-modifying enzymes. Microbiol Mol Biol Rev 57:138–163

Shiota N, Nagasawa A, Sakaki T, Yabusaki Y, Ohkawa H (1994) Herbicide-resistant tobacco plants expressing the fused enzyme between rat cytochrome p4501a1 (*cyp1a1*) and yeast nadphcytochrome p450 oxidoreductase. Plant Physiol 106:17–23

Sigareva M, Spivey R, Willits M, Kramer C, Chang YF (2004) An efficient mannose selection protocol for tomato that has no adverse effect on the ploidy level of transgenic plants. Plant Cell Rep 23:236–243

Simmonds J, Cass L, Routly E, Hubbard K, Donaldson P, Bancroft B, Davidson A, Hubbard S, Simmonds D (2004) Oxalate oxidase: a novel reporter gene for monocot and dicot transformations. Mol Breed 13:79–91

Singh Z, Sansavini S (1998) Transgenic transformation and fruit crop improvement. In: Janick J (ed) Plant Breeding Review, vol 16. Wiley, New York, USA, pp 87–134

Slater S, Mitsky TA, Houmiel KL, Hao M, Reiser SE, Taylor NB, Tran M, Valentin HE, Rodriguez DJ, Stone DA, Padgette SR, Kishore G, Gruys KJ (1999) Metabolic engineering of *Arabidopsis* and *Brassica* for poly (3-hydroxybutyrate-co-3-hydroxyvalerate) copolymer production. Nat Biotechnol 17:1011–1016

Smith AG, Marsh O, Elder GH (1993) Investigation of the subcellular location of the tetrapyrrole-biosynthesis enzyme coproporphyrinogen oxidase in higher plants. Biochem J 292:503–508

Sonnewald U, Ebneth M (2004) 2-deoxyglucose-6-phosphate (2-dog-6-p) phosphatase sequences as selection markers in plants. US Patent No 6,806,085: http://www.patentstorm.us/patents/6806085.html

Southern EM (1975) Detection of specific sequences among DNA fragments separated by gel electrophoresis. Mol Biol 98:503–517

Spertini D, Béliveau C, Bellemare G (1999) Screening of transgenic plants by amplification of unknown genomic DNA flanking T-DNA. Biotechniques 27:308–314

Sundar IK, Sakthivel N (2008) Advances in selectable marker genes for plant transformation. J Plant Physiol 165:1698–1716

Sreekala C, Wu L, Gu K, Wang D, Tian D, Yin Z (2005) Excision of a selectable marker in transgenic rice (*Oryza sativa* L.) using a chemically regulated Cre/loxP system. Plant Cell Rep 24:86–94

Stalker DM, McBride KE, Malyj L (1988) Herbicide resistance in transgenic plants expressing a bacterial detoxification gene. Science 242:419–423

Stano J, Kovacs P, Miieta K, Neubert K, Tintemann H, Koreova M (2002) Localization and measurement of extracellular plant galactosidases. Acta Histochem 104:441–444

Stein R, Gross W, Schnarrenberger C (1997) Characterization of a xylitol dehydrogenase and a d-arabitol dehydrogenase from the thermo- and acidophilic red alga *Galdieria sulphuraria*. Planta 202:487–493

Stewart CN (2001) The utility of green fluorescent protein in transgenic plants. Plant Cell Rep 20:376–382

Stewart CN (2006) Go with the glow: fluorescent proteins to light transgenic organisms. Trends Biotechnol 24:155–162

Stougaard J (1993) Substrate-dependent negative selection in plants using a bacterial cytosine deaminase gene. Plant J 3:755–761

Sugita K, Kasahara T, Matsunaga E, Ebinuma H (2000) A transformation vector for the production of marker-free transgenic plants containing a single copy transgene at high frequency. Plant J 22:461–469

Sugiura Y, Takita T, Umezawa H (1985) Bleomycin antibiotics: metal complex and their biological action. In: Sigel H, Sigel A (eds) Metal ions in biological systems: antibiotics and their complexes, vol 19. Marcel Dekker, New York, pp 81–108

Svab Z, Maliga P (1993) High-frequency plastid transformation in tobacco by selection for a chimeric *aadA* gene. Proc Natl Acad Sci USA 90:913–917

Svab Z, Harper EC, Jones JDG, Maliga P (1990) Aminoglycoside-3″-adenyltransferase confers resistance to spectinomycin and streptomycin in *Nicotiana tabacum*. Plant Mol Biol 14:197–205

Tachibana K, Watanabe T, Sekizawa T, Takematsu T (1986) Action mechanism of bialaphos II: accumulation of ammonia in plants treated with bialaphos. J Pest Sci 11:33–37

Takei K, Sakakibara H, Sugiyama T (2001) Identification of genes encoding adenylate isopentenyltransferase, a cytokinin biosynthesis enzyme, in *Arabidopsis thaliana*. J Biol Chem 276:26405–26410

Tang KX, Sun XF, Hu QN, Wu AZ, Lin CH, Lin HJ, Twyman RM, Christou P, Feng TY (2001) Transgenic rice plants expressing the ferredoxin-like protein (AP1) from sweet pepper show enhanced resistance to *Xanthomonas oryzae* pv. *oryzae*. Plant Sci 160:1035–1042

Taylor CB (1997) Promoter fusion analysis: an insufficient measure of gene expression. Plant Cell 9:273–275

Teeri TH, Lehvaslaiho H, Franck M, Uotila J, Heino P, Palva ET, Van Montagu M, Herrera-Estrella L (1989) Gene fusions to lacZ reveal new expression patterns of chimeric genes in transgenic plants. EMBO J 8:343–350

Tewari-Singh N, Sen J, Kiesecker H, Reddy VS, Jacobsen HJ, Guha-Mukherjee S (2004) Use of a herbicide or lysine plus threonine for non-antibiotic selection of transgenic chickpea. Plant Cell Rep 22:576–583

Thompson C, Dunwell JM, Johnstone CE, Lay V, Ray J, Schmitt M, Watson H, Nisbet G (1995) Degradation of oxalic acid by transgenic canola plants expressing oxalate oxidase. Euphytica 85:169–172

Thomson JG, Yau Y-Y, Blanvillain R, Chiniquy D, Thilmony R, Ow DW (2009) ParA resolvase catalyzes site-specific excision of DNA from the *Arabidopsis* genome. Transgenic Res 18:237–248

Tinoco ML, Vianna GR, Abud S, Souza PIM, Rech EL, Aragão FJL (2006) Radiation as a tool to remove selective marker genes from transgenic soybean plants. Biol Planta 50:146–148

Treacy BK, Hattori J, Homme IP, Barbour E, Boutilier K, Baszezynski CL, Huang B, Johnson DA, Miki BL (1997) *Bnm1*, a *Brassica* pollen-specific gene. Plant Mol Biol 34:603–611

Trulson A, Julia B, Carl J (1997) A method for visually selecting transgenic plant cells or tissues by carotenoid pigmentation. WO/1997/014807

Ulanov A, Widholm JM (2007) Effect of the expression of cyanamide hydratase on metabolites in cyanamide-treated soybean plants kept in the light or dark. J Exp Bot 58:4319–4332

Umbarger HE (1978) Amino acid biosynthesis and its regulation. Annu Rev Biochem 47:533–606

Ursin VM (1996) Aldehyde dehydrogenase selectable markers for plant transformation. US Patent No 5,633,153: http://www.patentstorm.us/patents/5633153.html

US Food and Drug Administration (1998) Guidance for industry: use of antibiotic resistance marker genes in transgenic plants: http://vm.cfsan.fda.gov/~dms/opa-armg.html

van Bel AJE, Hibberd J, Prüfer D, Knoblauch M (2001) Novel approach in plastid transformation. Curr Opin Biotechnol 12:144–149

Van Den Elzen PJM, Townsend J, Lee KY, Bedbrook JR (1985) A chimaeric hygromycin resistance gene as a selectable marker in plant cells. Plant Mol Biol 5:299–302

Van Slogteren GMS, Hooykaas PJJ, Planqué K, De Groot B (1982) The lysopinedehydrogenase gene used as a marker for the selection of octopine crown gall cells. Plant Mol Biol 1:133–142

Verweire D, Verleyen K, De Buck S, Claeys M, Angenon G (2007) Marker-free transgenic plants through genetically programmed auto-excision. Plant Physiol 145:1220–1231

Vilaine F, Casse-Delbart F (1987) Independent induction of transformed roots by the TL and TR regions of the Ri plasmid of agropine type *Agrobacterium rhizogenes*. Mol Gen Genet 206:17–23

Vincentz M, Caboche M (1991) Constitutive expression of nitrate reductase allows normal growth and development of *Nicotiana plumbaginifolia* plants. EMBO J 10:1027–1035

Wakasa Y, Ozawa K, Takaiwa F (2007) *Agrobacterium*-mediated transformation of a low glutelin mutant of 'Koshihikari' rice variety using the mutated-acetolactate synthase gene derived from rice genome as a selectable marker. Plant Cell Rep 26:1567–1573

Waldron C, Murphy EB, Roberts JL, Gustafson GD, Armour SL, Malcolm SK (1985) Resistance to hygromycin B. Plant Mol Biol 5:103–108

Wallis JG, Dziewanowska K, Guerra DJ (1996) Genetic transformation with the suli gene: a highly efficient selectable marker for *Solanum tuberosum* L. Cv. Russet burbank. Mol Breed 2:283–290

Wang H, Li Y, Xie L, Xu P (2003) Expression of a bacterial aroa mutant, aroA-M1, encoding 5-enolpyruvylshikimate-3-phosphate synthase for the production of glyphosate-resitant tobacco plants. J Plant Res 116:455–460

Wang Y, Chen B, Hu Y, Li J, Lin Z (2005) Inducible excision of selectable marker gene from transgenic plants by the Cre/lox site-specific recombination system. Transgenic Res 14:605–614

Wang JX, Zhao FY, Xu P (2006) Use of aroA-M1 as a selectable marker for *Brassica napus* transformation. Crop Sci 46:706–711

Wang J, Li Y, Liang C (2008) Recovery of transgenic plants by pollen-mediated transformation in *Brassica juncea*. Transgenic Res 17:417–424

Warwick S, Miki B (2004) Herbicide resistance. In: Nagata T, Lorz H, Widholm JM (eds) Biotechnology in agriculture and forestry, vol 54. Springer, New York, USA, pp 273–296

Weeks J, Koshiyama KY, Maier-Greiner U, Scheffner T, Anderson OD (2000) Wheat transformation using cyanamide as a new selective agent. Crop Sci 40:1749–1754

Wehrmann A, Van Vliet A, Opsomer C, Botterman J, Schulz A (1996) The similarities of *bar* and *pat* gene products make them equally applicable for plant engineers. Nat Biotechnol 14:1274–1278

Wenck A, Pugieux C, Turner M, Dunn M, Stacy C, Tiozzo A, Dunder E, Van Grinsven E, Khan R, Sigareva M, Wang WC, Reed J, Drayton P, Oliver D, Trafford H, Legris G, Rushton H, Tayab S, Launis K, Chang YF, Chen D-F, Mechers L (2003) Reef-coral proteins as visual, nondestructive reporters for plant transformation. Plant Cell Rep 22:244–251

Wohlleben W, Arnold W, Broer I, Hillemann D, Strauch E, Pühler A (1988) Nucleotide sequence of the phosphinothricin N-acetyltransferase gene from *Streptomyces viridochromogenes* Tü494 and its expression in *Nicotiana tabacum*. Gene Dev 70:25–37

Weinmann P, Gossen M, Hillen W, Bujard H, Gatz C (1994) A chimeric transactivator allows tetracycline-responsive gene expression in whole plants. Plant J 5:559–569

Xu C-J, Yang L, Chen KS (2005) Development of a rapid, reliable and simple multiplex PCR assay for early detection of transgenic plant materials. Acta Physiol Plant 27:283–288

Yamada T, Tozawa Y, Hasegawa H, Terakawa T, Ohkawa Y, Wakasa K (2005) Use of a feedback-insensitive α subunit of anthranilate synthase as a selectable marker for transformation of rice and potato. Mol Breed 14:363–373

Yang N-S, Steven GP, Paul C (1987) Detection of opines by colorimetric assay. Ann Biochem 160:342–345

Ye G-N, Hajdukiewicz PTJ, Broyles D, Rodriguez D, Xu CW, Nehra N, Staub JM (2001) Plastid-expressed 5-enolpyruvylshikimate-3-phosphate synthase genes provide high level glyphosate tolerance in tobacco. Plant J 25:261–270

Yemets A, Radchuk V, Bayer O, Bayer G, Pakhomov A, Baird WV, Blume YB (2008) Development of transformation vectors based upon a modified plant α-tubulin gene as the selectable marker. Cell Biol Int 32:566–570

Yoder J, Goldsbrough A (1994) Tranformation systems for generating marker-free transgenic plants. Biotechnology 12:263–267

You S-J, Liau C-H, Huang H-E, Feng T-Y, Prasad V, Hsiao H-H, Lu J-C, Chan M-T (2003) Sweet pepper ferredoxin-like protein (*pflp*) gene as a novel selection marker for orchid transformation. Planta 217:60–65

Yuan JS, Burris J, Stewart NR, Mentewab A, Stewart CN Jr (2007) Statistical tools for transgene copy number estimation based on real-time PCR. BMC Bioinform 8:S6

Zhang W, Subbarao S, Addae P, Shen A, Armstrong C, Peschke V, Gilbertson L (2003) Cre/lox-mediated marker gene excision in transgenic maize (*Zea mays* L.) plants. Theor Appl Genet 107:1157–1168

Zhang Y, Li H, Ouyang B, Lu Y, Ye Z (2006) Chemical-induced autoexcision of selectable markers in elite tomato plants transformed with a gene conferring resistance to lepidopteran insects. Biotechnol Lett 28:1247–1253

Zhang Z, Yang J, Collinge DB, Thordal-Christensen H (1996) Ethanol increases sensitivity of oxalate oxidase assays and facilitates direct activity staining in SDS gels. Plant Mol Biol Rep 14:266–272

Zhao F-Y, Li Y-F, Xu P-L (2006) *Agrobacterium*-mediated transformation of cotton (*Gossypium hirsutum* L. cv. Zhongmian 35) using glyphosate as a selectable marker. Biotechnol Lett 28:1199–1207

Zhao Y, Qian Q, Wang H-Z, Huang DN (2007) Co-transformation of gene expression cassettes via particle bombardment to generate safe transgenic plant without any unwanted DNA. In Vitro Cell Dev Biol Plant 43:328–334

Zheng SJ, Henken B, Sofiari E, Jacobsen E, Krens FA, Kik C (2001) Molecular characterization of transgenic shallots (*Allium cepa* L.) by adaptor ligation PCR (AL-PCR) and sequencing of genomic DNA flanking T-DNA borders. Transgenic Res 10:237–245

Zhou H, Arrowsmith J, Fromm M, Hironaka C, Taylor M, Rodriguez D (1995) Glyphosate-tolerant cp4 and gox gene as a selectable marker in wheat transformation. Plant Cell Rep 15:159–163

Zhu YJ, Agbayani R, McCafferty H, Albert HH, Moore PH (2005) Effective selection of transgenic papaya plants with the PMI/Man selection system. Plant Cell Rep 24:426–432

Zhu Z, Wu R (2008) Regeneration of transgenic rice plants using high salt for selection without the need for antibiotics or herbicides. Plant Sci 174:519–523

Ziemienowicz A (2001) Plant selectable markers and reporter genes. Acta Physiol Plant 23:363–374

Zubko E, Scutt C, Meyer P (2000) Intrachromosomal recombination between attp regions as a tool to remove selectable marker genes from tobacco transgenes. Nat Biotechnol 18:442–445

Zuo J, Niu QW, Moller SG, Chua NH (2001) Chemical-regulated, site-specific DNA excision in transgenic plants. Nat Biotechnol 19:157–161

Zuo J, Niub QW, Ikedab Y, Chua N-H (2002) Marker-free transformation: increasing transformation frequency by the use of regeneration-promoting genes. Curr Opin Biotechnol 13:173–180

Chapter 5
Levels and Stability of Expression of Transgenes

**Rajib Bandopadhyay, Inamul Haque, Dharmendra Singh,
and Kunal Mukhopadhyay**

5.1   Introduction

It is well known that in a given cell, at a particular time, only a fraction of the entire genome is expressed. Expression of a gene, nuclear, or organellar starts with the onset of transcription and ends in the synthesis of the functional protein. The regulation of gene expression is a complex process that requires the coordinated activity of different proteins and nucleic acids that ultimately determine whether a gene is transcribed, and if transcribed, whether it results in the production of a protein that develops a phenotype. The same also holds true for transgenic crops, which lie at the very core of insert design.

There are multiple checkpoints at which the expression of a gene can be regulated and controlled. Much of the emphasis of studies related to gene expression has been on regulation of gene transcription, and a number of methods are used to effect the control of gene expression. Controlling transgene expression for a commercially valuable trait is necessary to capture its value. Many gene functions are either lethal or produce severe deformity (resulting in loss of value) if over-expressed. Thus, expression of a transgene at a particular site or in response to a particular elicitor is always desirable.

Usually, the regions responsible for the initiation of transcription lie within the 5' region, upstream to the coding sequence of the gene. These are the promoter regions, defined as *cis*-acting (as they are on the same DNA strand that codes for the gene) nontranscribed elements, which provide sequences for the binding of various transcription initiation factors and RNA polymerase.

One can use a promoter that has known regulatory characteristics; for example, a promoter that is expressed throughout the plant tissue or only in vascular tissues, in

R. Bandopadhyay (✉)

Department of Biotechnology, Birla Institute of Technology, Mesra, Ranchi, Jharkhand 835215, India

Plant Genome Mapping Laboratory, University of Georgia, Athens, GA 30602, USA

e-mail: rajib_bandopadhyay@bitmesra.ac.in

C. Kole et al. (eds.), *Transgenic Crop Plants*,
DOI 10.1007/978-3-642-04809-8_5, © Springer-Verlag Berlin Heidelberg 2010

the leaf epidermis, seed endosperm or embryo, and so on. One can mix and match fragments of DNA and transcription factors to develop chimeric promoters that have the desired patterns and levels of gene expression.

In this chapter, we will discuss the basic aspects of designing genes for insertion and quantifying transgene expression, followed by the different types of promoters and their use in transgenic crops. Also, several factors responsible for high-level expression and stability in transgenic plants/crops will be discussed.

5.2   Gene Design for Insertion

Once a gene of choice has been targeted and cloned, it has to undergo several modifications before it can be effectively inserted into a plant. A promoter sequence is added for the gene to be expressed. Most promoters used for transgenic crop varieties have been "constitutive," i.e., causing gene expression throughout the life cycle of the plant in most of the tissues. The most commonly used constitutive promoter is CaMV 35S, from the cauliflower mosaic virus, which generally results in a high level of expression in most plants. Some promoters are more specific and are discussed in detail in Sect. 5.4.

Genes of interest are sometimes modified to achieve high level of expression. As plants prefer G-C rich regions, as compared to A-T rich bacterial genes, in order to overexpress bacterial genes in plants, A-T rich regions are to be substituted by G-C rich regions in such a way that the amino acid sequence of the protein remains unaltered (Evans et al. 2003). A selectable marker gene is inserted into the construct so as to identify the cells or tissues that have been successfully transformed (as discussed in Chap. 4). In some cases (e.g., resistance to pesticides), the transgene itself acts as a selectable marker. In other instances, a reporter gene is also inserted in the construct. A reporter gene is a coding sequence that upon expression in the transgenic plant provides conclusive evidence of genetic transformation. These reporter genes are very useful for transient expression experiments where the spatial and temporal activity of a promoter can be elucidated. The genes naturally exhibit an enzyme activity that does not exist in the host plant. Most common reporter genes are from bacteria, insects, or jellyfish as these organisms are so unrelated to angiosperms that their *cis*-regulatory elements are not functional in plants. Thus, when cloned into plant transformation vectors, a terminator sequence should also be fused downstream to the gene. Commonly used reporter genes are *CAT* (*E. coli*), β-*GUS* (*E. coli*), luciferase (firefly), green fluorescence protein (jellyfish), etc.

5.3   Quantification of Transgene Expression

It is necessary to know how a transgene is expressing in order to evaluate its effectiveness and level of expression in transgenic plants. The transgene copy number can greatly influence the expression level and genetic stability in the plant, and therefore, estimation of the transgene number is of prime importance (Bhat and Srinivasan 2002). Previously, Southern and Northern analyses were used for this purpose (Sabelli and Shewry 1995a, b). But, with time, other methods such as comparative genomic hybridization, fluorescence *in situ* hybridization, multiplex amplifiable probe hybridization, and microarray had been employed to determine the transgene copy number. All these methods are time consuming, laborious, and require large quantities of DNA. Moreover, nucleic acid hybridization-based techniques often involve the application of hazardous radioisotopes.

Recently, quantitative real-time PCR (qPCR) has proved to be an efficient method for transgene expression studies in plants. In qPCR, expression level is monitored per cycle of the reaction, comparing the fluorescent signal generated by the DNA or mRNA sample proportional to its initial quantity (Page and Minocha 2004). It has proved to be a more sensitive and rapid method, providing stringent evaluation through the use of SYBR Green I Fluorescent intercalating dye, which has the ability to detect a single gene among a number of genes in combination with highly specific gene primers. Till now, qPCR has been extensively applied to several transgenic crops such as maize (Ingham et al. 2001; Song et al. 2002; Shou et al. 2004; Assem and Hassan 2008), wheat (Li et al. 2004), rice (Yang et al. 2005), potato (Toplak et al. 2004), rapeseed (Weng et al 2004), tomato (Mason et al. 2001), tobacco (Miyamoto et al. 2000), cassava (Beltrán et al. 2009), and strawberry (Schaart et al. 2002) for analyzing transgene expression.

5.4 Promoters

As discussed earlier, the $5'$ upstream regions of a gene are not transcribed but provide sites for attachment of transcription initiation factors. The promoter itself contains many elements (short regions of a defined DNA sequence) for initiation factor attachment. The very basic of these elements is the TATA box, which is present about 25–30 bp upstream of the transcription start site and is primarily responsible for the correct positioning of RNA polymerase II. Many genes contain multiple operational TATA boxes, for example, three for *inrpk1* gene in *Ipomoea nil* (Bassett et al. 2004) and three for *phas* gene in *Phaseolus vulgaris* (Grace et al. 2004). It was thought earlier that the absence of TATA box is associated with constitutively expressing housekeeping genes, but recently TATA box was found to be absent in some inducible genes as well. Apart from TATA box, CAAT and GC boxes are also found to be present upstream; they too enhance the activity of RNA polymerase. Sequence elements like TATA boxes are also referred to as minimal or core promoter elements.

Along with the core promoter elements, other sequence elements are also found that provide sites for attachment of specific transcription factors or enhancer binding protein that trigger transcription of the gene (Alberts et al. 2002). These sequence elements are also called regulatory elements, enhancer binding elements, or simply enhancers. Enhancers are consensus DNA sequence motifs and are associated with levels, place, and timing of expression in response to internal or external (biotic or abiotic) factors. Enhancers can be located upstream, downstream, within coding regions or even in the intron sequences. One of the chief factors responsible for control of gene expression at the transcription level is the activation of enhancer sequences. A few *cis*-elements have the ability to silence or repress expression of the gene; these are called silencers. The activities of some of the plant promoters are summarized in Table 5.1. The table was generated using TGP, PlantCARE, NCBI, and Plant-Promoter databases for different genes specifically expressed in plants under the influence of suitable reporters/inducers, resulting in higher expression.

5.4.1 Types of Promoters and Their Applications in Transgenic Crops

5.4.1.1 Constitutive Promoters

Constitutive promoters maintain a constant level of activity. The cauliflower mosaic virus (CaMV) 35S promoter (derived from a DNA viral genome) is probably the most widely used plant promoter (Odell et al. 1985). Although "constitutive," many show differences in the level of expression in different tissues. Apart from delivering very high levels of expression in virtually all regions of the transgenic plant, the CaMV 35S promoter is easily obtainable for research purposes as plant transformation vector cassettes that allow for easy subcloning of the insert transgene of interest.

High levels of transgene expression can be achieved by the CaMV 35S promoter in both monocot and dicot plants (Benfey et al. 1990; Battraw and Hall 1990). The original full-size promoter (-941 to $+9$) has no significant difference in activity when compared to a -243 bp fragment. Interaction between the *cis*-acting elements within 343 bp upstream of the promoter results in high constitutive expression (Fang et al. 1989). However, tissue-specific individual elements have also been found (Benfey and Chua 1989). For the control of expression in specific tissues, two domains "Domain A" (-90 to $+8$) and "Domain B" (-343 to -90) are very important. Domain A is involved in expression in roots (Lam et al. 1989), while Domain B contains a conserved GATA motif, very much similar to the light responsive *cis* elements of light inducible promoters (Potenza et al. 2004). Even though CaMV 35S is a very strong promoter, it is strongly down-regulated in plant parasitic nematode feeding sites (Urwin et al. 1997). With the success of CaMV 35S promoter, other viral promoters have also been developed. They include the cassava vein mosaic virus (CsVMV; Verdaguer et al. 1996, 1998; Li et al. 2001), Australian banana streak virus (BSV; Schenk et al. 2001), mirabilis mosaic virus (MMV; Dey and Maiti 1999), and figwort mosaic virus (FMV; Maiti et al. 1997) promoters.

Table 5.1 Promoters used for transgene expression in plants

Promoter	Gene	Specificity	Reporter/inducer	Comment	References
At:SAG12_P1, P2	*sag12*, senescence-associated gene	transgenic *Arabidopsis* old leaves	mRNA, GUS activity/auxin, cytokinin, sugar	Essential for senescence-specific regulation of sag12	Noh and Amasino (1999)
Ms:PEPC7_P1, P2	*pepc7*, *pepc*	transgenic alfalfa plants	GUS activity/Rhizobium	very high levels of GUS activity	Pathirana et al. (1997)
Nt:COMTII_P1, P2, P3	*comtII*	transgenic tobacco leaves	GUS activity/chitin, glucan, TMV, pectin, wounding, MeJa	two to seven-fold increase in GUS activity	Toquin et al. (2003)
Ib:BMY1_P1, P2	*bmy1, amyb, beta-amy*	transgenic tobacco plants	GUS activity/sucrose (10%)	20–57-fold higher GUS activity	Maeo et al. (2001)
Dc:HCBT2_P1 P2, P3, P4	*hcbt2*	transgenic parsley protoplasts	GUS activity/elicitior	3.7–5.5-fold higher GUS activity	Yang et al. (1998)
Le:E4_P1, P2	*e4*	transgenic tomato plants	GUS activity/Ethylene	Increased 10–22-fold in unripe and 1,000-fold in ripened fruit	Montgomery et al. (1993a)
Zm:SBE1_P1, P2	*sbe1*	transgenic maize suspension endosperm cells	LUC activity/sucrose (9%)	two-fold greater LUC activity	Kim and Guiltinan (1999)
Vf:GRP3_P1, P2, P3	*grp3, glycine-rich early nodulin*	transgenic *Vicia hirsute* (5 week-old) nodules	GUS activity/Rhizobium	Essential for full promoter activity	Kuster et al. (1995)
Nt:CHN48_P1, P2	*chn48*	transgenic tobacco calli	LUC activity/elicitor	10–40-fold higher activity	Yamamoto et al. (1999)
At:P5CSA_P1, P2	*P5CSA_P1*	transgenic *Arabidopsis* plants	GUS activity/dehydration	five-fold increase in activity	Yoshiba et al. (1999)
Ib:SPOA1_P1, P2	*spoa1, gspo-a1*	transgenic tobacco plants	GUS activity/sucrose (3%)	tissue specific GUS staining	Ohta et al. (1991)
Pc:CMPG1b_P1	*cmpg1, elil7*	transgenic parsley suspension cells	GUS activity/elicitor	44-fold higher activity	Kirsch et al. (2001)
Pc:PR2_P1, P2	*pr2*	transgenic parsley protoplasts	GUS activity/elicitor	three- to eight-fold higher activity	van de Löcht et al. (1990)

(continued)

Table 5.1 (continued)

Promoter	Gene	Specificity	Reporter/inducer	Comment	References
Pc:PR1.1_P1, P2, P3	prl.1	transgenic parsley suspension cells	GUS activity/elicitor	7.5–12.1-fold higher activity	Rushton et al. (1996)
Pc:PR1.2_P1, P2	prl.2	transgenic parsley suspension cells	GUS activity/elicitor	5.2–15.2-fold higher activity	Rushton et al. (1996)
Ps:PSL_P1, P2, P3	psl	transgenic tobacco seeds	GUS activity, protein	High level of expression	de Pater et al. (1996)
Lg:LHCB2_P1, P2, P3, P4, P5	lhcb2, cabAB19	transgenic swollen duckweed seedlings	LUC activity/red light (2 min)	2–14-fold higher activity	Kehoe et al. (1994)
Np:LHCB1.2_P1, P2, P3, P4, P5	lhcb1.2	transgenic tobacco seedlings, transgenic Arabidopsis seedlings	GUS activity/far-red light (cont.), red light (pulses 3 min)	Medium 8–20-fold higher expression level	Cerdan et al. (2000)
Os:Amy1A_P1, P2, P3, P4	amy1A	transgenic rice embryos	GUS activity/Glucose, gibberellin A3	two- to six-fold lower expression with glucose, higher with gibberellin A3	Morita et al. (1998)
Os:Amy3D_P1	amy3D	transgenic rice embryos	GUS activity/glucose	six-fold lower Expression	Morita et al. (1998)
Zm:SH1_P1	sh1	transgenic tobacco plants	GUS activity/anaerobic conditions	High level of expression	Yang and Russell (1990)
At:CAB3_P1, P2, P3, P4	cab3	transgenic tobacco shoots (3–4 week-old), suspension cells	CAT activity/white light	Medium to high level of expression	Mitra et al. (1989)
Pc:WRKY1_P1, P2, P3	wrky1	transgenic parsley protoplasts	GUS activity/elicitor	2–50-fold higher activity	Eulgem et al. (1999)
Hv:LOXA_P1, P2, P3	loxA	transgenic barley leaves	GUS activity/methyl jasmonate (MeJA)	11.9–19.4-fold higher expression	Rouster et al. (1997)
Cr:CPR_P1	crp	transgenic tobacco plants	GUS activity/elicitor	Essential for controlled expression	Cardoso et al. (1997)
Nt:BGLUCANASE_P1, P2	gln2, gglb50	transgenic tobacco plants, transgenic tobacco plants (1 month-old)	GUS activity/tobacco mosaic virus, salicylic acid, ethephon (ethylene), water	2–10-fold higher activity	van de Rhee et al. (1993), Livne et al. (1997)

Nt:PR2D_P1, P2, P3	*pr2d*	transgenic tobacco plants	GUS activity/tobacco mosaic virus, salicylic acid, water	2–18-fold higher activity	Hennig et al. (1993)
Nt:PR2D_P2	*pr2b*	transgenic tobacco plants	GUS activity/tobacco mosaic virus, salicylic acid, water	five- to nine-fold higher activity	van de Rhee et al. (1993)
Np:GN1_P1	*gn1*	transgenic tobacco plants	GUS activity/SA, ethylene, water, elicitor, wounding	1.7–21-fold higher activity	Castresana et al. (1990)
At:CEL5_P1	*At1g22880*	transgenic *Arabidopsis* seedlings	GUS activity/auxin, ABA	Effective GUS staining activity	del Campillo et al. (2004)
As:PHYA3_P1, P2	*phyA3*	transgenic etiolated rice seedlings (2 day-old)	CAT activity/far-red light	five-fold higher activity	Bruce and Quail (1990)
St:GLUB_P1	*gluB8-1-3*	transgenic potato plants, transgenic tobacco plants	GUS activity/pathogen, tobacco mosaic virus, elicitor	2–12-fold higher activity	Mac et al. (2004)
Hv:GIII_P1	*GIII*	transgenic rice plants, transgenic rice calli	GUS activity/salicylic acid	Fragment length dependent GUS activity	Li et al. (2005)
Nt:PR1A_P1, P2	*pr1A*	transgenic tobacco plants	GUS activity/salicylic acid, tobacco mosaic virus	58–110-fold higher activity	Strompen et al. (1998), Uknes et al. (1993)
At:CYP85A1_P1	*cyp85A1*	transgenic *Arabidopsis* seedlings	GUS activity/Brassinolide	weak activity	Castle et al. (2005)
At:APX1_P1, P2, P3	*apx1*	transgenic *Arabidopsis* seedlings (2 week-old), (10 day-old)	GUS mRNA, GUS activity/ heat shock, methyl viologen, iron	Effective GUS staining	Storozhenko et al. (1998), Fourcroy et al (2004)
At:APX2_P1	*apx2*	transgenic *Arabidopsis* plants	LUC activity/hydrogen peroxide (H_2O_2), high light	ten-fold higher activity	Kimura et al. (2001)

(*continued*)

Table 5.1 (continued)

Promoter	Gene	Specificity	Reporter/inducer	Comment	References
At:APX2_P2	*apx1b*	transgenic tobacco mesophyll protoplasts	GUS activity/Heat Stress transcription Factor (HsfA2)	ten-fold higher activity	Schramm et al. (2006)
At:ATHB6_P1	*athb6*	transgenic *Arabidopsis* seedlings, transgenic *Arabidopsis* plants	GUS activity/drought, abscisic acid (ABA), salt	increased GUS staining with high level expression	Söderman et al. (1999)
At:CYP85A2_P1	*cyp85A2*	transgenic *Arabidopsis* seedlings	GUS activity/brassinolide	promoter activity down-regulated by brassinolide	Castle et al. (2005)
At:ELIP2_P1	*elip2*	transgenic *Arabidopsis* seedlings (10–14 day-old)	LUC activity/high light	100-fold increased expression	Kimura et al. (2001)
At:Fer1_P1	*fer1*	transgenic *Arabidopsis* plants, transgenic *Arabidopsis* cells	GUS activity/iron, senescence	17-fold derepression in response to 0.5 mM iron citrate	Tarantino et al. (2003)
At:Fer1_P2	*fer1*	transgenic *Arabidopsis* cells	GUS activity/iron	six-fold higher activity in response to 0.5 mM iron citrate	Petit et al. (2001)
Ca:Chi2_P1, P2, P3	*chi2*	transgenic tobacco	GUS activity/infection, mannitol, salt, NaCl, salicylic acid	1.5–4.5-fold higher activity	Hong and Hwang (2006)
Gm:Fer_P1, P2	*fer*	transgenic soybean leaves	LUC activity, GUS activity/ iron	iron dependent GUS activity	Wei and Theil (2000)
Hv:BLT101.1_P1, P2, P3	*blt101.1*	transgenic barley	GUS activity/low temperature	2.5-fold higher expression at low temperature	Brown et al. (2001)
Hv:BLT4.9_P1, P2, P3	*blt4.9*	transgenic barley	GUS activity/low temperature	2.5–6-fold higher expression at low temperature	Dunn et al. (1998)
POPLA:PAL1_P1	*pal1*	transgenic tobacco	GUS activity	Differential Gus activity	Gray-Mitsumune et al. (1999)

POPLA:PAL2_P1	pal2	transgenic tobacco, transgenic poplar	GUS activity	Differential Gus activity	Gray-Mitsumune et al. (1999)
Ta:GERMIN_P1	gf-2.8	transgenic tobacco seedlings, transgenic tobacco plants	GUS activity/heavy metal, cadmium, copper, cobalt, wounding, TMV	Gus activity detected after induction	Berna and Bernier (1999)
Zm:Fer1_P1	fer1	transgenic maize cells (BMS, Black Mexican Sweet)	GUS activity/iron	2.5–8-fold increase in activity	Petit et al. (2001)
At:FPS2_P1	fps2	transgenic Arabidopsis plants, transgenic Arabidopsis protoplasts	GUS activity	Differential Gus activity	Cunillera et al. (2000)
At:PAL1_P1, P2	pal1	transgenic Arabidopsis plants (3 week-old), transgenic tobacco plants (6 leaf stage)	GUS activity, GUS mRNA, PAL mRNA/ wounding, HgCl2, white light, elicitor, H2O2, mitomycin C (MMC)	30% increase in GUS activity	Ohl et al. (1990)
At:PDF1.2_P1	pdf1.2	transgenic Arabidopsis seedlings (10 day-old), transgenic Arabidopsis plants (4 week-old), transgenic tobacco seedlings (2 week-old), (6 week-old)	GUS activity/jasmonic acid (JA), pathogen, elicitor, methyl jasmonic acid (MeJA), paraquat, methyl viologen, rose bengal, TMV, ethylene	9–25-fold higher activity	Manners et al. (1998)
At:RAD54_P1	atrad54	transgenic Arabidopsis seedlings (1 week-old)	GUS mRNA/gamma irradiation	GUS activity detected after gamma irradiation	Osakabe et al. (2006)
At:THI1_P1, P2	thi1	transgenic Arabidopsis plants (14 day-old)	GUS activity/white light, flooding, salt, sugar deprivation	two to six-fold higher activity	Ribeiro et al. (2005)

(continued)

Table 5.1 (continued)

Promoter	Gene	Specificity	Reporter/inducer	Comment	References
Gm:SCAM4_P1	cam-4	transgenic Arabidopsis seedlings (2–4 day-old), transgenic Arabidopsis leaf protoplasts, transgenic Arabidopsis plants (4 week-old)	GUS mRNA, GUS activity/ salt, glycol chitin, Ca2+-ionophore A23187, elicitor, pathogen	3–15-fold higher activity	Park et al. (2004)
Le:LAT52_P1	lat52	transgenic tobacco mature pollen	GUS activity, LUC activity	Essential for full promoter activity	Bate and Twell (1998)
Zm:GAPC4_P1	gapc4, gpc4	transgenic tobacco leaves	GUS activity/anaerobic conditions	seven-fold higher activity	Geffers et al. (2001)
Nt:SAR8.2B_P1, P2, P3	sar8.2b	transgenic Arabidopsis plants (3 week-old)	GUS activity/Salicylic acid	4–31-fold higher GUS activity	Song et al. (2002)
Nt:PMT1A_P1	pmt1a	transgenic tobacco suspension cells	GUS activity/ (MeJA), ethephon	10–15-fold higher activity	Xu and Timko (2004)
Nt:G10_P1	g10, tobacco late pollen gene g10	transgenic tobacco pollen and leaves	GUS activity	Essential for full promoter activity	Rogers et al. (2001)
Ps:DRR206D_P1	drr206-d, pi206	transgenic tobacco plants (6-leaf stage)	GUS activity/Mitomycin C, actinomycin D, etoposide,H_2O_2, elicitor	Promoter only induced by natural tobacco Pathogen	Choi et al. (2001)
Nt:AP24_P1, P2	ap24	transgenic tobacco seedlings, transgenic tobacco plants	GUS activity/ethylene, NaCl, absicisic acid	2.5–13-fold higher Activity	Raghothama et al (1997)
At:OPR1_P1	opr1	transgenic Arabidopsis seedlings (2-week old)	GUS activity/Methyl jasmonate (MeJa), senescence	2–3.5-fold higher activity	He and Gan (2001)
Sc:OSML13_P1 Sc:OSML81_P1	osml13 osml81	transgenic potato plants	GUS activity/ABA, NaCl, SA, wounding, fungal infection, cold	Differential GUS activity	Zhu et al. (1995)
Sc:CI21A_P1, P2	ci21A	transgenic potato plants	GUS activity/low temperature (cold)	1.7–7-fold higher activity	Schneider et al. (1997)

There are a few limitations in the use of virus-derived promoters: first, the potential risk to human health from the genes of infective plant viruses (Hodgson 2000), and second, the ability of plant cells to recognize the inserted sequences of nonplant origin and inactivate them via "transcriptional gene silencing." Silencing is less common in promoters of plant origins.

Strong constitutive promoters of plant origin have also been isolated and used for the development of transgenic plants. Actin, a fundamental cytoskeleton component of the cell, is expressed in almost all the cells of a plant. The Act2 promoter, developed from *Arabidopsis* showed strong expression in all other parts except the seed coat, hypocotyl, ovary, and pollen sac (An et al. 1996). Similarly, rice Act1 promoter has also been developed (Zhang et al. 1991). Ubiquitins, a highly conserved protein family, are linked to many important cellular functions like chromatin structure and DNA repair. Maize ubiquitin 1 promoter (pUbi) has been successfully used for plant transformation of monocots (Weeks et al. 1993; Gupta et al. 2001) and has shown high levels of expression in actively dividing cells. Transgenic plants developed using Ubi.U4 promoter from *Nicotiana sylvestris* showed a three-fold higher activity when compared to the CaMV-based promoters. The ubiquitin-derived promoters perform very well in metabolically and mitotically active cells.

The constitutive action of a promoter has many drawbacks. Expression (or overexpression) of the transgene at a place where it is not expressed or expressed at a wrong time can have severe consequences on the growth and development of the plant. It can lead to enhanced susceptibility to some pathogens (Berrocal-Lobo et al. 2002) or decreased growth (Bowling et al. 1997). Another concern is the development of resistance by target insects against overexpressed toxins like, e.g., *Bt* toxin (Huang et al. 1999). For these reasons, it is ideal to strategically develop promoters that are "switched on" precisely when they are needed.

5.4.1.2 Nonconstitutive (Tissue-Enhanced) Promoters

Development of a new trait or value-addition of a previously existing one by genetic engineering requires the development of transgenes that are under control and expressed in a tissue-specific, developmental, or inducible manner. This will conserve energy and circumvent the drawbacks associated with constitutive expression to a large extent. It is more realistic to call these promoters tissue enhanced rather than tissue specific as their expression may not be confined to a specific tissue or plant part. Tissue-enhanced gene expression pattern is achieved as a result of several factors executed at various levels of gene control. Also, the more distant $5'$-*cis* acting enhancer element may be eliminated during isolation of the promoter, and there may not be effective functional interaction between the promoter *cis*-elements with the heterologous *trans*-acting factors present in the transgenic host plant. Thus, the development of such promoters can be very complex and difficult. Because of these complexities, it is preferable to use promoters from homologous or closely related plant taxa, and knowledge about the functionality of both homologous and heterologous promoters in target crop plants is essential.

Roots

These promoters are of high interest as they can be used for multiple applications. Expression of proteins responsible for resistance against drought and salt tolerance, resistance to bacterial (or fungal) pathogens or nematodes (Atkinson et al. 2003), and phytoremediation (Grichko et al. 2000) can improve crop yield. Although root-specific promoters have been isolated in plants (Yamamoto et al. 1991; Liu and Ekramoddoullah 2003), other stress-related studies have identified many candidate genes and their promoters. In maize, bacterial promoters are more popular (Qing et al. 2009). *Agrobacterium rhizogenes* causes hairy root disease in dicots. The promoters for rooting loci genes (*rol*) present in the root-inducing (Ri) plasmids are largely studied because of their root-mediated transformation and expression of the transgenes. Most important of the rol promoters is the rolD promoter, which has been much utilized (Stearns et al. 2005; Jayaraj et al. 2008) and extensively used in nitrogen assimilation studies (Fraisier et al. 2000; Fei et al. 2003). Very high levels of rolD promoter activity have been reported earlier (Elmayan and Tepfer 1995), but recently, in a comparative study, it was reported that *Arabidopsis* ubiquitin promoter (UBQ3) has the highest expression in roots (Wally et al. 2008). The domain A (-90 bp upstream) of CaMV 35S also shows root-specific activity (Benfey and Chua 1989; Benfey et al. 1990).

TobRB7, a putative membrane channel aquaporin, is another valuable plant based root-specific promoter isolated from tobacco (Yamamoto et al. 1991). Root-specific activity of this promoter was observed within 2 days of germination. Recently, a novel gene has been isolated from tomato, having very high expression levels in roots (*SlREO*); the 2.4-kb region representing the *SlREO* promoter sequence showed strict root specificity (Jones et al. 2008).

Root Nodules

Root nodules are formed as a result of an endosymbiotic association between *Rhizobium* and other species of bacteria with leguminous host plants. Within the nodule, the bacteroids fix atmospheric nitrogen that is used by the plant and in return, receive carbon substrates from the plant. This type of symbiosis is well studied. Leghemoglobin is an oxygen-binding protein synthesized in the nodule. The expression of leghemoglobin coincides with the nitrogen fixation in the nodule. Thus, this promoter can be used in nodules to increase nitrogen assimilation. The leghemoglobin promoter *glb3* from *Sesbania rostrata* was expressed in *Lotus corniculatus* and tobacco-harboring chimeric *glb3-uidA* (gus) gene fusions (Szabados et al. 1990; Szczyglowski et al. 1996). A 1.9-kb fragment of the *glb3* 5′-upstream region was found to direct high level of nodule-specific β-glucuronidase (GUS) activity in *L. corniculatus* that is restricted to the *Rhizobium*-infected cells of the nodules. In tobacco (a nonleguminous plant), the activity was restricted primarily to the roots and to phloem cells of the stem and petiole vascular system. A deletion analysis revealed that the region between −429 and −48 bp relative to the ATG was effective for nodule-specific expression.

Tubers

Tubers are storage organs in roots and are staple food source in many countries of the world. Improvement of tuber nutrition value, resistance toward infectious disease and pesticides can be manifested by using tuber-specific expression of transgenes. Patatins are glycoproteins that are one of the major products found in potato tuber. These are tuber- specific and can be induced by sucrose (Jefferson et al. 1990). The patatin promoters *Pat1* and *Pat2* were used to overexpress transgenes from the minipathway, of bacterial origin, to drive the synthesis of β-carotene (Provitamin A) in vitamin-A-deficient tubers of potato (Diretto et al. 2007). To enhance the metabolism of the environmental contaminants in tubers the rat P450 monooxygenase gene (*CYP1A1*) was overexpressed in tubers of potato, also under the control of patatin promoter (Yamada et al. 2002). High transgene expression was seen in developing tubers, and the amount of residual herbicides was much lower than that in nontransgenic plants, indicating that the transgenic plant metabolized and detoxified the herbicides. The processing quality of potato products (fries and chips) was increased by overexpressing transgenes in potato tuber under the control of TSSR (tuber-specific and sucrose-responsive) sequence from potato class I patatin promoter (Zhu et al. 2008). It was also demonstrated that tuber-specific expression of the native and slightly modified *MYB* transcription factor gene *StMtf1*(M) activates the phenylpropanoid biosynthetic pathway. The transgenic potato tubers contained four-fold increased levels of caffeoylquinates, including chlorogenic acid while also accumulating various flavonols and anthocyanins (Rommens et al. 2008). An 800-bp 5′ upstream sequence of the granule-bound starch synthase (*GBSS*) gene from potato was highly expressed in stolons and tubers (Visser et al. 1991), where the activities of the transgene in these two organs were 3–25-fold higher than the expression of the CaMV-*GUS* gene. The *GBSS* gene promoter was also used to obtain tuber-specific high expression of *AmA1*, a nonallergenic seed albumin gene from *Amaranthus hypochondriacus* in potato (Chakraborty et al. 2000). Two promoters, sporamin and β-amylase, have been well characterized in sweet potato (Maeo et al. 2001). The sweet potato sporamin promoter was found to control the expression of the *E. coli appA* gene in transgenic potato, which encoded a bifunctional enzyme exhibiting both acid phosphatase and phytase activities (Hong et al. 2008). Phytase expression levels in transgenic potato tubers were stable over several cycles of propagation. The study demonstrated that the sporamin promoter can effectively direct high-level recombinant protein expression in potato tubers. Moreover, overexpression of phytase in transgenic potato offers an ideal feed additive for improving phytate-Phosphorous digestibility in monogastric animals along with improvement of tuber yield, enhanced Phosphorous acquisition from organic fertilizers, and has a potential for phytoremediation.

Leaves

The light received in the environment can be roughly categorized as UV, visible, and far red. Three classes of photoreceptors have been identified in higher plants: red light and far-red light absorbing phytochromes (PHYs), blue-light receptors, and UV-light receptors. In *Arabidopsis*, five members compose the PHY family of photoreceptors (PHY A-E) and at least three different blue light photoreceptors have been identified [cryptochromes (CRYs), NPH1, and NPL1] (Martínez-Hernández et al. 2002). These photoreceptors, with association of other molecular systems (transcription factors), control the expression of many genes at the transcriptional and post-transcriptional level. Two important transcription factors are basic Leucine zipper factor HY5 (Oyama et al. 1997) and bHLH factor PIF3 (Martínez-García et al. 2000).

The photosynthesis-associated nuclear genes (*PhANGS*), like the chlorophyll a/b-binding proteins (Cab) and the small subunit of Rubisco (RbcS), contain a number of *cis*-acting elements, the transcription of which is controlled by light. Some of the motifs like G, I, and GTI boxes are found in the promoter regions of many light-regulated genes (Giuliano et al. 1988; Green et al. 1988; Menkens et al. 1995). The LS5-LS7 region from the *Lemna gibba Cab19* gene (Kehoe et al. 1994) and the CGF-1 factor-binding site from the *Arabidopsis CAB2* gene (Anderson and Kay 1995) contain the GATA and GT-1 sequences; still these two regions are unable to activate transcription, thus suggesting that additional regulatory elements are involved. This has led to the general hypothesis that light-responsive elements (LREs) are formed by the aggregation of different transcription factors. It has also been shown that artificial sequences composed of paired combinations of tetrameric repeats of G- and GATA boxes or GT1- and GATA-boxes, but not multimers of a single motif, function as LREs (Puente et al. 1996). Monocot rbcS promoters have different *cis*-acting elements and have different patterns of spatial expression than dicots. The C3 rbcS is specifically expressed in mesophyll cells, while the C4 rbcS is expressed in bundle sheath cells, and not in mesophyll cells (Nomura et al. 2000; Patel and Berry 2008). Overexpression of *Arabidopsis* phytocrome A (PHYA), under the control of rbcS promoter, in commercially important rice varieties produced an increased number of panicles per plant (Garg et al. 2006). In an attempt to obtain high-level production of intact *Acidothermus cellulolyticus* endoglucanase (E1) in transgenic tobacco plants using the constitutive (Mac) as well as light-inducible tomato Rubisco small subunit promoter (RbcS-*3C*), it was observed that RbcS-3 promoter was more favorable for E1 expression in transgenic plants than the Mac promoter (Dai et al. 2005). Moreover, by replacing *RbcS*-3C UTL with AMV RNA4 UTL, E1 production was enhanced more than two-fold. In a comparative study of the expression pattern of heterologous RbcS, RbcS3CP (0.8 kbp) from tomato, SRS1P (1.5 kbp) from soybean, and CaMV 35S in apple, it was found that the activity of SRS1P promoter was strictly dependent on light, whereas that of the RbcS-3C promoter appeared not to be so (Gittins et al. 2000). Later rolCP and CoYMVP were used for expression in vegetative tissues of apple; the CoYMV promoter was slightly more active than the rolC promoter, although expression was at a lower level than the CaMV 35S promoter (Gittins et al. 2003). The results indicated that both promoters could be suitable to drive the expression of transgenes to combat pests and diseases of apple that are dependent on interaction with the phloem.

The Cab proteins are highly expressed in green tissues and are often associated with other proteins to form the light-harvesting complex (Lhc). The expression pattern of the *Cab* gene in plants is different from that of the RbcS under certain physiological conditions as response to light quality and diurnal rhythm is different between these two genes (Ha and An 1988). Upon analysis of regulatory elements of *Cab-E* gene from *Nicotiana plumbaginifolia*, three positive and one negative *cis*-acting elements that influence photoregulation were found and of the three positive promoters two (PRE1 and PRE2) confer maximum level of photoregulation (Castresana et al. 1988). Tobacco plants when transformed with a chimeric gene encoding the A1 subunit of cholera toxin regulated by wheat Cab-1 promoter greatly reduced susceptibility to the bacterial pathogen *Pseudomonas tabaci* (Beffa et al. 1995).

Both RbcS and Cab are members of a multigene family and are expressed at very high level in green tissues (especially leaves), but many genes within the family contribute to the total protein content. Thus, the level of transgene expression is potentially dependent on the gene promoter used, so a strong green tissue-specific promoter from a single gene family will be most valuable.

Flowers

A substantial economic market has developed for cut flowers. Floral-specific promoters are therefore important for use in engineering transgenic flower varieties that may enhance vase life, visually appealing character of the flowers (reviewed by Mol et al. 1999) along with fragrance of interest and resistance to pests (Dolgov et al. 1995). The UEP1 promoter from Chrysanthemum when fused with a reporter gene (*GUS*) and transformed back into Chrysanthemum showed very high levels of expression in ray florets and three-fold lower expression in disk florets (Annadana et al. 2002). The activity of UEP1 promoter in ray florets is limited to petal tissues and does not extend into the tube of the petal or the sexual whorls of the floret. The promoter had 50-fold higher expression when compared with double CaMV-based promoters in petal tissues of ray florets (Annadana et al. 2002). This study also showed that CER6 promoter, associated with the wax biosynthesis pathway, had very high expression in ray florets, but the expression was much variable when compared to the UEP1 promoter.

Flavonoids are common color pigments in flowers and also perform many other functions including signaling and UV-protection. Engineering of the flavonoid biosynthetic pathway has led to the development of blue carnations (Holton 1995) and blue roses (Katsumoto et al. 2007). Chalcone synthase (CHS) genes as well as promoters have been studied extensively. The French bean CHS15 promoter showed expression in flowers and root tips of transformed tobacco plants (Faktor et al. 1996). In flowers, expression was confined to the pigmented part of petals and was induced in a transient fashion. Floral and root-specific expression required two conserved motifs, G-box and H-box, located near the TATA box. To evaluate the tissue-specific role of these motifs, a 39-bp DNA fragment containing the two motifs was prepared and fused with minimal promoters of CHS15 and CaMV 35S along with a marker gene (*GUS*). Tobacco plants were transformed and it was observed that the 39-bp polymer confers, upon both minimal promoters, a high level of expression that follows the typical tissue-specific expression pattern (Faktor et al. 1997). A chromoplast-specific carotenoid-associated gene (*OgCHRC*) and its promoter (Pchrc) was isolated from an orchid species (*Oncidium*), which showed very high and had flower-specific expression (Chiou et al. 2008).

Pistils

Pistil comprises the female part of the flower and includes stigma, style, and ovary. Identification of ovule-specific promoters is useful for the genetic engineering of crops with a variety of desirable traits, such as genetically engineered parthenocarpy, female sterility, or seedless fruits. The *SK2* gene from *Solanum tuberosum* encodes a pistil-specific endochitinase; the promoter from this gene was fused with a reporter (*GUS*) and when transformed back into potato, high-level expression specific to pistil was observed (Ficker et al. 1997). The 2.4-kb 5′-flanking region of the pistil-specific thaumatin gene (*PsTL1*) from Japanese pear, when transformed in tobacco, showed high expression in pistil, low in anther, and no detectable expression in the floral organs or the leaves. The promoter for *Arabidopsis AGL11* gene, when transformed back into *Arabidopsis,* showed high expression in the center of the young ovary, while expression was not seen in vegetative plant tissues, sepals, petals, or androecium (Nain et al. 2008).

Pollen/Anther

Anther as well as pollen-specific expression can be classified into "early" and "late" phases. The "early" phase comprises genes that are expressed during anther development and sporophytic tissue formation, while "late" phase involves expression during gametophyte generation and pollen formation/maturation. The 122-bp 5′ region of a tapetum-specific gene (*TA29*) isolated from tobacco programmed tapetum-specific expression as seen by fusing this promoter with a reporter (Koltunow et al. 1990). The expression increased in the developing anther and decreased as the microspores began to mature into pollen. The *TA29* promoter, fused with RNase (*barnase*), has been used to develop nuclear male-sterile plants (Mariani et al. 1990). In a comparative analysis, expression patterns of Bp4 promoter from rapeseed and the NTM19 promoter from tobacco were studied in transgenic tobacco (Custers et al. 1997). The Bp4 promoter became active only after the first pollen mitosis and not in the microspores, while the NTM19 promoter turned out to be highly microspore specific and directed very high level of GUS expression to the unicellular microspores; more importantly both the promoters were expressed only in the male germline (Custers et al. 1997). In *indica* rice, promoter of OSIPA was active during the late stages of pollen development and remained active till anthesis, whereas OSIPK promoter was active at a low level in developing anther till the pollen matured. OSIPK promoter activity diminished before anthesis. Both the promoters showed a potential to target expression of the genes of interest in developmental stage-specific manner and could help engineer pollen-specific traits in transgenic crops (Gupta et al. 2007). The anther- and tapetum-specific gene *TomA108* was present in as single copy per haploid genome of tomato. The fusion of β-glucuronidase to the TomA108 promoter demonstrated that the promoter was highly active from early meiosis to free microspores production in tapetum of tobacco (Xu et al. 2006).

Recently, a gene from pea, *PsEND1*, showed very high and early expression in anther primordium cells. Later *PsEND1* expression became restricted to the epidermis, connective, endothecium, and middle layer, but it was never observed in tapetal cells or microsporocytes. On fusion of the PsEND1 promoter region to the cytotoxic *barnase* gene to induce specific ablation of the cell layers, where the PsEND1 was expressed it produced male-sterile plants in tobacco and tomato (Roque et al. 2007). The PsEND1-*barnase* gene is quite different from other chimeric genes previously used to obtain male-sterile plants. The tapetum-specific promoter produces the ablation of specific cell lines during the initial steps of the anther development, but this chimeric construct (PsEND1-*barnase*) arrests the microsporogenesis before differentiation of the microspore mother cells and so, no viable pollen grains are produced. This strategy represents an excellent alternative to generate genetically engineered male-sterile plants. The PsEND1 promoter has high potential to prevent undesirable horizontal gene flow in many plant species (Roque et al. 2007). Two anther-specific cDNAs (designated *GhACS1* and *GhACS2*) encoding acyl-CoA synthetases (ACSs) isolated from cotton flower cDNA library were seen to accumulate in developing anthers. *GUS* expression controlled under the GhACS1 promoter showed high and specific expression in primary sporogenous cells, pollen mother cells, microspores, and tapetal cells (Wang and Li 2009).

Compared to "early" phase genes a few "late" phase genes have also been characterized. The promoter of tomato *Lat52* gene showed pollen-specific activity when transformed to tomato, tobacco, and *Arabidopsis* plants. Its expression was also correlated with the onset of microspore mitosis and increased progressively until anthesis (Twell et al. 1990). The elements necessary for expression in transgenics were present within 600 bp of the 5′ flanking region. The promoter sequence of *BAN215-6* gene from Chinese cabbage (*Brassica campestris*) showed high similarity with the *Lat52* gene (Kim et al. 1997). Expression studies, by *Agrobacterium*-mediated transformation of tobacco plants, revealed that 383 bp of the *BAN215-6* promoter region was sufficient for the anther-specific expression. The expression level was increased during anther development, reaching highest levels in mature pollens (Kim et al. 1997). The promoter of a maize pectin methytransferase gene (*ZmC5*) was found to be expressed specifically in late pollen development when transformed to tobacco plants (Wakeley et al. 1998). By genome walking PCR, a novel β-mannase gene (*LeMAN5*) was discovered in tomato, which is involved in cell wall disassembly and degrading mannan polymers. The 5′-upstream region of this endo-β-mannanase gene contained four copies of the pollen-specific *cis*-acting elements POLLEN1LELAT52 (*AGAAA*). The expression of the putative LeMAN5 promoter region (−543 to +38) in transgenic *Arabidopsis* was detected in mature pollen, sporangia, discharged pollen, and elongating pollen tubes (Filichkin et al. 2004).

Fruit

Fruits are one of the best delivery vehicles for value-added nutrients and other characters like increasing shelf-life, development of oral vaccines, etc. and there has always been a need for fruit-enhanced gene expression. The promoters of fruit-specific genes, especially fruit ripening genes, have been sought after. The ACC (1-aminocyclopropane-1-carboxylate) oxidase gene, the *E8* gene, and polygalacturonase (PG) genes are all fruit-ripening-specific promoters and have been characterized from apple and tomato (Montgomery et al. 1993a,b; Nicholass et al. 1995; Atkinson et al. 1998). The ACC oxidase gene is induced by application of ethylene, and fragments of 1,966 and 1,159 bp of the 5′ region showed both fruit and ripening specificity, whereas for the *PG* gene promoter, fragments of 1,460 and 532 bp conferred ripening-specific expression in transgenic tomato fruit (Atkinson et al. 1998). The promoter of the *E8* gene of tomato is by far the most important fruit-ripening-specific promoter. It has been successfully applied in a number of instances including enhancement of aroma of tomato by expressing *Clarkia breweri* S-linalool synthase gene (*LIS*) (Lewinsohn et al. 2001), fruit-specific expression of viral proteins (Sandhu et al. 2000), and cholera toxin gene (*CTB*) in an effort to make edible vaccines (He et al. 2008). The tomato *PG* gene is also associated with fruit ripening and its promoter was successfully employed to overexpress a bacterial phytoene synthase gene resulting in increased carotenoid content (Fraser et al. 2002) and a lemon basil α-zingiberene synthase gene (*ZIS*) in tomato fruit to increase both mono- and sesqui-terpene contents (Davidovich-Rikanati et al. 2008).

Seeds

Like fruits, seeds are also an excellent vehicle to pack transgenic products. Seed-specific transgenic technology can be used to enhance nutrient quality, production of pharmaceutical compounds, edible vaccines, etc. The genes expressed at very high level in the seeds are seed storage proteins and these have become the target of choice. Promoters for dicots as well as monocots have been extensively studied and several seed-specific elements have been characterized. The promoter region of soybean β-conglycinin was expressed in the embryo during the mid to late stages of seed development (Chen et al. 1989). The 2.4-kb upstream region of the sunflower *Helianthinin* gene (*HaG3-A*) also conferred high embryo-specific expression in transgenic *Arabidopsis* (Nunberg et al. 1994). The 0.8-kb fragment of the 5′ β-phaseolin gene of French bean (*Phaseolus vulgaris*) showed strong, temporally regulated, and embryo-specific expression in transgenic tobacco plants (Bustos et al. 1989). The expression pattern of the promoter fragment (1,108 bp) of the α-globulin gene in cotton was studied in transgenic cotton, *Arabidopsis*, and tobacco. Expression was initiated during the torpedo stage of seed development in tobacco, *Arabidopsis*, and during cotyledon expansion stage in cotton. The activity increased sharply until embryo maturation in all the three species. Expression was not detected in stem, leaf, root, pollen, or floral bud of transgenic cotton, thus confirming the high seed specificity of the promoter (Sunilkumar et al. 2002).

For monocots, several seed-specific promoters have been used successfully to incorporate many traits. The promoter region of the endosperm-specific protein hordein (D and B hordein) from barley has been well characterized in transgenic rice, barley, and wheat (Furtado et al. 2008, 2009). Six promoters (GluA-1, GluA-2, GluA-3, GluB-3, GluB-5, GluC) of seed storage glutenin genes were isolated from rice and their expression potential was checked in transgenic rice. The GluA-1, GluA-2, and GluA-3 promoters directed expression in the outer portion of the endosperm, while GluB-5 and GluC promoters directed expression in the whole endosperm. The GluB-3 promoter directed expression solely in aleurone and sub-aleurone layers, while maximum activity was pertained to the GluC promoter (Qu et al. 2008). Recently, edible vaccines are being made in transgenic rice against house dust mite allergy (Yang et al. 2008) and Japanese cedar pollen allergen (Yang et al. 2007) under the control of GluB-1 promoter and cholera toxin B subunit under the control of wheat Bx17 promoter containing an intron of the rice *act1* (Oszvald et al. 2008). The zein promoters from maize have been used for many applications. Transgenic maize with enhanced provitamin A content in the kernel was developed by endosperm-specific expression of the bacterial genes (*crtB* and *crtI*) under the control of a "super γ-zein promoter" (Aluru et al. 2008). Increase of total carotenoids was up to 34-fold with a preferential accumulation of β-carotene in the maize endosperm. The *phyA2* from *Aspergillus niger* was successfully expressed in maize seeds using the maize embryo specific globulin-1 promoter. The transgenic seeds showed a 50-fold increase in phytase activity (Chen et al. 2008). The developed maize hybrids had improved phosphorus availability for pig and poultry feed.

5.5   Factors Affecting Stability and Level of Transgene Expression

5.5.1   SAR/MAR Effect on Transgene Expression

Transgenic plants often display the chromosomal position effect, which results because of transgene integration events taking place within euchromatin, producing irregular and mixed expression. The pre-existing chromatin structure at the site of integration ultimately determines the expression level, acting either as an enhancer or as a silencer (Taddei et al. 2004). The chromosomal position effect can be prevented if the transgene is flanked by matrix attachment regions (MARs) also known as scaffold attachment regions (SARs) which are DNA elements that bind to the nuclear matrix (Mirkovitch et al. 1984; Allen et al. 2000). The location of MARs within transcription regulatory elements suggests that MARs may serve to bring these DNA sequences in proximity to the scaffold, thereby promoting enhancer and promoter activity by facilitating interaction with transcription factors (Nardozza et al. 1996). This inference is supported by loop domain model studies in which different expression profiles were observed on comparative analysis of transgenes that were flanked by MARs and those lacking it. Transgenes lacking MARs are influenced much by the surrounding chromatin structure; their expression levels are also dependent on local chromatin state. Transgenes that are flanked by MARs act independent of local chromatin state; thus, multiple copies of MAR-flanked transgene insertion might proportionally increase expression (Gasser and Laemmli 1986; Stief et al. 1989).

During gene activation chromatin structure becomes relaxed and the DNA is more accessible to DNase I. MARs that flank the chromatin loop domain function as the boundaries to differentiate active from inactive chromatin (Martienssen 2003). Later, on the basis of this inference, comparative study of the higher order chromatin structure and their accessibility to DNase I was performed in *Arabidopsis* and maize nuclei resulting in 45-Kb and 25-Kb domains, respectively, (Paul and Ferl 1998). It was reported that transgenic plants containing the synthetic MAR (sMAR) sequences derived from the MAR 3' end of the immunoglobulin heavy chain (IgH) enhancer, exhibited high levels of expression compared to transgenic plants that lacked the sMARs (Nowak et al. 2001). A diversity of promoters and MAR sequences has been used to analyze transgene expression. Mankin et al. (2003) analyzed the effects of a MAR, from the tobacco *RB7* gene on transgene expression from six different promoters in stably transformed tobacco cell cultures. The presence of MARs flanking the transgene increased expression of constructs based on the constitutive CaMV 35S, NOS (nopaline synthase 5' region), and OCS (octopine synthase 5' region) promoters (Mankin et al. 2003). Expression from a heat-shock induced promoter also increased five- to nine-folds, and MARs did not cause expression in the absence of heat shock (Schöffl et al. 1993). The effect of MAR fragments from tobacco gene transformed to two hybrid poplar clones and in tobacco plants was analyzed and found that MARs increased expression approximately ten- and two-fold, respectively, 1 month after cocultivation with *Agrobacterium*. Apart from gene expression, increased frequency of kanamycin resistance was also reported in poplar shoots (Han et al. 1997).

Different studies on MAR function in plant transgene expression provided interesting conclusions. The effect of MAR on transgene expression is analyzed only after the integration of transgene construct within plant genome. Enhancement of transgene expression has been reported in MARs containing stably transformed plant cell lines of soybean Gmhsp 17.6.L (Schöffl et al. 1993), yeast ARS1 (Allen et al. 1993; Vain et al. 1999), tobacco and rice Rb7 (Allen et al. 1996), tomato HSC80 (Chinn and Comai 1996), bean phaseolin (van der Geest et al. 1994), maize Adh1, Mha1 (Brouwer et al. 2002), and *Arabidopsis* ARS (Liu and Tabe 1998).

In a nutshell, MARs are not highly conserved but possess AT-rich DNA motifs of 100–3,000 bp containing binding sites for DNA topoisomerase II, DNA helicase, and DNA polymerase and thus are involved in structural organization of the genome. The loops created by MARs are topologically independent units of gene regulation and were found to facilitate the transcription of genes by changing topology along with less-condensed chromatin structure. The transgene constructs containing MARS are observed to create its own chromatin domain favorable for transcription; thus, MARS can reduce variability of transgene expression and increase level of expression.

5.5.2 Effect of 5′ and 3′ UTR Regions

The use of a specific promoter, with or without one or more enhancers, does not necessarily guarantee the desired level of gene expression in plants. In addition to the desired transcription levels, other factors such as improper splicing, polyadenylation, and nuclear export can affect accumulation of both mRNA and the protein of interest. Therefore, methods of increasing RNA stability and translational efficiency through mechanisms of post-transcriptional regulation are needed in the transgenic approach.

With regard to post-transcriptional regulation, it has been demonstrated that certain 5′ and 3′ untranslated regions (UTRs) of eukaryotic mRNAs play a major role in translational efficiency and RNA stability. For example, the 5′ and 3′ UTRs of tobacco mosaic virus (TMV) and alfalfa mosaic virus (AMV) coat protein mRNAs can enhance gene expression 5.4-fold and three-fold, respectively, in tobacco plants (Zeyenko et al. 1994). The 5′ and 3′ UTRs of the maize alcohol dehydrogenase-1 gene (*adh1*) are required for efficient translation in hypoxic protoplasts (Bailey-Serres and Dawe 1996; Hulzink et al. 2002).

Experiments with various 5′ UTR leader sequences demonstrate that various structural features of a 5′ UTR can be correlated with levels of translational efficiency. It was reported that 5′ UTR elements are required for the high-level expression of pollen *ACT1* gene in *Arabidopsis* (Vitale et al. 2003). During the process of initiation of translation 40S ribosomal subunit enters at 5' end of the mRNA and moves linearly until it reaches the first AUG codon, whereupon a 60S ribosomal subunit attaches and the first peptide bond is formed. Certain 5' UTR contain AUG codons in mRNA, which interact with 40S ribosomal subunit resulting in a weak context in terms of initiation codon, thus decreasing the rate of translation (Kozak 1991; Lee et al. 2009; Luttermann and Meyers 2009). Additionally, the 5' UTR nucleotide sequences flanking the AUG initiation site on the mRNA have an impact on translational efficiency. If the framework of the flanking 5' UTR is not favorable, part of the 40S ribosomal subunit fails to recognize the translation start site such that the rate of polypeptide synthesis will be slowed down (Kozak 1991; Pain 1996). Secondary structures of 5' UTRs (e.g., hairpin formation) also obstruct the movement of 40S ribosomal subunits during their scanning process and therefore negatively impact the efficiency of translation (Kozak 1986; Sonenberg and Pelletier 1988). The relative GC content of a 5' UTR sequence was shown to be the stability indicator of the potential secondary structure, high GC content indicated instability (Kozak 1991), and long UTRs exhibit a large number of inhibitory secondary structures. The translational efficiency of any given 5' UTR is highly dependent upon its particular structure and optimization of the leader sequence, which has been shown to increase gene expression as a direct result of improved translation initiation efficiency. Furthermore, significant increase in gene expression has been produced by addition of leader sequences from plant viruses or heat-shock genes (Datla et al. 1993).

In addition to 5' UTR sequences, 3' UTR sequences of mRNAs also influence in gene expression and known to control nuclear export, polyadenylation status, subcellular targeting, and rates of translation and degradation of mRNA from RNases. In particular, 3' UTRs contain one or more inverted repeats that can fold into stem-loop structures, which act as a barrier to exoribonucleases, and interact with RNA-binding proteins known to promote RNA stability (Gutiérrez et al. 1999). However, certain elements found within 3' UTR were reported to be RNA destabilizing, one such example occurring in plants is the DST element which can be found in small auxin up RNAs (SAURs) (Gil and Green 1996). A further destabilizing feature of some 3' UTRs is the presence of AUUUA pentamers (Ohme-Takagi et al. 1993).

The 3' UTRs were demonstrated to play a significant role in gene expression of several maize genes. Specifically, a 200-bp 3' sequence is responsible for suppression of light induction of maize small m3 subunit of the ribulose-1, 5-biphosphate carboxylase gene (*rbc/m3*) in mesophyll cells (Viret et al. 1994). Monde et al. (2000) observed that the *pet D3'*-UTR stem loop secondary structure was not able to form RNA-protein complex, essential for translational activity and thus acted as weak terminator required for RNA maturation. One 3' UTR frequently used in genetic engineering of plants is derived from nopaline synthase gene (3' *nos*) (Wyatt et al. 1993).

In certain plant viruses, such as alfalfa mosaic virus (AMV) and tobacco mosaic virus (TMV), the highly structured 3' UTRs are essential for replication and can be folded into either a linear array of stem-loop structures, which contain several high-affinity coat protein binding sites or a tRNA-like site recognized by RNA-dependent RNA polymerases (Olsthoorn et al. 1999).

5.5.3   Effect of Introns

Introns are the intragenic regions that are not translated into proteins. These noncoding portions are present in pre-mRNA and further removed by splicing to yield mature RNA. Introns contain acceptor and donor sites at either end as well as a branch point site, which is required for proper splicing by the spliceosome. The number and the length of introns vary widely among species and among genes within the same species. Introns with alternative splicing may introduce greater variability in protein sequences translated from a single gene. Introns also enhance the level of transgene expression in plants (Callis et al. 1987).

Recent studies provided several examples of introns, whose impact on expression is larger than that of the promoter from the same gene. Many genes with fully functional promoter are not essentially expressed at all but require an intron for their expression. A study in *Arabidopsis* showed that *PRF2* intron is required for full expression of a PRF2 promoter and the β-glucoronidase (*GUS*) and also to convert PFR5:*GUS* fusion from a reproductive to vegetative pattern (Jeong et al. 2006). Introns can increase the expression level through their enhancer element, an alternative promoter activity, or it can be independent of their conventional enhancer elements, i.e., intron-mediated enhancement (IME). The second intron of *Arabidopsis* agamous gene (*AG*) is a well-characterized enhancer-containing intron that can function in both orientations to force the expression of a reporter from a minimal promoter. The *Arabidopsis AG, STK, FLC* introns and wheat *VRN-1* intron act as enhancers. All these introns are large in size providing sufficient room for controlling elements and allow the establishment of stable chromatin conformation required for appropriate expression (Rose 2008).

The studies conducted by Morello et al. (2002, 2006) revealed the role of intron as an alternative promoter in rice. Presences of introns in promoterless genes drive weak expression; these introns are considered to contain promoters that are responsible for expression. The first intron acts as the alternative promoter as observed in *Ostub16* and *OsCDPK2* in rice, *PpAct1* and *PpAct5* in *Physcomitrella patens* and sesame, and *FAD2* in *Arabidopsis* (Kim et al. 2006; Weise et al. 2006).

IME of gene expression in plants indicates that the insertion of one or more introns in a gene construct results in increased accumulation of mRNA and protein relative to similar fusions that lack introns (Mascarenhas et al. 1990). The deletion studies of different introns such as maize *Adh1, Sh1* first intron, rice *Ostub A1* first intron, *Arabidopsis TRP1* first intron and *PRF2* intron1 revealed that no specific sequences were absolutely required and no conserved motif was found between enhancing introns (Rose 2008). Sequence analog studies showed enhancement can be restored by substituting the U/GC-rich region of intron with similar sequence analog from another part of intron (Rose 2002). The mutation studies in *Arabidopsis* *TRP1* intron1 and maize *Sh1* intron1 revealed that IME is destroyed by simultaneous elimination of branch-points and the 5' splice site, further indicating that splicing machinery is required for IME (Rose 2002). Additionally, the positions of introns also influence IME on gene expression. The most prominent is the location of the intron within the gene, i.e., the introns present in 5' UTR of the rice *rubi3* gene was shown to enhance expression (Lu et al. 2008). The other significant position of intron is near the starting of the gene (Rose 2004; Chung et al. 2006). Presently, there are several examples of introns (e.g., first intron of *OsTua2, OsTua3, OsTub4,* and *OsTub6*) that can greatly influence both the amount and the actual size of the expression, attributing different patterns of expression to the different intron iso-types, thus generating the intron-dependent spatial expression (IDSE) profile (Gianì

Table 5.2 Introns affecting transgene expression in different plants

Intron	Specificity	Remark	Reference
COX 5c-1	Arabidopsis	Increased GUS expression level	Curi et al. (2005)
COX 5c-2			
Ubi7	Potato	Ten-fold higher expression	Garbarino et al. (1995)
Adh1	Maize	40–100-fold increase in expression	Callis et al. (1987)
Act1	Arabidopsis	High level of reproductive tissue expression	Vitale et al. (2003)
RBCS2	Chlamydomonas reinhardtii	Stable high-level expression	Lumbreras et al. (1998)
reg A3	Volvox carteri	Required for *regA* expression	Stark et al. (2001)
regA5			
STK	Arabidopsis	Intron-mediated promoter expression in ovules and septum	Kooiker et al. (2005)
VRN-1	Wheat	Essential for promoter activity	Fu et al. (2005)
Ostub 16	Rice	Required for maximum promoter activity	Morello et al. (2002)
OsCDPK 2	Rice	Required for promoter activity	Morello et al. (2006)
PpAct 1	Physcomitrella patens	11–18-fold higher expression	Weise et al. (2006)
PpAct 5			
PpAct 7			
Sh1	Maize	10–1,000-fold enhanced expression	Maas et al. (1990, 1991)
Gap A1	Maize	Required for full promoter activity	Donath et al. (1995)
Actin 3rd intron	Maize	IME	Luehrsen and Walbot (1991)
Hsp81	Maize	IME	Sinibaldi and Mettler (1992)
Act 1	Rice	IME	McElroy et al. (1990)
tpi	Rice	Required for promoter activity	Xu et al. (1994)

et al. 2009). The specificity of introns acts as enhancer, and alternative promoter, or mediates enhancement on the basis of their numbers and position in a gene construct. Some of the introns with defined specificity have been summarized in Table 5.2.

5.5.4 Role of Transcription Factors

Transcription factors are sequence-specific DNA-binding proteins that interact with the promoter regions of the target genes and modulate the rate of initiation of mRNA synthesis by RNA polymerase II (Gantet and Memelink 2002). The role of transcription factors in transgene expression is studied by overexpression and antisense technology. For highly conserved transcription factors such as MADS-box or the Myb-like transcription factors, generation of antisense plant is difficult since the target requires an antisense RNA homology of over 50 bp, which is not preferred. Also, due to the presence of highly conserved regions, the specificity of antisense RNA is significantly reduced (Cannon et al. 1990). High-level expression of a transcription factor in a transgenic plant cell might favor the binding of the transcription factor to low affinity binding sites and result in activation of gene expression from noncognate promoters.

To study the effect of transcription factor on transgene, the steroid-binding domain of the glucocorticoid receptor is fused to a plant transcription factor. The absence of ligand represses nuclear localization and DNA-binding activities of transcription factor. After induction, repression is relieved and active protein can rapidly enter the nucleus and exert its transcription factor function. A glucocorticoid-responsive GAL4-VP16 fusion protein has been used to induce the activation of a luciferase reporter gene in transgenic *Arabidopsis* and tobacco plants, either by growing the plants on nutrient agar containing dexamethasone or by spraying the plants with the inducing compound (Aoyama and Chua 1997).

The *Arabidopsis* transparent testa glabra (*ttg*) mutant plants are not able to produce trichomes, anthocyanins, and seed coat pigment but generate excess root hairs. Production of trichomes and anthocyanins could be restored by overexpression of the maize transcription factor *R* in a constitutive and inducible manner (Lloyd et al. 1994). An interaction study carried out between transcription factor (*myb305*) and its promoter-binding site in *PAL2* in transgenic tobacco plant revealed that when leaves were inoculated with a PVX-construct expressing *Myb305* reporter gene, expression increased (Sablowski et al. 1995). Thus, the ectopic expression of *Myb305* in infected tissue incites the higher expression of *GUS* reporter gene in transgenic tobacco plant with nonmutant *PAL2* promoter element.

Synthetic transcription factors are an assembly of multiple zinc finger domains designed to achieve better regulation of gene expression. It is estimated that *Arabidopsis* contains 85 genes that encode zinc finger transcription factors (Riechmann and Ratcliffe 2000). Such synthetic zinc finger transcription factors (*TFsZF*) can be custom designed for binding to any DNA sequence (Segal and Barbas 2001).

Furthermore, the addition of herpes simplex virus VP16 activation domain to the polydactyl six-zinc finger protein 2C7 increased the expression more than 450-fold in transgenic plants (Liu et al. 1997). Later, Van Eenennaam et al. (2004) constructed five, three-finger zinc finger protein (ZFP) DNA-binding domains which tightly bound to 9-bp DNA sequences located on either the promoter or the coding region of the *Arabidopsis GMT* gene. When these ZFPs were fused to a maize *opaque-2* nuclear localization signal and the maize *C1* activation domain, four out of the five resulting ZFP-TFs were able to up-regulate the expression of the *GMT* gene in leaf protoplast transient assays. The seed-specific expression of these ZFP-TFs was reported to produce heritable increase in seed α-tocopherol level in subsequent generations of transgenic *Arabidopsis*.

The transcription factors, *R* and *C1*, interact to regulate anthocyanin biosynthesis in the maize kernel (Grotewold et al. 2000). In a recent study, it was reported that ectopic expression of a conifer *Abscisic Acid Insensitive 3* (*ABI3*) transcription factor induced high-level synthesis of recombinant human α-L-iduronidase gene in transgenic tobacco leaves (Kermode et al. 2007). Transgenic rice with *DREB 1s/CBF* or *OsDREB 1A/1B* transcription factor interact specifically with *DRE/CRT* or *OsDRE cis*-acting elements and control the expression of many stress-inducible genes (Ito et al. 2005). In continuation, Zhao et al. (2009) reported on the role of transcription factors on abiotic stress where the expression of yeast *YAP1* gene in transgenic *Arabidopsis* resulted in increased salt tolerance. The *YAP1* contains a basic leucine zipper domain similar to that of Jun (Moye-Rowley et al. 1989), which is a component of mammalian *AP-1* transcription factor complexes. Nuclear *YAP1* regulates the expression of up to 70 genes that are related to oxidative stress caused by high salinity (Zhao et al. 2009).

5.5.5   Effect of DNA Acetylation and Methylation

The interaction of histones with DNA plays an important role in chromatin remodeling and consequently the activation or repression of gene expression (Tian et al 2005). Intrinsic histone acetyltransferases (HATs) and histone deacetylases (HDs, HDAs, HDACs) drive acetylation and deacetylation, respectively, thus providing a mechanism for reversibly modulating chromatin structure and transcriptional regulation (Jenuwein and Allis 2001). Hyperacetylation relaxes chromatin structure and activates gene expression, whereas hypoacetylation induces chromatin compaction and gene repression. Histone acetylation and deacetylation are reversible and therefore play a significant role in transcriptional regulation associated with developmental programs and environmental conditions. These include day-length (Tian et al. 2003), flowering (He et al. 2003), osmotic and oxidative stress (Brunet et al 2004, De Nadal et al 2004), and cell aging (Imai et al 2000).

Acetylation neutralizes the lysine residues on the amino terminal tails of the histones, thereby neutralizing the positive charges of histone tails and decreasing their affinity to bind DNA. HATs are often associated with proteins forming coactivator complexes, stabilizing the chromatin in an open conformation and transcriptionally active state. These complexes are targeted to promoters by specific transcription factors, allowing the RNApol II holoenzyme to access the promoter DNA sequence, which results in activation of transcription and increased gene expression. Histone deacetylases (HDAC) ameliorate the affinity of histones for DNA as deacetylation of histone tails result in stronger interaction between the basic histone tails and DNA. HDACs are often associated with other proteins that are associated with chromatin condensation and repression of transcription. These corepressor complexes promote heterochromatin formation, blocking access of RNApol II, thus resulting in repression of transcription.

Arabidopsis has 18 members of putative histone deacetylase family (Pandey et al. 2002). Among them *AtHDA6* is responsible for silencing transgenes (Murfett et al. 2001), whereas *AtHD1* is reported to be a global transcriptional regulator throughout the development of *Arabidopsis* (Tian et al. 2003). The analysis of microarray data revealed that gene activation is associated with increased levels of site-specific histone acetylation, whereas gene repression does not correlate with the changes in histone acetylation or histone methylation. Many of the HDACs found in plants are *Rpd3*, *HD2*, *SIR2*, and their homologs (Chen and Tian 2007).

DNA methylation is known to play a role in plant gene silencing (Ng and Bird 1999). Methyl CpG-binding protein (MeCP2) was reported to be involved in the recruitment of HDAC to methylated DNA through a corepressor complex, which results in gene silencing. Earlier studies showed that hemimethylation results in inhibition of transient gene expression, whereas nonmethylated gene expressed normally (Weber et al. 1990). In one of the studies, a mutation isolated via a transgene reactivation screen in *Arabidopsis*, mom1, was thought to act downstream of DNA methylation signals in controlling silencing because it did not confer obvious methylation changes (Amedeo et al. 2000). In recent studies, Shibuya et al. (2009) reported that the *pMADS3* gene in petunia, specifically expressed in the stamen and carpels of developing flower, showed ectopic expression after introduction of intron 2. This is known as ect-pMADS3 phenomenon and is due to transcriptional activation based on RNA-directed DNA methylation (RdDM) occurring in a particular CG in a putative *cis*-element in *pMADS3* intron 2. The CG methylation was maintained over generations, along with *pMADS3* ectopic expression, even in the absence of RNA triggers. Transcriptional or post-transcriptional gene silencing was expected; instead, upregulated gene expression was observed (Shibuya et al. 2009).

Recently, the new Amplicon-plus targeting technology (APTT) has been developed to overcome the problems of post-transcriptional gene silencing and lower accumulation of transgenic protein. This technology uses a novel combination of techniques, i.e., expression of a mutated PTGS suppressor, P1/HC-Pro, with PVX (potato virus X vector) amplicon encoding a highly-labile L1 protein of canine oral papillomavirus (COPV L1). Appreciable amount of protein accumulation was achieved by targeting the L1 to various cellular compartments, by creating a fusion between the protein of interest and different targeting peptides. Additionally, a scalable "wound-and-agrospray" inoculation method has been developed that allows high-throughput *Agrobacterium* inoculation of *Nicotiana tabacum* to facilitate large-scale application of this technology (Azhakanandam et al. 2007).

5.6   Conclusions

Genetic transformation of crops has opened a new dimension to increase production that benefits both producers and consumers. Its effect can be best utilized in less developed or developing countries where crop yield is severely affected by biotic and abiotic stress. Also, value addition of existing nutrients along with production of novel nutraceuticals will help alleviate nutrition-related deficiencies in famine-stricken countries. Apart from enhancing food value in crop species, transgenic technology can be used to develop visual marker systems to monitor crops and carry out fine scale studies of agricultural crops. Despite the hostility against genetically modified crops in Eastern Europe, many countries in Asia and North America have accepted transgenic crops. In the present scenario, some of the factors responsible for the control of transgene expression at different levels have been summarized in Fig. 5.1. The primary challenge lies with the detailed under-standing of the underlying mechanism involved in gene expression, and there is a pressing need to study gene expression, especially its regulation.

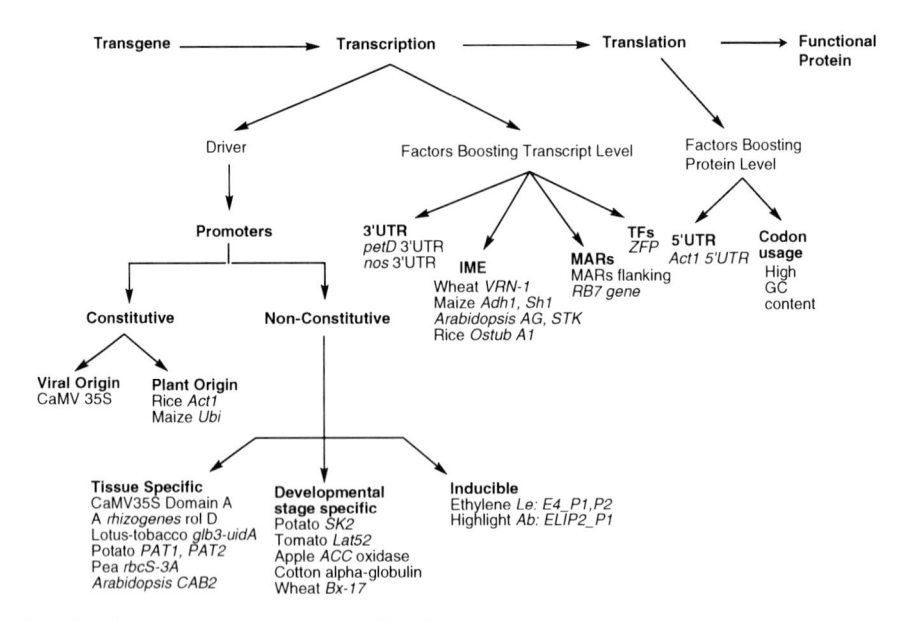

Fig. 5.1 Transgene expression and stability factors

Acknowledgements RB is thankful to Prof. A.H. Paterson of PGML, UGA, for hosting as a BOYSCAST Fellow, and gratefully acknowledges the financial support from the Department of Science and Technology, GOI (SR/BY/L-08/2007).

References

Alberts B, Johnson A, Lewis J, Raff M, Roberts K, Walter P (2002) Molecular biology of the cell. Garland Sci, New York, USA

Allen GC, Hall G, Michalowski S, Newman W, Spiker S, Weissinger AK, Thompson WF (1996) High-level transgene expression in plant cells: effects of a strong scaffold attachment region from tobacco. Plant Cell 8:899–913

Allen GC, Hall GE Jr, Childs LC, Weissinger AK, Spiker S, Thompson WF (1993) Scaffold attachment regions increase reporter gene expression in stably transformed plant cells. Plant Cell 5:603–613

Allen GC, Spiker S, Thompson WF (2000) Use of matrix attachment regions (MARs) to minimize transgene silencing. Plant Mol Biol 43:361–376

Aluru M, Xu Y, Guo R, Wang Z, Li S, White W, Wang K, Rodermel S (2008) Generation of transgenic maize with enhanced provitamin A content. J Exp Bot 59:3551–3562

Amedeo P, Habu Y, Afsar K, Mittelsten Scheid O, Paszkowski J (2000) Disruption of the plant gene MOM releases transcriptional silencing of methylated genes. Nature 405:203–206

An YQ, McDowell JM, Huang S, McKinney EC, Chambliss S, Meagher RB (1996) Strong, constitutive expression of the *Arabidopsis* ACT2/ACT8 actin subclass in vegetative tissues. Plant J 10:107–121

Anderson SL, Kay SA (1995) Functional dissection of circadian clock- and phytochrome-regulated transcription of the *Arabidopsis CAB2* gene. Proc Natl Acad Sci USA 92: 1500–1504

Annadana S, Beekwilder MJ, Kuipers G, Visser PB, Outchkourov N, Pereira A, Udayakumar M, De Jong J, Jongsma MA (2002) Cloning of the chrysanthemum *UEP*1 promoter and comparative expression in florets and leaves of *Dendranthema grandiflora*. Transgenic Res 11:437–445

Aoyama T, Chua NH (1997) A glucocorticoid-mediated transcriptional induction system in transgenic plants. Plant J 11:605–612

Assem SK, Hassan OS (2008) Real time quantitative PCR analysis of transgenic maize plants produced by *Agrobacterium* mediated transformation and particle bombardment. J Appl Sci Res 4:408–414

Atkinson HJ, Urwin PE, McPherson MJ (2003) Engineering plants for nematode resistance. Annu Rev Phytopathol 41:615–639

Atkinson RG, Bolitho KM, Wright MA, Iturriagagoitia-Bueno T, Reid SJ, Ross GS (1998) Apple ACC-oxidase and polygalacturonase: ripening-specific gene expression and promoter analysis in transgenic tomato. Plant Mol Biol 38:449–460

Azhakanandam K, Weissinger SM, Nicholson JS, Qu R, Weissinger AK (2007) Amplicon-plus targeting technology (APTT) for rapid production of a highly unstable vaccine protein in tobacco plants. Plant Mol Biol 63:393–404

Bailey-Serres J, Dawe RK (1996) Both 5′ and 3′ sequences of maize *adh1* mRNA are required for enhanced translation under low-oxygen conditions. Plant Physiol 112:685–695

Bassett CL, Nickerson ML, Farrell RE Jr, Harrison M (2004) Multiple transcripts of a gene for a leucine-rich repeat receptor kinase from morning glory (*Ipomoea nil*) originate from different TATA boxes in a tissue-specific manner. Mol Genet Genomics 271:752–760

Bate N, Twell D (1998) Functional architecture of a late pollen promoter: pollen-specific transcription is developmentally regulated by multiple stage-specific and co-dependent activator elements. Plant Mol Biol 37:859–869

Battraw MJ, Hall TC (1990) Histochemical analysis of CaMV 35S promoter-β-glucuronidase gene expression in transgenic rice plants. Plant Mol Biol 15:527–538

Beffa R, Szell M, Meuwly P, Pay A, Vögeli-Lange R, Métraux JP, Neuhaus G, Meins F Jr, Nagy F (1995) Cholera toxin elevates pathogen resistance and induces pathogenesis-related gene expression in tobacco. EMBO J 14:5753–5761

Beltrán J, Jaimes H, Echeverry M, Ladino Y, López D, Duque MC, Chavarriaga P, Tohme J (2009) Quantitative analysis of transgenes in cassava plants using real-time PCR technology. In Vitro Cell Dev Biol Plant 45:48–56

Benfey PN, Chua NH (1989) Regulated genes in transgenic plants. Science 244:174–181

Benfey PN, Ren L, Chua NH (1990) Tissue-specific expression from CaMV 35S enhancer subdomains in early stages of plant development. EMBO J 9:1677–1684

Berna A, Bernier F (1999) Regulation by biotic and abiotic stress of a wheat germin gene encoding oxalate oxidase, a H_2O_2-producing enzyme. Plant Mol Biol 39:539–549

Berrocal-Lobo M, Molina A, Solano R (2002) Constitutive expression of ETHYLENE-RESPONSE-FACTOR1 in *Arabidopsis* confers resistance to several necrotrophic fungi. Plant J 29:23–32

Bhat SR, Srinivasan S (2002) Molecular and genetic analyses of transgenic plants: considerations and approaches. Plant Sci 163:673–681

Bowling SA, Clarke JD, Liu Y, Klessig DF, Dong X (1997) The *cpr5* mutant of *Arabidopsis* expresses both NPR1-dependent and NPR1-independent resistance. Plant Cell 9:1573–1584

Brouwer C, Bruce W, Maddock S, Avramova Z, Dowen B (2002) Suppression of transgene silencing by matrix attachment regions in maize: A dual role for the maize 5'*ADH1* matrix attachment region. Plant Cell 14:2251–2264

Brown AP, Dunn MA, Goddard NJ, Hughes MA (2001) Identification of a novel low-temperature-response element in the promoter of the barley (*Hordeum vulgare* L.) gene *blt101.1*. Planta 213:770–780

Bruce WB, Quail PH (1990) *cis*-Acting elements involved in photoregulation of an oat phytochrome promoter in rice. Plant Cell 2:1081–1089

Brunet A, Sweeney LB, Sturgill JF, Chua KF, Greer PL, Lin Y, Tran H, Ross SE, Mostoslavsky R, Cohen HY, Hu LS, Cheng HL, Jedrychowski MP, Gygi SP, Sinclair DA, Alt FW, Greenberg ME (2004) Stress-dependent regulation of FOXO transcription factors by the SIRT1 deacetylase. Science 303:2011–2015

Bustos MM, Guiltinan MJ, Jordano J, Begum D, Kalkan FA, Hall TC (1989) Regulation of β-glucuronidase expression in transgenic tobacco plants by an A/T-rich, cis-acting sequence found upstream of a French bean β-phaseolin gene. Plant Cell 1:839–853

Callis J, Fromm M, Walbot V (1987) Introns increase gene expression in cultured maize cells. Genes Dev 1:1183–1200

Cannon M, Platz J, O'Leary M, Sookdeo C, Cannon F (1990) Organ-specific modulation of gene expression in transgenic plants using antisense RNA. Plant Mol Biol 15:39–47

Cardoso MI, Meijer AH, Rueb S, Machado JA, Memelink J, Hoge J (1997) A promoter region that controls basal and elicitor-inducible expression levels of the NADPH:cytochrome P450 reductase gene (*Cpr*) from *Catharanthus roseus* binds nuclear factor GT-1. Mol Gen Genet 256:674–681

Castle J, Szekeres M, Jenkins G, Bishop GJ (2005) Unique and overlapping expression patterns of Arabidopsis *CYP85* genes involved in brassinosteroid C-6 oxidation. Plant Mol Biol 57:129–140

Castresana C, de Carvalho F, Gheysen G, Habets M, Inzé D, Van Montagu M (1990) Tissue-specific and pathogen-induced regulation of a Nicotiana plumbaginifolia β-1, 3-glucanase gene. Plant Cell 2:1131–1143

Castresana C, Garcia-Luque I, Alonso E, Malik VS, Cashmore AR (1988) Both positive and negative regulatory elements mediate expression of a photoregulated CAB gene from *Nicotiana plumbaginifolia*. EMBO J 7:1929–1936

Cerdan PD, Staneloni RJ, Ortega J, Bunge MM, Rodriguez-Batiller MJ, Sanchez RA, Casal JJ (2000) Sustained but not transient phytochrome A signaling targets a region of an *Lhcb1*2* promoter not necessary for phytochrome B action. Plant Cell 12:1203–1211

Chakraborty S, Chakraborty N, Datta A (2000) Increased nutritive value of transgenic potato by expressing a nonallergenic seed albumin gene from *Amaranthus hypochondriacus*. Proc Natl Acad Sci USA 97:3724–3729

Chen R, Xue G, Chen P, Yao B, Yang W, Ma Q, Fan Y, Zhao Z, Tarczynski MC, Shi J (2008) Transgenic maize plants expressing a fungal phytase gene. Transgenic Res 17:633–643

Chen ZJ, Tian L (2007) Roles of dynamic and reversible histone acetylation in plant development and polyploidy. Biochim Biophys Acta 1769:295–307

Chen ZL, Naito S, Nakamura I, Beachy RN (1989) Regulated expression of genes encoding soybean beta-conglycinins in transgenic plants. Dev Genet 10:112–122

Chinn AM, Comai L (1996) The heat shock cognate 80 gene of tomato is flanked by matrix attachment regions. Plant Mol Biol 32:959–968

Chiou YC, Wu K, Yeh KW (2008) Characterization and promoter activity of chromoplast specific carotenoid associated gene (*CHRC*) from *Oncidium* Gower Ramsey. Biotechnol Lett 30:1861–1866

Choi JJ, Klosterman SJ, Hadwiger LA (2001) A comparison of the effects of DNA-damaging agents and biotic elicitors on the induction of plant defense genes, nuclear distortion, and cell death. Plant Physiol 125:752–762

Chung BY, Simons C, Firth AE, Brown CM, Hellens RP (2006) Effect of 5'UTR introns on gene expression in *Arabidopsis thaliana*. BMC Genomics 7:120

Cunillera N, Boronat A, Ferrer A (2000) Spatial and temporal patterns of GUS expression directed by 5' regions of the *Arabidopsis thaliana* farnesyl diphosphate synthase genes *FPS1* and *FPS2*. Plant Mol Biol 44:747–758

Curi GC, Chan RL, Gonzalez DH (2005) The leader intron of *Arabidopsis thaliana* genes encoding cytochrome *c* oxidase subunit 5c promotes high-level expression by increasing transcript abudance and translation efficiency. J Exp Bot 56:2563–2571

Custers JB, Oldenhof MT, Schrauwen JA, Cordewener JH, Wullems GJ, van Lookeren Campagne MM (1997) Analysis of microspore-specific promoters in transgenic tobacco. Plant Mol Biol 35:689–699

Dai Z, Hooker BS, Quesenberry RD, Thomas SR (2005) Optimization of *Acidothermus cellulolyticus* endoglucanase (E1) production in transgenic tobacco plants by transcriptional, post-transcription and post-translational modification. Transgenic Res 14:627–643

Datla RSS, Bekkaoui F, Hammerlindl JK, Pilate G, Dunstan DI, Crosby WL (1993) Improved high-level constitutive foreign gene expression in plants using an AMV RNA4 untranslated leader sequence. Plant Sci 94:139–149

Davidovich-Rikanati R, Lewinsohn E, Bar E, Iijima Y, Pichersky E, Sitrit Y (2008) Overexpression of the lemon basil alpha-zingiberene synthase gene increases both mono- and sesquiterpene contents in tomato fruit. Plant J 56:228–238

De Nadal E, Zapater M, Alepuz PM, Sumoy L, Mas G, Posas F (2004) The MAPK *Hog1* recruits *Rpd3* histone deacetylase to activate osmoresponsive genes. Nature 427:370–374

de Pater S, Pham K, Klitsie I, Kijne J (1996) The 22 bp W1 element in the pea lectin promoter is necessary and, as a multimer, sufficient for high gene expression in tobacco seeds. Plant Mol Biol 32:515–523

del Campillo E, Abdel-Aziz A, Crawford D, Patterson SE (2004) Root cap specific expression of an endo-β-1, 4-D-glucanase (cellulase): a new marker to study root development in Arabidopsis. Plant Mol Biol 56:309–323

Dey N, Maiti IB (1999) Structure and promoter/leader deletion analysis of mirabilis mosaic virus (MMV) full-length transcript promoter in transgenic plants. Plant Mol Biol 40:771–782

Diretto G, Al-Babili S, Tavazza R, Papacchioli V, Beyer P, Giuliano G (2007) Metabolic engineering of potato carotenoid content through tuber-specific overexpression of a bacterial mini-pathway. PLoS ONE 2:e350

Dolgov SV, Mityshkina TU, Rukavtsova EB, Buryanov YI, Vainstein A, Weiss D (1995) Production of transgenic plants of *Chrysanthemum morifolium* Ramat with the *Bacillus thuringiensis* delta endotoxin. Acta Hortic 420:46–47

Donath M, Mendel R, Cerff R, Martin W (1995) Intron-dependent transient expression of the maize *GapA1* gene. Plant Mol Biol 28:667–676

Dunn MA, White AJ, Vural S, Hughes MA (1998) Identification of promoter elements in a low-temperature-responsive gene (*blt*4.9) from barley (*Hordeum vulgare* L.). Plant Mol Biol 38:551–564

Elmayan T, Tepfer M (1995) Evaluation in tobacco of the organ specificity and strength of the *rolD* promoter, domain A of the 35S promoter and the $35S^2$ promoter. Transgenic Res 4: 388–396

Eulgem T, Rushton PJ, Schmelzer E, Hahlbrock K, Somssich IE (1999) Early nuclear events in plant defence signalling: rapid gene activation by WRKY transcription factors. EMBO J 18:4689–4699

Evans DE, Coleman JOD, Kearns A (2003) Plant cell culture. BIOS Scientific Publ, Taylor and Francis, London, UK

Faktor O, Kooter JM, Dixon RA, Lamb CJ (1996) Functional dissection of a bean chalcone synthase gene promoter in transgenic tobacco plants reveals sequence motifs essential for floral expression. Plant Mol Biol 32:849–859

Faktor O, Loake G, Dixon RA, Lamb CJ (1997) The G-box and H-box in a 39 bp region of a French bean chalcone synthase promoter constitute a tissue-specific regulatory element. Plant J 11:1105–1113

Fang RX, Nagy F, Sivasubramaniam S, Chua NH (1989) Multiple cis regulatory elements for maximal expression of the cauliflower mosaic virus 35S promoter in transgenic plants. Plant Cell 1:141–150

Fei H, Chaillou S, Hirel B, Mahon JD, Vessey JK (2003) Overexpression of a soybean cytosolic glutamine synthetase gene linked to organspecific promoters in pea plants grown in different concentrations of nitrate. Planta 216:467–474

Ficker M, Wemmer T, Thompson RD (1997) A promoter directing high level expression in pistils of transgenic plants. Plant Mol Biol 35:425–431

Filichkin SA, Leonard JM, Monteros A, Liu PP, Nonogaki H (2004) A novel endo-β-mannanase gene in tomato *LeMAN5* is associated with anther and pollen development. Plant Physiol 134:1080–1087

Fourcroy P, Vansuyt G, Kushnir S, Inzé D, Briat JF (2004) Iron-regulated expression of a cytosolic ascorbate peroxidase encoded by the *APX1* gene in Arabidopsis seedlings. Plant Physiol 34:605–613

Fraisier V, Gojon A, Tillard P, Daniel-Vedele F (2000) Constitutive expression of a putative high-affinity nitrate transporter in *Nicotiana plumbaginifolia*: Evidence for post-transcriptional regulation by a reduced nitrogen source. Plant J 23:489–496

Fraser PD, Romer S, Shipton CA, Mills PB, Kiano JW, Misawa N, Drake RG, Schuch W, Bramley PM (2002) Evaluation of transgenic tomato plants expressing an additional phytoene synthase in a fruit-specific manner. Proc Natl Acad Sci USA 99:1092–1097

Fu D, Szucs P, Yan L, Helguera M, Skinner J, Hayes P, Dubcovsky J (2005) Large deletions in the first intron of the *VRN-1* vernalization gene are associated with spring growth habit in barley and polyploid wheat. Mol Genet Genomics 273:54–65

Furtado A, Henry RJ, Pellegrineschi A (2009) Analysis of promoters in transgenic barley and wheat. Plant Biotechnol J 7:240–253

Furtado A, Henry RJ, Takaiwa F (2008) Comparison of promoters in transgenic rice. Plant Biotechnol J 6:679–693

Gantet P, Memelink J (2002) Transcription factors: tools to engineer the production of pharmacologically active plant metabolites. Trends Pharmacol Sci 23:563–569

Garbarino JE, Oosumi T, Belknap WR (1995) Isolation of a polyubiquitin promoter and its expression in transgenic potato plants. Plant Physiol 109:1371–1378

Garg AK, Sawers RJ, Wang H, Kim JK, Walker JM, Brutnell TP, Parthasarathy MV, Vierstra RD, Wu RJ (2006) Light-regulated overexpression of an *Arabidopsis* phytochrome A gene in rice alters plant architecture and increases grain yield. Planta 223:627–636

Gasser SM, Laemmli UK (1986) The organization of chromatin loops: characterization of a scaffold attachment site. EMBO J 5:511–518

Geffers R, Sell S, Cerff R, Hehl R (2001) The TATA box and a *Myb* binding site are essential for anaerobic expression of a maize *GapC4* minimal promoter in tobacco. Biochim Biophys Acta 1521:120–125

Gianì S, Altana A, Camanoni P, Morello L, Breviario D (2009) In trangenic rice, α- and β-tubulin regulatory sequences control GUS amount and distribution through intron mediated enhancement and intron dependent spatial expression. Transgenic Res 18:151–162

Gil P, Green PJ (1996) Multiple regions of the *Arabidopsis SAUR-AC1* gene control transcript abundance: the 3′ untranslated region functions as an mRNA instability determinant. EMBO J 15:1678–1686

Gittins JR, Pellny TK, Biricolti S, Hiles ER, Passey AJ, James DJ (2003) Transgene expression in the vegetative tissues of apple driven by the vascular-specific *rolC* and CoYMV promoters. Transgenic Res 12:391–402

Gittins JR, Pellny TK, Hiles ER, Rosa C, Biricolti S, James DJ (2000) Transgene expression driven by heterologous ribulose-1, 5-bisphosphate carboxylase/oxygenase small-subunit gene promoters in the vegetative tissues of apple (*Malus pumila* mill.). Planta 210:232–240

Giuliano G, Pichersky E, Malik VS, Timko MP, Scolnik PA, Cashmore AR (1988) An evolutionarily conserved protein binding sequence upstream of a plant light-regulated gene. Proc Natl Acad Sci USA 85:7089–7093

Grace ML, Chandrasekharan MB, Hall TC, Crowe AJ (2004) Sequence and spacing of TATA box elements are critical for accurate initiation from the β- phaseolin promoter. J Biol Chem 279:8102–8110

Gray-Mitsumune M, Molitor EK, Cukovic D, Carlson JE, Douglas CJ (1999) Developmentally regulated patterns of expression directed by poplar *PAL* promoters in transgenic tobacco and poplar. Plant Mol Biol 39:657–669

Green PJ, Yong MH, Cuozzo M, Kano-Murakami Y, Silverstein P, Chua NH (1988) Binding site requirements for pea nuclear protein factor GT-1 correlate with sequences required for light-dependent transcriptional activation of the *rbcS-3A* gene. EMBO J 7:4035–4044

Grichko VP, Filby B, Glick BR (2000) Increased ability of transgenic plants expressing the bacterial enzyme ACC deaminase to accumulate Cd, Co, Cu, Ni, Pb, and Zn. J Biotechnol 81:45–53

Grotewold E, Sainz MB, Tagliani L, Hernandez JM, Bowen B, Chandler VL (2000) Identification of the residues in the Myb domain of maize C1 that specify the interaction with the bHLH cofactor R. Proc Natl Acad Sci USA 97:13579–13584

Gupta P, Raghuvanshi S, Tyagi AK (2001) Assessment of the efficiency of various gene promoters *via* biolistics in leaf and regenerating seed callus of millets, *Eleusine coracana* and *Echinochloa crusgalli*. Plant Biotechnol 18:275–282

Gupta V, Khurana R, Tyagi AK (2007) Promoters of two anther-specific genes confer organ-specific gene expression in a stage-specific manner in transgenic systems. Plant Cell Rep 26:1919–1931

Gutiérrez RA, MacIntosh GC, Green PJ (1999) Current perspectives on mRNA stability in plants: multiple levels and mechanisms of control. Trends Plant Sci 4:429–438

Ha SB, An G (1988) Identification of upstream regulatory elements involved in the developmental expression of the *Arabidopsis thaliana cab1* gene. Proc Natl Acad Sci USA 85:8017–8021

Han KH, Ma C, Strauss SH (1997) Matrix attachment regions (MARs) enhance transformation frequency and transgene expression in poplar. Transgenic Res 6:415–420

He Y, Gan S (2001) Identical promoter elements are involved in regulation of the *OPR1* gene by senescence and jasmonic acid in *Arabidopsis*. Plant Mol Biol 47:595–605

He Y, Michaels SD, Amasino RM (2003) Regulation of flowering time by histone acetylation in *Arabidopsis*. Science 302:1751–1754

He ZM, Jiang XL, Qi Y, Luo DQ (2008) Assessment of the utility of the tomato fruit-specific *E8* promoter for driving vaccine antigen expression. Genetica 133:207–214

Hennig J, Dewey RE, Cutt JR, Klessig DF (1993) Pathogen, salicylic acid and developmental dependent expression of a β-1, 3-glucanase/GUS gene fusion in transgenic tobacco plants. Plant J 4:481–493

Hodgson J (2000) Scientists avert new GMO crisis. Nat Biotechnol 18:13

Holton TA (1995) Modification of flower colour via manipulation of P450 gene expression in transgenic plants. Drug Metabol Drug Interact 12:359–368

Hong JK, Hwang BK (2006) Promoter activation of pepper class II basic chitinase gene, *CAChi2*, and enhanced bacterial disease resistance and osmotic stress tolerance in the *CAChi2*-over-expressing *Arabidopsis*. Planta 223:433–448

Hong YF, Liu CY, Cheng KJ, Hour AL, Chan MT, Tseng TH, Chen KY, Shaw JF, Yu SM (2008) The sweet potato sporamin promoter confers high-level phytase expression and improves organic phosphorus acquisition and tuber yield of transgenic potato. Plant Mol Biol 67:347–361

Huang F, Buschman LL, Higgins RA, McGaughey WH (1999) Inheritance to *Bacillus thuringiensis* toxin (Dispel ES) in European corn borer. Science 284:965–967

Hulzink RJM, de Groot PFM, Croes AF, Quaedvlieg W, Twell D, Wullems GJ, van Herpen MMA (2002) The 5'-untranslated region of the *ntp303* gene strongly enhances translation during pollen tube growth, but not during pollen maturation. Plant Physiol 129:342–353

Imai S, Johnson FB, Marciniak RA, McVey M, Park PU, Guarente L (2000) Sir2: an NAD-dependent histone deacetylase that connects chromatin silencing, metabolism, and aging. Cold Spring Harb Symp Quant Biol 65:297–302

Ingham DJ, Beer S, Money S, Hansen G (2001) Quantitative real-time PCR assay for determining transgene copy number in transformed plants. Biotechniques 31:132–140

Ito Y, Katsura K, Maruyama K, Taji T, Kobayashi M, Seki M, Shinozaki K, Yamaguchi-Shinozaki K (2005) Functional analysis of rice DREB1/CBF-type transcription factors involved in cold-responsive gene expression in transgenic rice. Plant Cell Physiol 47:141–153

Jayaraj J, Devlin R, Punja Z (2008) Metabolic engineering of novel ketocarotenoid production in carrot plants. Transgenic Res 17:489–501

Jefferson R, Goldsbrough A, Bevan M (1990) Transcriptional regulation of a patatin-1 gene in potato. Plant Mol Biol 14:995–1006

Jenuwein T, Allis CD (2001) Translating the histone code. Science 293:1074–1080

Jeong YM, Mun JH, Lee I, Woo JC, Hong CB, Kim SG (2006) Distinct roles of the first introns on the expression of Arabidopsis profilin gene family members. Plant Physiol 140:196–209

Jones MO, Manning K, Andrews J, Wright C, Taylor IB, Thompson AJ (2008) The promoter from *SlREO*, a highly-expressed, root-specific *Solanum lycopersicum* gene, directs expression to cortex of mature roots. Funct Plant Biol 35:1224–1233

Katsumoto Y, Fukuchi-Mizutani M, Fukui Y, Brugliera F, Holton TA, Karan M, Nakamura N, Yonekura-Sakakibara K, Togami J, Pigeaire A, Tao GQ, Nehra NS, Lu CY, Dyson BK, Tsuda S, Ashikari T, Kusumi T, Mason JG, Tanaka Y (2007) Engineering of the rose flavonoid biosynthetic pathway successfully generated blue-hued flowers accumulating delphinidin. Plant Cell Physiol 48:1589–1600

Kehoe DM, Degenhardt J, Winicov I, Tobin EM (1994) Two 10-bp regions are critical for phytochrome regulation of a *Lemna gibba Lhcb* gene promoter. Plant Cell 6:1123–1134

Kermode AR, Zeng Y, Hu X, Lauson S, Abrams SR, He X (2007) Ectopic expression of a conifer *Abscisic Acid Insensitive 3* transcription factor induces high-level synthesis of recombinant human a-L-iduronidase in transgenic tobacco leaves. Plant Mol Biol 63:763–776

Kim HU, Park BS, Jin YM, Chung TY (1997) Promoter sequences of two homologous pectin esterase genes from Chinese cabbage (*Brassica campestris* L. ssp. *pekinensis*) and pollen-specific expression of the *GUS* gene driven by a promoter in tobacco plants. Mol Cell 7:21–27

Kim KN, Guiltinan MJ (1999) Identification of *cis*-acting elements important for expression of the starch-branching enzyme I gene in maize endosperm. Plant Physiol 121:225–236

Kim MJ, Kim H, Shin JS, Chung CH, Ohlrogge JB, Suh MC (2006) Seed-specific expression of sesame microsomal oleic acid desaturase is controlled by combinatorial properties between negative *cis*-regulatory elements in the *SeFAD2* promoter and enhancers in the 5′-UTR intron. Mol Genet Genomics 276:351–368

Kimura M, Yoshizumi T, Manabe K, Yamamoto YY, Matsui M (2001) *Arabidopsis* transcriptional regulation by light stress via hydrogen peroxide-dependent and -independent pathways. Genes Cells 6:607–617

Kirsch C, Logemann E, Lippok B, Schmelzer E, Hahlbrock K (2001) A highly specific pathogen-responsive promoter element from the immediate- early activated *CMPG1* gene in *Petroselinum crispum*. Plant J 26:217–227

Koltunow AM, Truettner J, Cox KH, Wallroth M, Goldberg RB (1990) Different temporal and spatial gene expression patterns occur during anther development. Plant Cell 2:1201–1224

Kooiker M, Airoldi CA, Losa A, Manzotti PS, Finzi L, Kater MM, Colombo L (2005) BASIC PENTACYSTEINE1, a GA binding protein that induces conformational changes in the regulatory region of the homeotic *Arabidopsis* gene *SEEDSTICK*. Plant Cell 17:722–729

Kozak M (1986) Point mutations define a sequence flanking the AUG initiator codon that modulates translation by eukaryotic ribosomes. Cell 44:283–292

Kozak M (1991) Structural features in eukaryotic mRNAs that modulate the initiation of translation. J Biol Chem 266:19867–19870

Kuster H, Quandt HJ, Broer I, Perlick AM, Puhler A (1995) The promoter of the *Vicia faba* L. VfENOD-GRP3 gene encoding a glycine-rich early nodulin mediates a predominant gene expression in the interzone II-III region of transgenic *Vicia hirsuta* root nodules. Plant Mol Biol 29:759–772

Lam E, Benfey PN, Gilmartin PM, Fang RX, Chua NH (1989) Site-specific mutations alter *in vitro* factor binding and change promoter expression pattern in transgenic plants. Proc Natl Acad Sci USA 86:7890–7894

Lee YY, Cevallos RC, Jan E (2009) An upstream open reading frame regulates translation of *gadd34* during cellular stresses that induce *eif2α* phosphorylation. J Biol Chem 284: 6661–6673

Lewinsohn E, Schalechet F, Wilkinson J, Matsui K, Tadmor Y, Nam KH, Amar O, Lastochkin E, Larkov O, Ravid U, Hiatt W, Gepstein S, Pichersky E (2001) Enhanced levels of the aroma and flavor compound S-linalool by metabolic engineering of the terpenoid pathway in tomato fruits. Plant Physiol 127:1256–1265

Li YF, Zhu R, Xu P (2005) Activation of the gene promoter of barley beta-1, 3-glucanase isoenzyme GIII is salicylic acid (SA)-dependent in transgenic rice plants. J Plant Res 118:215–221

Li Z, Hansen JL, Liu Y, Zemetra RS, Berger PH (2004) Using realtime PCR to determine copy number in wheat. Plant Mol Biol Rep 22:179–188

Li Z, Jayasankar S, Gray DJ (2001) Expression of a bifunctional green fluorescent protein (GFP) fusion marker under the control of three constitutive promoters and enhanced derivatives in transgenic grape (*Vitis vinifera*). Plant Sci 160:877–887

Liu JJ, Ekramoddoullah AK (2003) Root-specific expression of a western white pine PR10 gene is mediated by different promoter regions in transgenic tobacco. Plant Mol Biol 52:103–120

Liu JW, Tabe LM (1998) The influences of two plant nuclear matrix attachment regions (MARs) on gene expression in transgenic plants. Plant Cell Physiol 39:115–123

Liu Q, Segal DJ, Ghiara JB, Barbas CF 3rd (1997) Design of polydactyl zinc-finger proteins for unique addressing within complex genomes. Proc Natl Acad Sci USA 94:5525–5530

Livne B, Faktor O, Zeitoune S, Edelbaum O, Sela I (1997) TMV-induced expression of tobacco β-glucanase promoter activity is mediated by a single, inverted, GCC motif. Plant Sci 130:159–169

Lloyd AM, Schena M, Walbot V, Davis RW (1994) Epidermal cell fate determination in *Arabidopsis*: patterns defined by a steroid-inducible regulator. Science 266:436–439

Lu J, Sivamani E, Azhakanandam K, Samadder P, Li X, Qu R (2008) Gene expression enhancement mediated by the 5′ UTR intron of the rice *rubi3* gene varied remarkably among tissues in transgenic rice plants. Mol Genet Genomics 279:563–572

Luehrsen KR, Walbot V (1991) Intron enhancement of gene expression and the splicing efficiency of introns in maize cells. Mol Gen Genet 225:81–93

Lumbreras V, Stevens DR, Purton S (1998) Efficient foreign gene expression in *Chlamydomonas reinhardtii* mediated by an endogenous intron. Plant J 14:441–447

Luttermann C, Meyers G (2009) The importance of inter- and intramolecular base pairing for translation reinitiation on a eukaryotic bicistronic mRNA. Genes Dev 23:331–344

Maas C, Laufs J, Grant S, Korfhage C, Werr W (1991) The combination of a novel stimulatory element in the first exon of the maize *Shrunken-1* gene with the following intron 1 enhances reporter gene expression up to 1000-fold. Plant Mol Biol 16:199–207

Maas C, Schaal S, Werr W (1990) A feedback control element near the transcription start site of the maize shrunken gene determines promoter activity. EMBO J 9:3447–3452

Mac A, Krzymowska M, Barabasz A, Hennig J (2004) Transcriptional regulation of the gluB promoter during plant response to infection. Cell Mol Biol Lett 9:843–853

Maeo K, Tomiya T, Hayashi K, Akaike M, Morikami A, Ishiguro S, Nakamura K (2001) Sugar-responsible elements in the promoter of a gene for beta-amylase of sweet potato. Plant Mol Biol 46:627–637

Maiti IB, Ghosh SK, Gowda S, Kiernan J, Shepherd RJ (1997) Promoter/leader deletion analysis and plant expression vectors with the figwort mosaic virus (FMV) full length transcript (FLt) promoter containing single or double enhancer domains. Transgenic Res 6:143–156

Mankin SL, Allen GC, Phelan T, Spiker S, Thompson WF (2003) Elevation of transgene expression level by flanking matrix attachment regions (MAR) is promoter dependent: a study of the interactions of six promoters with the RB7 3′ MAR. Transgenic Res 12:3–12

Manners MJ, Penninckx IAMA, Vermaere K, Kazan K, Brown RL, Morgan A, Maclean DJ, Curtis MD, Cammue BPA, Broekaert WF (1998) The promoter of the plant defensin gene *PDF1.2* from *Arabidopsis* is systemically activated by fungal pathogens and responds to methyl jasmonate but not to salicylic acid. Plant Mol Biol 38:1071–1080

Mariani C, De Beuckeleer M, Treuttner J, Goldberg RB (1990) Induction of male sterility in plants by a chimaeric ribonuclease gene. Nature 347:737–741

Martienssen RA (2003) Maintenance of heterochromatin by RNA interference of tandem repeats. Nat Genet 35:213–214

Martínez-García JF, Huq E, Quail PH (2000) Direct targeting of light signals to a promoter element-bound transcription factor. Science 288:859–863

Martínez-Hernández A, López-Ochoa L, Argüello-Astorga G, Herrera-Estrella L (2002) Functional properties and regulatory complexity of a minimal *RBCS* light-responsive unit activated by phytochrome, cryptochrome, and plastid signals. Plant Physiol 128:1223–1233

Mascarenhas D, Mettler IJ, Pierce DA, Lowe HW (1990) Intron mediated enhancement of heterologous gene expression in maize. Plant Mol Biol 15:913–920

Mason G, Provero P, Vaira AM, Accotto GP (2001) Estimating the number of integrations in transformed plants by quantitative real time PCR. BMC Biotechnol 2:20

McElroy D, Zhang W, Cao J, Wu R (1990) Isolation of an efficient actin promoter for use in rice transformation. Plant Cell 2:163–171

Menkens AE, Schindler U, Cashmore AR (1995) The Gbox: a ubiquitous regulatory DNA element in plantsbound by the GBF family of bZIP proteins. Trends Biochem Sci 20:506–510

Mirkovitch J, Mirault ME, Laemmli UK (1984) Organisation of the higher-order chromatin loop: specific DNA attachment sites on nuclear scaffold. Cell 39:223–232

Mitra A, Choi HK, An G (1989) Structural and functional analyses of *Arabidopsis thaliana* chlorophyll a/b-binding protein (*cab*) promoters. Plant Mol Biol 12:169–179

Miyamoto T, Nakamura T, Nagao I, Obokata J (2000) Quantitative analysis of transiently expressed mRNA in particle-bombarded tobacco seedlings. Plant Mol Biol Rep 18:101–107

Mol J, Cornish E, Mason J, Koes R (1999) Novel coloured flowers. Curr Opin Biotechnol 10:198–201

Monde RA, Greene JC, Stern DB (2000) The sequence and secondary structure of the 3'UTR affect 3'-end maturation, RNA accumulation, and translation in tobacco chloroplasts. Plant Mol Biol 44:529–542

Montgomery J, Goldman S, Deikman J, Margossian L, Fischer RL (1993a) Identification of an ethylene-responsive region in the promoter of a fruit ripening gene. Proc Natl Acad Sci USA 90:5939–5943

Montgomery J, Pollard V, Deikman J, Fischer RL (1993b) Positive and negative regulatory regions control the spatial distribution of polygalacturonase transcription in tomato fruit pericarp. Plant Cell 5:1049–1062

Morello L, Bardini M, Cricri M, Sala F, Breviario D (2006) Functional analysis of DNA sequences controlling the expression of the rice *OsCDPK2* gene. Planta 223:479–491

Morello L, Bardini M, Sala F, Breviario D (2002) A long leader intron of the *Ostub16* rice β-tubulin gene is required for high-level gene expression and can autonomously promote transcription both *in vivo* and *in vitro*. Plant J 29:33–44

Morita A, Umemura T, Kuroyanagi M, Futsuhara Y, Perata P, Yamaguchi J (1998) Functional dissection of a sugar-repressed alpha-amylase gene (*RAmy1A*) promoter in rice embryos. FEBS Lett 13:81–85

Moye-Rowley WS, Harshman KD, Parker CS (1989) Yeast *YAP1* encodes a novel form of the jun family of transcriptional activator proteins. Genes Dev 3:283–292

Murfett J, Wang XJ, Hagen G, Guilfoyle TJ (2001) Identification of *Arabidopsis* histone deacetylase HDA6 mutants that affect transgene expression. Plant Cell 13:1047–1061

Nain V, Verma A, Kumar N, Sharma P, Ramesh B, Kumar PA (2008) Cloning of an ovule specific promoter from *Arabidopsis thaliana* and expression of β-glucuronidase. Indian J Exp Biol 46:207–211

Nardozza TA, Quigley MM, Getzenberg RH (1996) Association of transcription factors with the nuclear matrix. J Cell Biochem 61:467–477

Ng HH, Bird A (1999) DNA methylation and chromatin modification. Curr Opin Genet Dev 9:158–163

Nicholass FJ, Smith CJ, Schuch W, Bird CR, Grierson D (1995) High levels of ripening-specific reporter gene expression directed by tomato fruit polygalacturonase gene-flanking regions. Plant Mol Biol 28:423–435

Noh YS, Amasino RM (1999) Identification of a promoter region responsible for the senescence-specific expression of *SAG12*. Plant Mol Biol 41:181–194

Nomura M, Katayama K, Nishimura A, Ishida Y, Ohta S, Komari T, Miyao-Tokutomi M, Tajima S, Matsuoka M (2000) The promoter of *rbcS* in a C3 plant (rice) directs organ-specific, light-dependent expression in a C4 plant (maize), but does not confer bundle sheath cell-specific expression. Plant Mol Biol 44:99–106

Nowak W, Gawlowska M, Jarmolowski A, Augustyniak J (2001) Effect of nuclear matrix attachment regions on transgene expression in tobacco plant. Acta Biochim Pol 48:637–646

Nunberg AN, Li Z, Bogue MA, Vivekananda J, Reddy AS, Thomas TL (1994) Developmental and hormonal regulation of sunflower helianthinin genes: proximal promoter sequences confer regionalized seed expression. Plant Cell 6:473–486

Odell JT, Nagy F, Chua NH (1985) Identification of DNA sequences required for activity of the cauliflower mosaic virus 35S promoter. Nature 313:810–812

Ohl S, Hedrick SA, Chory J, Lamb CJ (1990) Functional properties of a phenylalanine ammonia-lyase promoter from *Arabidopsis*. Plant Cell 2:837–848

Ohme-Takagi M, Taylor CB, Newman TC, Green PJ (1993) The effect of sequences with high AU content on mRNA stability in tobacco. Proc Natl Acad Sci USA 90:11811–11815

Ohta S, Hattori T, Morikami A, Nakamura K (1991) High-level expression of a sweet potato sporamin gene promoter: β-glucuronidase (*GUS*) fusion gene in the stems of transgenic tobacco plants is conferred by multiple cell type-specific regulatory elements. Mol Gen Genet 225:369–378

Olsthoorn RC, Mertens S, Brederode FT, Bol JF (1999) A conformational switch at the 3′ end of a plant virus RNA regulates viral replication. EMBO J 18:4856–4864

Osakabe K, Abe K, Yoshioka T, Osakabe Y, Todoriki S, Ichikawa H, Hohn B, Toki S (2006) Isolation and characterization of the *RAD54* gene from *Arabidopsis thaliana*. Plant J 48:827–842

Oszvald M, Kang TJ, Tomoskozi S, Jenes B, Kim TG, Cha YS, Tamas L, Yang MS (2008) Expression of cholera toxin B subunit in transgenic rice endosperm. Mol Biotechnol 40:261–268

Oyama T, Shimura Y, Okada K (1997) The Arabidopsis *HY5* gene encodes a bZIP protein that regulates stimulus-induced development of root and hypocotyl. Genes Dev 11:2983–2995

Page AF, Minocha SC (2004) Analysis of gene expression in transgenic plants. In: Pena L (ed) Methods in molecular biology. Transgenic plants: methods and protocols, vol 286. Humana Press, New Jersey, USA, pp 291–311

Pain VM (1996) Initiation of protein synthesis in eukaryotic cells. Eur J Biochem 236:747–771

Pandey R, Muller A, Napoli CA, Selinger DA, Pikaard CS (2002) Analysis of histone acetyltransferase and histone deacetylase families of *Arabidopsis thaliana* suggests functional diversification of chromatin modification among multicellular eukaryotes. Nucleic Acids Res 30:5036–5055

Park HC, Kim ML, Kang YH, Jeon JM, Yoo JH, Kim MC, Park CY, Jeong JC, Moon BC, Lee JH, Yoon HW, Lee SH, Chung WS, Lim CO, Lee SY, Hong JC, Cho MJ (2004) Pathogen- and NaCl-induced expression of the SCaM-4 promoter is mediated in part by a GT-1 Box that interacts with a GT-1-like transcription factor. Plant Physiol 135:2150–2161

Patel M, Berry JO (2008) Rubisco gene expression in C4 plants. J Exp Bot 59:1625–1634

Pathirana MS, Samac DA, Roeven R, Yoshioka H, Vance CP, Gantt JS (1997) Analyses of phosphoenolpyruvate carboxylase gene structure and expression in alfalfa nodules. Plant J 12:293–304

Paul AL, Ferl RJ (1998) Higher order chromatin structures in maize and *Arabidopsis*. Plant Cell 10:1349–1359

Petit JM, van Wuytswinkel O, Briat JF, Lobréaux S (2001) Characterization of an iron-dependent regulatory sequence involved in the transcriptional control of *AtFer1* and *ZmFer1* plant ferritin genes by iron. J Biol Chem 276:5584–5590

Potenza C, Aleman L, Sengupta-Gopalan C (2004) Targeting transgene expression in research, agricultural, and environmental applications: Promoters used in plant transformation. In Vitro Cell Dev Biol Plant 40:1–22

Puente P, Wei N, Deng XW (1996) Combinatorial interplay of promoter elements constitutes the minimal determinants for light and developmental control of gene expression in *Arabidopsis*. EMBO J 15:3732–3743

Qing DJ, Lu HF, Li N, Dong HT, Dong DF, Li YZ (2009) Comparative profiles of gene expression in leaves and roots of maize seedlings under the conditions of the salt stress and the removal of the salt stress. Plant Cell Physiol 50:889–903

Qu LQ, Xing YP, Liu WX, Xu XP, Song YR (2008) Expression pattern and activity of six glutelin gene promoters in transgenic rice. J Exp Bot 59:2417–2424

Raghothama KG, Maggio A, Narasimhan ML, Kononowicz AK, Wang G, D'Urzo MP, Hasegawa PM, Bressan RA (1997) Tissue-specific activation of the osmotin gene by ABA, C_2H_4 and NaCl involves the same promoter region. Plant Mol Biol 34:393–402

Ribeiro DT, Farias LP, de Almeida JD, Kashiwabara PM, Ribeiro AF, Silva-Filho MC, Menck CF, Van Sluys MA (2005) Functional characterization of the *thi1* promoter region from *Arabidopsis thaliana*. J Exp Bot 56:1797–1804

Riechmann JL, Ratcliffe OJ (2000) A genomic perspective on plant transcription factors. Curr Opin Plant Biol 3:423–434

Rogers HJ, Bate N, Combe J, Sullivan J, Sweetman J, Swan C, Lonsdale DM, Twell D (2001) Functional analysis of *cis*-regulatory elements within the promoter of the tobacco late pollen gene *g10*. Plant Mol Biol 45:577–585

Rommens CM, Richael CM, Yan H, Navarre DA, Ye J, Krucker M, Swords K (2008) Engineered native pathways for high kaempferol and caffeoylquinate production in potato. Plant Biotechnol J 6:870–886

Roque E, Gómez MD, Ellul P, Wallbraun M, Madueño F, Beltrán JP, Cañas LA (2007) The *PsEND1* promoter: a novel tool to produce genetically engineered male-sterile plants by early anther ablation. Plant Cell Rep 26:313–325

Rose AB (2002) Requirements for intron-mediated enhancement of gene expression in *Arabidopsis*. RNA 8:1444–1453

Rose AB (2004) The effect of intron location on intron-mediated enhancement of gene expression in *Arabidopsis*. Plant J 40:744–751

Rose AB (2008) Intron-mediated regulation of gene expression. In: Reddy ASN, Golovkin M (eds) Nuclear pre-mRNA processing in plants: current topics in microbiology and immunology, vol 326. Springer, Berlin, pp 277–290

Rouster J, Leah R, Mundy J, Cameron-Mills V (1997) Identification of a methyl jasmonate-responsive region in the promoter of a lipoxygenase 1 gene expressed in barley grain. Plant J 11:513–523

Rushton PJ, Torres JT, Parniske M, Wernert P, Hahlbrock K, Somssich IE (1996) Interaction of elicitor-induced DNA-binding proteins with elicitor response elements in the promoters of parsley PR1 genes. EMBO J 15:5690–5700

Sabelli PA, Shewry PR (1995a) Gene characterization by southern analysis. In: Jones H (ed) Methods in molecular biology. Plant gene transfer and expression protocols, vol 49. Humana Press, New Jersey, USA, pp 161–180

Sabelli PA, Shewry PR (1995b) Northern analysis and nucleic acid probes. In: Jones H (ed) Methods in molecular biology. Plant gene transfer and expression protocols, vol 49. Humana Press, New Jersey, USA, pp 213–228

Sablowski RWM, Baulcombe DC, Bevan M (1995) Expression of a flower-specific Myb protein in leaf cells using a viral vector causes ectopic activation of a target promoter. Proc Natl Acad Sci USA 92:6901–6905

Sandhu JS, Krasnyanski SF, Domier LL, Korban SS, Osadjan MD, Buetow DE (2000) Oral immunization of mice with transgenic tomato fruit expressing respiratory syncytial virus-F protein induces a systemic immune response. Transgenic Res 9:127–135

Schaart JG, Salentijn EMJ, Krens FA (2002) Tissue-specific expression of the β-glucuronidase reporter gene in transgenic strawberry (*Fragaria* × *ananassa*) plants. Plant Cell Rep 21:313–319

Schenk PM, Remans T, Sági L, Elliott AR, Dietzgen RG, Swennen R, Ebert PR, Grof CPL, Manners JM (2001) Promoters for pregenomic RNA of banana streak badnavirus are active for transgene expression in monocot and dicot plants. Plant Mol Biol 47:399–412

Schneider A, Salamini F, Gebhardt C (1997) Expression patterns and promoter activity of the cold-regulated gene *ci21A* of potato. Plant Physiol 113:335–345

Schöffl F, Schröder G, Kliem M, Rieping M (1993) An SAR-sequence containing 395 bp-DNA fragment mediates enhanced, gene-dosage-correlated expression of a chimaeric heat shock gene in transgenic tobacco plants. Transgenic Res 2:93–100

Schramm F, Ganguli A, Kiehlmann E, Englich G, Walch D, von Koskull-Döring P (2006) The heat stress transcription factor HsfA2 serves as a regulatory amplifier of a subset of genes in the heat stress response in *Arabidopsis*. Plant Mol Biol 60:759–772

Segal DJ, Barbas CF (2001) Custom DNA-binding proteins come of age: Polydactyl zinc-finger proteins. Curr Opin Biotechnol 12:632–637

Shibuya K, Fukushima S, Takatsuji H (2009) RNA-directed DNA methylation induces transcriptional activation in plants. Proc Natl Acad Sci USA 106:1660–1665

Shou H, Frame BR, Whitham SA, Wang K (2004) Assessment of transgenic maize events produced by particle bombardment or *Agrobacterium*-mediated transformation. Mol Breed 13:201–208

Sinibaldi RM, Mettler IJ (1992) Intron splicing and intronmediated enhanced expression in monocots. In: Cohn WE, Moldave K (eds) Progress in nucleic acid research and molecular biology, vol 42. Academic Press, New York, USA, pp 229–257

Söderman E, Hjellstrom M, Fahleson J, Engstrom P (1999) The HD-Zip gene *ATHB6* in *Arabidopsis* is expressed in developing leaves, roots and carpels and up-regulated by water deficit conditions. Plant Mol Biol 40:1073–1083

Sonenberg N, Pelletier J (1988) Internal initiation of translation of eukaryotic mRNA directed by a sequence derived from poliovirus RNA. Nature 334:320–325

Song P, Cai CQ, Skokut M, Kosegi BD, Petolino JF (2002) Quantitative real-time PCR as a screening tool for estimating transgene copy number in WHISKERS-derived transgenic maize. Plant Cell Rep 20:948–954

Stark K, Kirk DL, Schmitt R (2001) Two enhancers and one silencer located in the introns of *regA* control somatic cell differentiation in *Volvox carteri*. Genes Dev 15:1449–1460

Stearns JC, Shah S, Greenberg BM, Dixon DG, Glick BR (2005) Tolerance of transgenic canola expressing 1-aminocyclopropane-1-carboxylic acid deaminase to growth inhibition by nickel. Plant Physiol Biochem 43:701–708

Stief A, Winter DM, Stratling WH, Sippel AE (1989) A nuclear DNA attachment element mediates elevated and position-independent gene activity. Nature 341:343–345

Storozhenko S, De Pauw P, Van Montagu M, Inzé D, Kushnir S (1998) The heat-shock element is a functional component of the Arabidopsis *APX1* gene promoter. Plant Physiol 118:1005–1014

Strompen G, Gruner R, Pfitzner UM (1998) An *as-1*-like motif controls the level of expression of the gene for the pathogenesis-related protein 1a from tobacco. Plant Mol Biol 37:871–883

Sunilkumar G, Connell JP, Smith CW, Reddy AS, Rathore KS (2002) Cotton alpha-globulin promoter: isolation and functional characterization in transgenic cotton, *Arabidopsis*, and tobacco. Transgenic Res 11:347–359

Szabados L, Ratet P, Grunenberg B, de Bruijn FJ (1990) Functional analysis of the *Sesbania rostrata* leghemoglobin *glb*3 gene 5'-upstream region in transgenic *Lotus corniculatus* and *Nicotiana tabacum* plants. Plant Cell 2:973–986

Szczyglowski K, Potter T, Stoltzfus J, Fujimoto SY, de Bruijn FJ (1996) Differential expression of the *Sesbania rostrata* leghemoglobin *glb3* gene promoter in transgenic legume and non-legume plants. Plant Mol Biol 31:931–935

Taddei A, Hediger F, Neumann FR, Gasser SM (2004) The function of nuclear architecture: a genetic approach. Annu Rev Genet 38:305–345

Tarantino D, Petit JM, Lobreaux S, Briat JF, Soave C, Murgia I (2003) Differential involvement of the IDRS *cis*-element in the developmental and environmental regulation of the *AtFer1* ferritin gene from *Arabidopsis*. Planta 217:709–716

Tian L, Fong PM, Wang JJ, Wei NE, Jiang H, Doerge RW, Chen ZJ (2005) Reversible histone acetylation and deacetylation mediate genome-wide, promoter-dependent and locus-specific changes in gene expression during plant development. Genetics 169:337–345

Tian L, Wang J, Fong MP, Chen M, Cao H, Gelvin SB, Chen ZJ (2003) Genetic control of developmental changes induced by disruption of Arabidopsis histone deacetylase 1 (*AtHD1*) expression. Genetics 165:399–409

Toplak N, Okršlar V, Stanič-Racman D, Gruden K, Žel J (2004) A high-throughput method for quantifying transgene expression in transformed plants with real-time PCR analysis. Plant Mol Biol Rep 22:237–250

Toquin V, Grausem B, Geoffroy P, Legrand M (2003) Structure of the tobacco caffeic acid *O*-methyltransferase (COMT) II gene: identification of promoter sequences involved in gene inducibility by various stimuli. Plant Mol Biol 52:495–509

Twell D, Yamaguchi J, McCormick S (1990) Pollen-specific gene expression in transgenic plants: coordinate regulation of two different tomato gene promoters during microsporogenesis. Development 109:705–713

Uknes S, Dincher S, Friedrich L, Negrotto D, Williams S, Thompson-Taylor H, Potter S, Ward E, Ryals J (1993) Regulation of pathogenesis-related protein-1a gene expression in tobacco. Plant Cell 5:159–169

Urwin PE, Møller SG, Lilley CJ, McPherson MJ, Atkinson HJ (1997) Continual green-fluorescent protein monitoring of cauliflower mosaic virus 35S promoter activity in nematode-induced feeding cells in *Arabidopsis thaliana*. Mol Plant Microbe Interact 10:394–400

Vain P, Worland B, Kohli A, Snape JW, Christou P, Allen GC, Thompson WF (1999) Matrix attachment regions increase transgene expression levels and stability in transgenic rice plants and their progeny. Plant J 18:233–242

van de Löcht U, Meier I, Hahlbrock K, Somssich IE (1990) A 125 bp promoter fragment is sufficient for strong elicitor-mediated gene activation in parsley. EMBO J 9:2945–2950

van de Rhee MD, Lemmers R, Bol JF (1993) Analysis of regulatory elements involved in stress-induced and organ-specific expression of tobacco acidic and basic β-1, 3-glucanase genes. Plant Mol Biol 21:451–461

van der Geest AHM, Hall GE Jr, Spiker S, Hall TC (1994) The beta-phaseolin gene is flanked by matrix attachment regions. Plant J 6:413–423

Van Eenennaam AL, Li G, Venkatramesh M, Levering C, Gong X, Jamieson AC, Rebar EJ, Shewmaker CK, Case CC (2004) Elevation of seed α-tocopherol levels using plant-based transcription factors targeted to an endogenous locus. Metab Eng 6:101–108

Verdaguer B, de Kochko A, Beachy RN, Fauquet C (1996) Isolation and expression in transgenic tobacco and rice plants, of the cassava vein mosaic virus (CsVMV) promoter. Plant Mol Biol 31:1129–1139

Verdaguer B, de Kochko A, Fux CI, Beachy RN, Fauquet C (1998) Functional organization of the cassava vein mosaic virus (CsVMV) promoter. Plant Mol Biol 37:1055–1067

Viret JF, Mabrouk Y, Bogorad L (1994) Transcriptional photoregulation of cell-type-preferred expression of maize *rbcS*-m3: 3′ and 5′ sequences are involved. Proc Natl Acad Sci USA 91:8577–8581

Visser RGF, Stolte A, Jacobsen E (1991) Expression of a chimaeric granule-bound starch synthase-GUS gene in transgenic potato plants. Plant Mol Biol 17:691–699

Vitale A, Wu RJ, Cheng Z, Meagher RB (2003) Multiple conserved 5′ elements are required for high-level pollen expression of the *Arabidopsis* reproductive actin *ACT1*. Plant Mol Biol 52:1135–1151

Wakeley PR, Rogers HJ, Rozycka M, Greenland AJ, Hussey PJ (1998) A maize pectin methylesterase-like gene, ZmC5, specifically expressed in pollen. Plant Mol Biol 37:187–192

Wally O, Jayaraj J, Punja ZK (2008) Comparative expression of β-glucuronidase with five different promoters in transgenic carrot (*Daucus carota* L.) root and leaf tissues. Plant Cell Rep 27:279–287

Wang XL, Li XB (2009) The *GhACS1* gene encodes an acyl-CoA synthetase which is essential for normal microsporogenesis in early anther development of cotton. Plant J 57:473–486

Weber H, Ziechmann C, Graessmann A (1990) In vitro DNA methylation inhibits gene expression in transgenic tobacco. EMBO J 9:4409–4415

Weeks JT, Anderson OD, Blechl AE (1993) Rapid production of multiple independent lines of fertile transgenic wheat (*Triticum aestivum*). Plant Physiol 102:1077–1084

Wei J, Theil EC (2000) Identification and characterization of the iron regulatory element in the ferritin gene of a plant (soybean). J Biol Chem 275:17488–17493

Weise A, Rodriguez-Franco M, Timm B, Hermann M, Link S, Jost W, Gorr G (2006) Use of *Physcomitrella patens* actin 5′ regions for high transgene expression: importance of 5′ introns. Appl Microbiol Biotechnol 70:337–345

Weng H, Pan A, Yang L, Zhang C, Liu Z, Zhang D (2004) Estimating number of transgene copies in transgenic rapeseed by real-time PCR assay with *HMG I/Y* as an endogenous reference gene. Plant Mol Biol Rep 22:289–300

Wyatt RE, Ainley WM, Nagao RT, Conner TW, Key JL (1993) Expression of the Arabidopsis AtAux2–11 auxin-responsive gene in transgenic plants. Plant Mol Biol 22:731–749

Xu B, Timko MP (2004) Methyl jasmonate induced expression of the tobacco putrescine *N*-methyltransferase genes requires both G-box and GCC-motif elements. Plant Mol Biol 55:743–761

Xu SX, Liu GS, Chen RD (2006) Characterization of an anther- and tapetum-specific gene and its highly specific promoter isolated from tomato. Plant Cell Rep 25:231–240

Xu Y, Yu H, Hall TC (1994) Rice triosephosphate isomerase gene 59 sequence directs β-glucuronidase activity in transgenic tobacco but requires an intron for expression in rice. Plant Physiol 106:459–467

Yamada T, Ishige T, Shiota N, Inui H, Ohkawa H, Ohkawa Y (2002) Enhancement of metaboliz-ing herbicides in young tubers of transgenic potato plants with the rat *CYP1A1* gene. Theor Appl Genet 105:515–520

Yamamoto S, Suzuki K, Shinshi H (1999) Elicitor-responsive, ethylene- independent activation of GCC box-mediated transcription that is regulated by both protein phosphorylation and dephos-phorylation in cultured tobacco cells. Plant J 20:571–579

Yamamoto YT, Taylor CG, Acedo GN, Cheng CL, Conkling MA (1991) Characterization of *cis*-acting sequences regulating root specific gene expression in tobacco. Plant Cell 3:371–382

Yang L, Ding J, Zhang C, Jia J, Weng H, Liu W, Zhang D (2005) Estimating the copy number of transgenes in transformed rice by real-time quantitative PCR. Plant Cell Rep 23:759–763

Yang L, Kajiura H, Suzuki K, Hirose S, Fujiyama K, Takaiwa F (2008) Generation of a transgenic rice seed-based edible vaccine against house dust mite allergy. Biochem Biophys Res Comm 365:334–339

Yang L, Suzuki K, Hirose S, Wakasa Y, Takaiwa F (2007) Development of transgenic rice seed accumulating a major Japanese cedar pollen allergen (Cry j 1) structurally disrupted for oral immunotherapy. Plant Biotechnol J 5:815–826

Yang NS, Russell D (1990) Maize sucrose synthase-1 promoter directs phloem cell-specific expression of *Gus* gene in transgenic tobacco plants. Proc Natl Acad Sci USA 87:4144–4148

Yang Q, Grimmig B, Matern U (1998) Anthranilate N-hydroxycinnamoyl/benzoyltransferase gene from carnation: rapid elicitation of transcription and promoter analysis. Plant Mol Biol 38:1201–1214

Yoshiba Y, Nanjo T, Miura S, Yamaguchi-Shinozaki K, Shinozaki K (1999) Stress-responsive and developmental regulation of $\Delta^2(1)$-pyrroline-5-carboxylate synthetase 1 (P5CS1) gene expres-sion in *Arabidopsis thaliana*. Biochem Biophys Res Commun 11:766–772

Zeyenko VV, Ryabova LA, Gallie DR, Spirin AS (1994) Enhancing effect of the 3′-untranslated region of tobacco mosaic virus RNA on protein synthesis in vitro. FEBS Lett 354:271–273

Zhang W, McElroy D, Wu R (1991) Analysis of rice *Act1* 5′ region activity in transgenic rice plants. Plant Cell 3:1155–1165

Zhao J, Guo S, Chen S, Zhang H, Zhao Y (2009) Expression of Yeast *YAP1* in transgenic *Arabidopsis* results in increased salt tolerance. J Plant Biol 52:56–64

Zhu B, Chen THH, Li PH (1995) Activation of two osmotin-like protein genes by abiotic stimuli and fungal pathogen in transgenic potato plants. Plant Physiol 108:929–937

Zhu Q, Song B, Zhang C, Ou Y, Xie C, Liu J (2008) Construction and functional characteristics of tuber-specific and cold-inducible chimeric promoters in potato. Plant Cell Rep 27:47–55

Chapter 6
Silencing as a Tool for Transgenic Crop Improvement

Pudota B Bhaskar and Jiming Jiang

6.1 Introduction

RNA silencing, also known as post-transcriptional gene silencing (PTGS) or RNA interference (RNAi), is a form of RNA degradation believed to be an important defense against foreign nucleic acids (Waterhouse et al. 2001). It was initially discovered in plants and was thought to function as part of a defense mechanism against viruses (Ratcliff et al. 1997). Subsequently, it was shown to be a common gene-silencing mechanism occurring in all eukaryotes, including plants and animals. The term RNAi was coined for the phenomenon when it was observed in the nematode *Caenorhabditis elegans* (Fire et al. 1998). However, this phenomenon of RNAi (PTGS) had actually been reported previously in transgenic petunia but was referred to as cosuppression, because transformation with a sense chalcone synthase transgene suppressed the expression of both the transgene and the endogenous gene (Napoli et al. 1990). It is now widely accepted that dsRNA is the effective trigger of PTGS/RNAi in plants and that this process operates by sequence-specific degradation (Kusaba 2004). Several milestones related to RNAi-based silencing are summarized in Table 6.1. In plants, cosuppression, PTGS, and virus-induced gene silencing (VIGS), all describe a homology-dependent gene-silencing phenomenon that involves what is more broadly known as RNAi. The science of RNAi broadly includes a few different and diverse RNA-silencing pathways that alter the expression levels of specific genes in plants, mediate the amplification and mobile signal mechanisms in RNAi pathways, and yield RNA-mediated DNA methylation (Baulcombe 2004; Lippman and Martienssen 2004).

P.B Bhaskar (✉)
Department of Horticulture, University of Wisconsin, 1575 Linden Drive, Madison, WI 53706, USA
e-mail: pudota1@wisc.edu

C. Kole et al. (eds.), *Transgenic Crop Plants*,
DOI 10.1007/978-3-642-04809-8_6, © Springer-Verlag Berlin Heidelberg 2010

Table 6.1 Time-line showing the breakthroughs related to RNAi-based silencing technology

Year	Breakthrough	Publication
1990	Cosuppression of purple color in plants (*Petunia*)	Napoli et al. (1990)
1998	A concept of using double-stranded RNA for triggering silencing	Fire et al. (1998)
2000	First successful intron-based hairpin RNA construct for silencing	Smith et al. (2000)
2002	First successful attempt of using RNAi for crop improvement	Liu et al. (2002)
2005	RNAi shown to improve the quality aspects by organ-specific silencing	Davuluri et al. (2005)
2006	A concept of using RNAi for improving nematode resistance was demonstrated	Huang et al. (2006)
2007	First artificial microRNAs used for gene silencing for virus resistance	Qu et al. (2007)
2007	RNAi was demonstrated to be successfully applied for the insect resistance in crops	Baum et al. (2007), Mao et al. (2007)
2008	amiRNAs shown to trigger gene silencing in a crop plant, rice	Warthmann et al. (2008)

In this chapter, we will focus on the RNAi-mediated gene-silencing method available for the development of transgenic crop plants, with a focus on usable and deployable crop improvements. Readers interested in a more extensive deliberation on RNAi mechanism may consult a number of recent reviews (Waterhouse and Helliwell 2003; Vazquez 2006; Matzke et al. 2007; Ramachandran and Chen 2008; Eamens et al. 2008). Another gene-silencing-based method that has proven successful for crop improvement is tilling (Henikoff et al. 2004). However, desirable phenotypes obtained using this approach are not transgenic (a non-GM method) and hence not discussed in the current chapter.

6.2   Procedures for Development of RNAi-Based Transgenic Gene-Silencing Lines

RNA silencing is a homology-based process that is triggered by double-stranded RNA (*dsRNA*) and eventually leads to suppression of gene expression. Initially, sense or antisense RNA strands were used to mediate PTGS, most often with modest effects on gene expression (Waterhouse et al. 2001). Through cleavage by endonucleases called Dicers, dsRNAs are efficiently converted into small RNAs (~21–24 nt), which are then used to direct a sequence-specific degradation of cognate single-stranded RNAs (Vazquez 2006). Considerable research has been conducted to determine the most efficient silencing construct. Intron-containing hairpin RNA (hpRNA)-based vectors have been proven to be highly efficient for plant RNAi-based gene silencing (Smith et al. 2000). In a hpRNA vector, the target gene is cloned as an inverted repeat spaced with an intron and is driven by either a strong whole plant promoter, such as the 35S CaMV (dicots) or the maize ubiquitin1 (monocots) or alternatively, an organ-specific silencing promoter. A spacer fragment between the arms of the inverted repeat is useful for increasing the stability of the vector in *Escherichia coli*, and using a splicable intron as a spacer has been shown to dramatically increase the frequency of strong silencing phenotypes (Smith et al. 2000). Typically, target-sequence inserts of 300–650 nt have been reported to provide reliably strong and frequent silencing in many crop plants (Helliwell et al. 2002; Matthew 2009).

Several types of RNAi-based vectors that make use of *Agrobacterium tumefaciens*-based plant transformation are available to the public and are being widely used. Predominantly, vectors developed by Waterhouse and colleagues at CSIRO, Australia, are supported by Gateway technology [TM] and facilitate easy incorporation of target sequence in the sense and antisense direction with an intron between them (*http://www.pi.csiro.au/rnai*). Another set of RNAi vectors are available through the *Arabidopsis* Biological Resource center (ABRC, *http://www.arabidopsis.org*) and were donated by the Functional Genomics of Plant Chromatin Consortium (*http://www.chromdb.org*). Despite similar designs, these vectors differ in terms of selectable markers, type of promoters, and cloning strategies. Detailed information about the choices of different RNAi vectors is available (Matthew 2009; Preuss and Pikkard 2003).

RNAi is also affected by the transformation method. The most effective and heritable silencing has been achieved through stable transformation by *Agrobacterium or* particle bombardment (Waterhouse and Helliwell 2003). One drawback of this system is that if PTGS in the whole organism is desired, then stably transformed plants carrying these constructs must be generated. Nevertheless, stable transformation of RNAi constructs has currently been used as a tool for the genetic improvement in a variety of crops (Mansoor et al. 2006; Eamens et al. 2008). Currently, we have a much great understanding of the endogenous gene-silencing mechanism, providing knowledge that can be used to develop precisely targeted gene-silencing approaches.

6.3 Crop Improvements with Silencing Tools

6.3.1 RNAi for Resistance to Diseases and Pests

6.3.1.1 RNAi for Resistance to Viruses

Although the mechanism was not clear at that time, the effects of gene silencing in plants were first used in efforts to develop resistance to diseases, particularly those caused by viruses (Powell-Abel et al. 1986). This "pathogen-derived resistance" (PDR) was achieved by transforming plants with either genes or genetic fragments derived from the pathogen with the aim of blocking a specific step in the life or infection cycle of the pathogen. Most of the strategies used for PDR were shown to be mediated by RNA, rather than protein, and led directly to the identification of PTGS – a phenomenon that is believed to be a form of antiviral defense (Voinnet 2001). An important finding recognized first in plants was that, once triggered, the silencing spreads throughout the organism by virtue of a gene-silencing signal, thus providing systemic rather than localized resistance (Voinnet et al. 1998). Unsurprisingly, virus-resistant transgenic plants are one of the first commercial applications resulting from gene-silencing technology.

The first demonstration that dsRNA mediates gene silencing in plants is the genetic study of Waterhouse et al. (1998). Transgenic plants were generated that expressed either the sense or antisense strand of a gene of potato virus Y (PVY). Both transgenic lines of tobacco were susceptible to PVY infection. However, progeny resulted from crosses between these susceptible tobacco lines showed resistance to PVY by generation of dsRNA. This suggests that two complementary RNAs transcribed from unlinked loci were able to anneal in the nucleus and induce a gene-specific suppressive state (Sharp 1999). This experiment first successfully demonstrated that dsRNA molecules are potent inducers of RNA silencing (Waterhouse et al. 1998).

Wang et al. (2000) first applied the deliberate use of RNA silencing for virus protection in the important cereal crop species barley. Barley yellow dwarf virus (BYDV) is a virus of global importance, as it infects and reduces yields of several crop species worldwide. An RNAi construct targeting the $5'$ end of this virus was transformed into barley, and the lines obtained showed complete immunity to BYDV. The transgenic lines were field tested and have been commercially released. During the last several years, efforts to control various viruses infecting several crop plants have been reported. These include RNAi approaches to control single-stranded DNA viruses (Geminiviruses Pooggin et al. 2003) or RNA viruses (Poty viruses, Waterhouse et al. 1998). In some cases, a simultaneous silencing of diverse plant viruses was achieved by designing a single RNAi construct that targets multiple distinct viruses (Missiou et al. 2004). Viruses have been the obvious targets for RNAi technologies, as most viruses have single-stranded RNA genomes. Currently transgenic lines of several crop plants have been field tested or commercially released and continue to show very strong resistance to several plant viruses. For a complete, detailed list of the types of crops transformed for resistance to viruses and about the performance of virus-resistant transgenic crop plants, refer to a recent review by Fuchs and Gonsalves (2007).

6.3.1.2 RNAi for Resistance to Parasitic Nematodes

Plant-parasitic nematodes, in particular root-knot nematodes, are the most economically devastating group of plant-parasitic nematodes worldwide, attacking nearly all food and fiber crops grown. The inadequacy of current control methods provides an opportunity for transgenic approaches to make an important contribution to an integrated pest management strategy. One of the recent approaches to GM-mediated nematode resistance is host-induced RNAi gene silencing.

Plants can be engineered to produce dsRNAs that silence essential genes in the nematode. This dsRNA, or its siRNAs, would then be delivered from the plant to the nematode through ingestion of the plant cytoplasm. Once the siRNAs are inside the nematode, the RNAi process would inactivate the gene targeted by the dsRNA (Bakhetia et al. 2005).

This logic has been put into use for the first time in a recent report in which the goal was to engineer a host plant to become resistant to root-knot nematodes. Huang et al. (2006) targeted the parasitism gene, 16D10 from the nematode *Meliodogyne incognita* that encodes a small peptide necessary for the infection. This gene/peptide is secreted by the nematode into the plant roots and is thought to have an important role in the early signaling that occurs during feeding-site formation playing an important role in the plant-parasite infection. Expression of dsRNA directed against this gene (using a 35S promoter) resulted in *Arabidopsis thaliana* plants with a 70–90% reduction in the number of nematode eggs in the host plant. In other words, host plants showed resistance to multiplication of the nematodes. This range of resistance extended to four different types of root-knot nematodes. The range of resistance was unique and had not been previously obtained by any natural root-knot nematode resistance genes. This work is a good illustration of how fundamental RNAi mechanism might lead to engineering crop plants for nematode resistance. However, the method is still in the discovery phase involving only model plants, but this unique method might in the near future emerge as a viable and flexible means of developing novel and durable nematode-resistant crops for this devastating pathogen and others.

6.3.1.3 RNAi for Resistance to Insects

Transgenic expression of *Bacillus thuringenesis* (*Bt*) toxin in crop plants has proven to be a great success for pest control in several crops. However, many important pests are not susceptible to *Bt*-protection, and there is a danger that some crop pests might develop resistance to *Bt*. In integrated pest management, there is always a search for alternative and potentially complementary control strategies, particularly for agents that are more robust and/or broadly applicable (Gordon and Waterhouse 2007).

Recently, RNAi-mediated resistance has been exploited to control insect pests via the *in planta* expression of a dsRNA (Baum et al. 2007; Mao et al. 2007). This is in fact the first demonstration that the lessons learned from the use of RNAi in model organisms can be applied to real-life biological processes to obtain gains in controlling crop pests. The observations that ingested dsRNA can silence genes in both nematodes and *Drosophila* lead to the possibility of applying this technology to control crop insect pests. In this method, RNAi is induced in insects after ingestion of plant-expressed hairpin RNA.

This concept was demonstrated through managing a coleopteran insect pest. The Western corn rootworm (*Diabrotica virgifera*) is one of the most devastating pests in North America. The USDA estimates that the corn rootworm causes

US$1 billion in lost revenue each year, which equals $800 million in yield losses. Transgenic corn plants were engineered to express dsRNAs that target a western corn rootworm V-ATPase gene. V-ATPases are found in the plasma membrane of many organelles, such as endosomes, lysosomes, and vesicles, playing crucial roles in the function of these organelles. It was hypothesized that disruption of this enzyme is detrimental to the insects; thus, RNAi was directed to silence this gene. The transgenic plants expressing lethal insect dsRNAs were challenged with rootworm larvae and showed significant root protection compared with the nontransgenic control plants (Baum et al. 2007). No negative agronomic effects were seen in multiple generations of these transgenic plants.

Another successful RNAi strategy was reported by Mao et al. (2007) to improve resistance against a notorious lepidopteran insect pest, *Helicoverpa armigera,* commonly called cotton bollworm. First a cytochrome P450 gene was identified from cotton boll worm that acts as an antidote allowing the pest to resist the naturally occurring toxin, gossypol, produced by cotton plants. Transgenic tobacco and *Arabidopsis* plants were generated expressing dsRNAs against the bollworm cytochrome P450 gene. When larvae were fed leaves from these transgenic plants, the expression of the gene decreased and larval growth was retarded. It was recently reported that the engineered cotton plants showed partial resistance to cotton bollworm pest, as expected (Price and Gatehouse 2008).

This new method of in planta RNAi against feeding insects seems to have potential for future pest control strategies. A wide range of potential targets among various crop pests can be identified and targeted for suppression of gene expression to achieve increased resistance in plants.

6.3.2 RNAi to Enhance Quality Traits

6.3.2.1 Decaffeinated Coffee

The first example of using RNAi to improve the quality aspects of any crop plant was demonstrated in coffee. A cup of coffee, on average, contains 150 mg of caffeine, which can cause health problems for many people worldwide (Ogita et al. 2005). Consequently, decaffeinated coffee is preferred by buyers who are sensitive to caffeine and accounts for about 10% of the world coffee market (Ogita et al. 2005). The solvent extraction process used currently to chemically reduce the caffeine levels of coffee beans may leave undesired components in the decaffeinated beans. Therefore, coffee plants that produce caffeine-free beans have always been an objective. Currently, it takes ~20 years to develop a coffee variety with reduced levels of caffeine. Ogita et al. (2003) addressed this problem with an RNAi strategy. Coffee plants were transformed with RNAi constructs to silence the theobromine synthase gene in the caffeine biosynthetic pathway. The transgenic plants obtained showed a 70% reduction in caffeine content compared to nontransformed coffee plants, and no phenotypic abnormalities were reported. While transgenic "decaf" lines have yet to be commercialized, this research has provided the first successful example of metabolic engineering of the alkaloids for quality improvement in crop plants.

6.3.2.2 Reduction of Toxic Gossypol in Cotton

One of the recent, dramatic applications of RNAi has been the elimination of the toxic compound gossypol from cottonseeds by Sunilkumar et al. (2006). This study clearly demonstrated the feasibility of a targeted RNAi-based approach to solve an age-old problem of cottonseed toxicity and provided an avenue to exploit the considerable quantities of protein and oil available in the global cottonseed output. Gossypol and related terpenoids are present throughout the cotton plant in the glands of foliage, floral organs, and bolls, as well as in the roots. These terpenoid compounds protect the cotton plants from both insects and pathogens and are essential for the survival of cotton under normal agricultural conditions, where it is exposed to a variety of pests and diseases, although its presence in the seed might be expendable (Townsend and Llewellyn 2007). A glanded-plant and glandless-seed trait does occur naturally in the native Australian cotton species *Gossypium sturtianum*. Gossypol-filled glands develop as the seeds germinate in order to provide the needed protection against pests and pathogens. Efforts to breed this trait into cultivated cotton were not successful, mainly due to considerable genome differences between two species. A natural glandless mutant of cotton was identified in the 1950s, and several breeding programs were launched to transfer this glandless trait into commercial cotton cultivars. However, glandless cotton varieties were a commercial failure due to their extraordinary susceptibility to insect pests since they constitutively lacked gossypol and protective terpenoids.

Remarkably, this long-standing goal of cotton geneticists was achieved through RNAi-mediated silencing that eliminates toxic gossypol from cottonseeds. This objective was achieved by silencing the δ-Candinene synthase. This enzyme catalyzes the first committed step involving in the cyclization of farnesyl diphosphate to δ-candinene (Chen et al. 1995), the compound from which gossypol and other sesquiterpenoid compounds are derived. The δ-cadinene synthase gene was silenced under the control of the cottonseed-specific ∝-globulin promoter. Seed from the transgenic cotton plants exhibited a significant reduction in gossypol content, whereas the cotton foliage, floral parts, and floral organs contained normal levels of gossypol. Transgenic plants with seed gossypol levels reduced to as low as 99% were stable and were maintained for three generations (Sunilkumar et al. 2006). Gossypol values in the seeds from some of the silenced lines were well below the limit deemed safe for human consumption by the United Nations Food and Agriculture Organization and World Health Organization. In this example, once again the use of an endogenous gene and a native promoter of cotton ensure appropriate spatial and temporal expression of the transgene.

6.3.2.3 Tearless Onions

Manipulation of plant secondary metabolic pathways can result in dramatic and simultaneous down- and up-regulation of products within that pathway and the even production of novel products. Onions *(Allium cepa)* synthesize a unique set of secondary sulfur metabolites. When onions are chopped, these metabolites are cleaved by the enzyme alliinase into their corresponding sulfenic acids and volatile sulfur compounds that give the respective flavors. One of the volatiles released is Lachrymatory Factor (LF), the chemical responsible for inducing tearing. In addition, it is hypothesized that LF production causes the absence of otherwise predicted sulfur volatiles, analogs of which in garlic (*A. sativum*) are known for their health attributes (Eady et al. 2008). Current "tearless" onion cultivars (e.g., Vidalia) are achieved through deficient uptake and partitioning of sulfur and/or growth in sulfur-deficient soils, but in so doing they accumulate fewer secondary sulfur compounds in the bulb, reducing their sensory and health qualities compared with more pungent high-sulfur cultivars.

Eady et al. (2008) made a healthier and tearless onion by reducing the levels of Lachrymatory Factor Synthase (LFS) and preventing the conversion of 1-propenyl sulfenic acid to the undesirable and irritating LF. By means of RNAi, LFS activity in onions was reduced by up to 1,544-fold. When these onions are chopped, significantly reduced levels of LF are produced. No phenotypic abnormalities were reported among the transgenic onion plants compared to the controls in greenhouse experiments. The authors also confirmed that RNAi silencing of LFS shifted sulfur metabolism away from tearing agents, giving rise to a cascade of predicted secondary compounds that had not been detected previously or only in trace amounts in onion (Minorsky 2008). The researchers hope to initiate formal taste evaluation trials of these transgenic, tearless onions following regulatory approval. These onions may add potential value to the future agrifood industry due to their desirable health promoting attributes.

6.3.2.4 Low-acrylamide French Fries and Potato Chips

Acrylamide is a toxic substance that is naturally produced in starchy foods as a result of high-temperature cooking, such as baking, grilling, or frying. In 2002, the Swedish National Food Administration reported alarmingly high levels of acrylamide in carbohydrate-rich heated foods (products from potato tubers, wheat flour, and coffee beans) (Tareke et al. 2002). Since acrylamide is considered as probably carcinogen for animals and humans, this finding resulted in worldwide concern. In potato, acrylamide is formed by a Maillard-type reaction among amino acids (Asparagine) and reducing sugars at high frying temperatures (Mottram et al. 2002). Several food companies recently have agreed to substantially reduce the acrylamide levels in fried potato products over the next 3–5 years. Thus novel methods to reduce the acrylamide levels in fried potato products have been a major goal for potato industry breeding programs.

Previous attempts to lower the acrylamide levels negatively affected color, texture, taste, and overall consumer palatability of the fried products. In some cases these required changes in current grower or processor practices, which limited their broader acceptance (Rommens et al. 2008). A most effective approach to reduce acrylamide levels by RNAi technology has recently been reported by Simplot^{TM} Company (Rommens et al. 2008). Potato plants were transformed with an all-native sequence RNAi-silencing construct (Rommens 2004; Yan et al. 2006) that targets two asparagine synthetase genes under the control of potato tuber specific promoters. Asparagine synthetase catalyzes the ATP-dependent conversion of aspartate into asparagines. The resulted intragenic plants produced tubers with very low levels of the acrylamide precursor asparagine. Green house experiments have shown that these lines contained a 20-fold reduction of asparagines in tubers. Chips and fries processed from these tubers remarkably showed as little as 5% accumulation of acrylamide compared to the controls. Surprisingly, this modification neither altered overall yield of the tubers grown under greenhouse conditions; nor the color, texture, or taste of the fried products. If silenced lines retain all the original agronomic characteristics under field conditions, the researchers hope all-native-DNA potato products with very low levels of acrylamide could be offered as a choice on the market.

6.4 Limitations of RNAi-Silencing Technology

A possible limitation of RNAi technology is the off-target effects of siRNA that might silence nontarget genes. Since RNAi is based on sequence recognition, targeting a gene by RNAi may give rise to the silencing of another gene that has short regions of similar sequence. This phenomenon is referred to as off-target silencing. However, no potential off-target effects were reported so far in plants (Mansoor et al. 2006; Xu et al. 2006). Transcript profiling has extensively been used in plant research and as of yet no off-target expression level changes have been noticed. Several reports recently confirm that RNAi in plants exhibits a high level of sequence specificity. Nevertheless, the possibility of off-target effects in plants cannot be ruled out and therefore needs careful attention. Caution is warranted in interpreting gene function and phenotype information resulting from RNAi experiments.

Other limitations of RNAi occur when there is a lack of efficacy or variable levels of silencing effects. Traditional DNA mutations (insertion or deletion) most often are irreversible and the effect on the function of the affected gene is generally predictable. By contrast, RNAi silencing can have widely varying effects depending on the target gene and the region of the transcript that is targeted. Sibling plants carrying identical RNAi constructs can produce varying phenotypes (Wang et al. 2005; Small 2007). There are multiple reasons for this variability that need to be considered when interpreting RNAi phenotypes. One should examine multiple independent lines to check for a reproducible phenotype and attempts should also be made to check that off-target effects are not affecting genes related to the target gene.

6.5 Future Directions

Only a few of the achievements through RNAi technology were discussed in this chapter. However, RNAi has been used for a variety of applications, including altering the flower color or obtaining novel colors by RNAi-mediated engineering of flavonoid biosynthetic pathways (Tanaka et al. 2005); developing cyanogen-free transgenic cassava, which is a major staple crop in sub-Saharan Africa (Jorgensen et al. 2005); and changing the pattern and quality of fatty acid composition of soybean by silencing the undesirable fatty acids (Flores et al. 2008). These examples, among others, demonstrate that targeted gene silencing can be used to modulate biosynthetic pathways in a specific tissue in order to obtain a desired phenotype, a feat that is often not possible by traditional breeding. These studies open the gateway to new frontiers in the use of genetic manipulation to enhance global food supply. Over the horizon, plant molecular biologists and plant breeders can see the possibility of using similar approaches to eliminate harmful compounds from plants that otherwise could serve as potential food sources, such as *Lathyrus sativus*, a hardy tropical/subtropical legume plant that naturally contains the neurotoxin β-*N*-oxalylamino-L-alanine (BOAA). This noneconomic crop is a potential target of RNAi silencing of the gene(s) responsible for the production of BOAA. A similar strategy could be applied to fava beans in order to eliminate various glycosides, undesirable compounds for human consumption. Tissue- or organ-specific silencing approaches are needed to achieve targeted gene silencing in particular plant cells with minimal interference to the normal plant lifecycle (Tang and Galili 2004).

The study of gene silencing has led to a revolution in the understanding of gene expression, as underlined by the recent award of a Nobel Prize on this concept (2006 Nobel Prize in Physiology/Medicine to Dr. Andrew Fire and Dr. Craig Mello on Gene Silencing). Because RNAi is a very efficient knockdown technology in plants, it is useful for genetic improvement in cultivars and crop plants with low transformation efficiencies, although transformation is still a challenge to face. That having been said, RNAi has clear advantages over insertional mutagenesis. The primary advantage is an ability to specifically target the gene of interest. As RNAi is a homology-dependent process, careful selection of a unique region of the target sequence can ensure that a specific gene family member is silenced, or targeting highly conserved sequence domains can silence multiple members of a gene family. In this way, redundancy is not limiting. RNAi can also be used to analyze the functions of essential genes. Variable levels of gene silencing can be achieved in different transgenic lines using the same RNAi construct, allowing selection of lines with a greater or lesser degree of silencing.

Since its identification several years ago, RNAi has become the technology of choice for plant scientists investigating gene function and manipulating plants for novel traits. Though most of the products developed using RNAi technologies are yet to hit the commercial line, there seems to be an enormous promise for future improvement especially quality traits in crop plants. The use of tissue-specific and inducible promoters should improve our ability to silence gene expression in desired target tissues and at the appropriate developmental stage, thus minimizing off-target effects. New approaches such as amiRNA (Warthmann et al. 2008) promise to bring more precision and predictability to the technology in the near future. Nevertheless, the bottleneck of public acceptance of crops derived through genetic modification should not be neglected, and remains a political and not technical challenge.

References

Bakhetia M, Charlton WL, Urwin PE, McPherson MJ, Atkinson HJ (2005) RNA interference and plant parasitic nematodes. Trends Plant Sci 10:362–367

Baulcombe D (2004) RNA silencing in plants. Nature 431:356–363

Baum JA, Bogaert T, Clinton W, Heck GR, Feldmann P, Ilagan O, Johnson S, Plaetinck G, Munyikwa T, Pleau M, Vaughn T, Roberts J (2007) Control of coleopteran insect pests through RNA interference. Nat Biotechnol 25:1322–1326

Chen XY, Chen Y, Heinstein P, Davisson VJ (1995) Cloning, expression and characterization of (+)-δ-candinene synthase: a catalyst for cotton phytoalexin biosynthesis. Arch Biochem Biophys 324:255–266

Davuluri GR, Tuinen AV, Fraser PD, Manfredonia A, Newman R, Burgess D, Brummell DA, King SR, Palys J, Uhlig J, Bramley PM, Pennings HMJ, Bowler C (2005) Fruit-specific RNAi-mediated suppression of DET1 enhances carotenoid and flavonoid content in tomatoes. Nat Biotechnol 23:890–895

Eady CC, Kamoi T, Kato M, Porter NG, Davis S, Shaw M, Kamoi A, Imai S (2008) Silencing onion lachrymatory factor synthase causes a significant change in the sulfur secondary metabolite profile. Plant Physiol 147:2096–2106

Eamens A, Wang MB, Smith NA, Waterhouse PM (2008) RNA silencing in plants: yesterday, today and tomorrow. Plant Physiol 147:456–468

Fire A, Xu S, Montgomery MK, Kostas SA, Driver SE, Mello CC (1998) Potent and specific genetic interference by double-stranded RNA in *Caenorhabditis elegans*. Nature 391:806–811

Flores T, Karpova O, Su X, Zeng P, Bilyeu K, Sleper DA, Nguyen HT, Zhang ZJ (2008) Silencing of Gm *FAD3* gene by siRNA leads to low α-linolenic acids (18:3) of *fad3* -mutant phenotype in soybean [*Glycine max* (Merr.)]. Transgenic Res 17:839–850

Fuchs M, Gonsalves D (2007) Safety of virus-resistant transgenic plants two decades after their introduction: lessons from realistic field risk assessment studies. Annu Rev Phytopathol 45:173–202

Gordon KH, Waterhouse PM (2007) RNAi for insect-proof plants. Nat Biotechnol 25:1231–1232

Helliwell CA, Wesley SV, Wielopolska AJ, Waterhouse PM (2002) High-throughput vectors for efficient gene silencing in plants. Funct Plant Biol 29:1217–1225

Henikoff S, Till BJ, Comai L (2004) TILLING: traditional mutagenesis meets functional genomics. Plant Physiol 135:630–636

Huang GZ, Allen R, Davis EL, Baum TJ, Hussey RS (2006) Engineering broad root-knot resistance in transgenic plants by RNAi silencing of a conserved and essential root-knot nematode parasitism gene. Proc Natl Acad Sci USA 103:14302–14306

Jorgensen K, Bak S, Busk PK, Sorensen C, Olsen CE, Puonti-Kaerlas J, Moller BL (2005) Cassava plants with depleted cyanogenic glucoside content in leaves and tubers. Distribution of cyanogenic glucosides, their site of synthesis and transport, and blockage of the biosynthesis by RNA interference technology. Plant Physiol 139:363–374

Kusaba M (2004) RNA interference in crop plants. Curr Opin Biotechnol 15:139–143

Lippman Z, Martienssen R (2004) The role of RNA interference in heterochromatic silencing. Nature 431:364–369

Liu Q, Singh SP, Green AG (2002) High-stearic and high-oleic cottonseed oils produced by hairpin RNA-mediated post-transcriptional gene silencing. Plant Physiol 129:1732–1743

Mansoor S, Amin I, Hussain M, Zafar Y, Briddon RW (2006) Engineering novel traits in plants through RNA interference. Trends Plant Sci 11:559–565

Mao YB, Cai WJ, Wang JW, Hong GJ, Tao XY, Wang LJ, Huang YP, Chen XY (2007) Silencing a cotton bollworm P450 monooxygenase gene by plant-mediated RNAi impairs larval tolerance of gossypol. Nat Biotechnol 25:1307–1313

Matthew L (2009) Hairpin RNAi in plants. In: Doran T, Helliwell C (eds) RNA interference: methods for plants and animals. CAB Int Publ, Wallingford, pp 1–25

Matzke M, Kanno T, Huettel B, Daxinger L, Matzke AJ (2007) Targets of RNA-directed DNA methylation. Curr Opin Plant Biol 10:512–519

Minorsky PV (2008) On the Inside. Plant Physiol 147:1761–1762

Missiou A, Kalantidis K, Boutla A, Tzortzakak S, Tabler M, Tsagris M (2004) Generation of transgenic potato plants highly resistant to Potato Virus Y (PVY) through RNA silencing. Mol Breed 14:185–197

Mottram DS, Wedzicha BL, Dodson AT (2002) Acrylamide is formed in the Maillard reaction. Nature 419:448–449

Napoli C, Lemieux C, Jogensen R (1990) Introduction of a chimeric chalcone synthase gene into petunia results in reversible co-suppression of homologous genes in trans. Plant Cell 2:279–289

Ogita S, Uefuji H, Morimoto M, Sano H (2005) Metabolic engineering of caffeine production. Plant Biotechnol 22:461–468

Ogita S, Uefuji H, Yamaguchi Y, Koizumi N, Sano H (2003) RNA Interference: producing decaffeinated coffee plants. Nature 423:823

Pooggin M, Shivaprasad PV, Veluthambi K, Thomas H (2003) RNAi targeting of DNA virus in plants. Nat Biotechnol 21:131–132

Powell-Abel P, Nelson RS, De B, Hoffmann N, Rogers SG (1986) Delay of disease development in transgenic plants that express the tobacco mosaic virus coat protein gene. Science 232:738–743

Preuss S, Pikkard CS (2003) Targeted gene silencing in plants using RNA interference. In: Engelke D (ed) RNA Interference (RNAi): nuts and bolts of siRNA technology. DNA Press, Philadelphia, PA, pp 23–36

Price DRG, Gatehouse JA (2008) RNAi-mediated crop protection against insects. Trends Biotechnol 26:393–400

Qu J, Ye J, Fang R (2007) Artificial microRNA-mediated virus resistance in plants. J Virol 81:6690–6699

Ramachandran V, Chen X (2008) Small RNA metabolism in Arabidopsis. Trends Plant Sci 13:368–374

Ratcliff F, Harrison BD, Baulcombe DC (1997) A similarity between viral defense and gene silencing in plants. Science 276:1558–1560

Rommens CM (2004) All-native DNA transformation: a new approach to plant genetic engineering. Trends Plant Sci 9:457–464

Rommens CM, Yan H, Swords K, Richael C, Ye J (2008) Low-acrylamide French fries and potato chips. Plant Biotechnol J 6:843–853

Sharp PA (1999) RNAi and double-strand RNA. Genes Dev 13:139–141

Small I (2007) RNAi for revealing and engineering plant gene functions. Curr Opin Biotechnol 18:148–153

Smith NA, Singh SP, Wang MB, Stoutjesdijk PA, Green AG, Waterhouse PM (2000) Gene expression: total silencing by intron-spliced hairpin RNAs. Nature 407:319–332

Sunilkumar G, Campbell LM, Puckhaber L, Stipanovic RD, Rathore KS (2006) Engineering cottonseed for use in human nutrition by tissue-specific reduction of toxic gossypol. Proc Natl Acad Sci USA 103:18054–18059

Tanaka Y, Nakamura N, Togami J (2005) Altering flower color in transgenic plants by RNAI-mediated engineering of flavonoid biosynthetic pathway. In: Barik S (ed) Methods in molecular biology: RNAi: design and application. Humana Press, Totowa, NJ, pp 245–257

Tang G, Galili G (2004) Using RNAi to improve plant nutritional value: from mechanism to application. Trends Biotechnol 22:463–469

Tareke E, Rydberg P, Karlsson S, Tornquist M (2002) Analysis of acrylamide, a carcinogen formed in heated foodstuffs. J Agric Food Chem 50:4998–5006

Townsend BJ, Llewellyn DJ (2007) Reduced terpene levels in cottonseed add food to fiber. Trends Biotechnol 25:239–241

Vazquez F (2006) Arabidopsis endogenous small RNAs: highways and byways. Trends Plant Sci 11:460–468

Voinnet O (2001) RNA silencing as a plant immune system against viruses. Trends Genet 17:449–459

Voinnet O, Vain P, Angell S, Baulcombe DC (1998) Systemic spread of sequence-specific transgene RNA degradation is initiated by localised introduction of ectopic promoterless DNA. Cell 95:177–187

Wang MB, Abbott DC, Waterhouse PM (2000) A single copy of virus-derived transgene encoding hairpin RNA gives immunity to barley yellow dwarf mosaic virus. Mol Plant Pathol 1:347–356

Wang T, Iyer LM, Pancholy R, Shi X, Hall TC (2005) Assessment of penetrance and expressivity of RNAi-mediated silencing of Arabidopsis phytoene desaturase gene. New Phytol 167:751–760

Warthmann N, Chen H, Ossowski S, Weigel D, Hervé P (2008) Highly specific gene silencing by artificial miRNAs in rice. PLoS ONE 3(3):e1829

Waterhouse PM, Graham MW, Wang MB (1998) Virus resistance and gene silencing in plants can be induced by simultaneous expression of sense and antisense RNA. Proc Natl Acad Sci USA 95:13959–13964

Waterhouse PM, Helliwell CA (2003) Exploring plant genomes by RNA-induced gene silencing. Nat Rev Genet 4:29–38

Waterhouse PM, Ming-Bo W, Lough T (2001) Gene silencing as an adaptive defense against viruses. Nature 411:834–842

Xu P, Zhang Y, Kang L, Roossinck MJ, Mysore KS (2006) Computational estimation and experimental verification of off-target silencing during posttranscriptional gene silencing in plants. Plant Physiol 142:429–440

Yan H, Chretien R, Ye J, Rommens CM (2006) New construct approaches for efficient gene silencing in plants. Plant Physiol 141:1508–1518

Chapter 7
Transgene Integration, Expression and Stability in Plants: Strategies for Improvements

Ajay Kohli, Berta Miro, and Richard M. Twyman

7.1 Introduction

The transfer of DNA into plants has been common practice for over 20 years, and transgenic plants are now a burgeoning industry. In 2007, over 114 million ha (282.4 million acres) of transgenic crops were grown commercially in 23 countries, the most prevalent traits being herbicide tolerance, pest resistance, or both traits stacked together (James 2007). In the laboratory, one encounters a vastly greater diversity of traits, including disease resistance, stress tolerance, nutritional improvement, modified development, and the use of plants to produce specific, high-value molecules, such as secondary metabolites, chemical precursors, antibodies, vaccine subunits, and industrial enzymes. It is notable that in the majority of cases, the purpose of gene transfer into plants is to achieve a specific, desirable phenotype. Plants that fail to live up to expectations are routinely discarded so that the best performers can be nurtured.

Despite the focus on phenotype, over the last decade there has been an increasing interest in creating transgenic plants to study the process of gene transfer itself (Kohli et al. 2003). On the academic side, it has been appreciated for many years that the structure of a transgene locus can have a major influence on the level and stability of transgene expression; thus, researchers have studied DNA integration mechanisms, particularly with regard to how transgenes interact with the plant's DNA repair and genome defense systems. On the applied side, the global adoption of transgenic crops and the development of transgenic plants producing pharmaceuticals and other important molecules have attracted the interest of regulatory authorities (Ramessar et al. 2008). The demand for robust risk assessment practices

A. Kohli (✉)

Plant Molecular Biology Laboratory, Plant Breeding Genetics and Biotechnology, International Rice Research Institute, DAPO-7777, Metro Manila, The Philippines

e-mail: a.kohli@cgiar.org

C. Kole et al. (eds.), *Transgenic Crop Plants*,

DOI 10.1007/978-3-642-04809-8_7, © Springer-Verlag Berlin Heidelberg 2010

means that transgenic plants have to be characterized in great detail, including information on the sequence, structure, organization, and genomic position of the transgenic locus. Recently, this has culminated in the first report of the genome sequence of a transgenic plant, including the analysis of the transgenic locus (Ming et al. 2008). The principles and practices of transgenic technology have come under scrutiny, leading to research focusing on the use and elimination of marker genes, the role of vector sequences that integrate along with the transgene, and the random nature of transgene integration events with regard to copy number, transgene orientation, and transgene rearrangements. Researchers, therefore, have practical as well as academic reasons for studying transgene integration and expression, and have developed new ways to analyze transgenic loci. Current research focuses on ways to better control the way DNA integrates into the plant genome.

In this chapter, we describe the methods used to study transgene locus structure and discuss evidence supporting current models of transgene integration for both *Agrobacterium*-mediated transformation and direct transfer methods. We discuss how transgene loci are organized and how this affects the level and stability of transgene expression from generation to generation. Finally, we look to the future by describing how recent research has advanced the state of the art in gene transfer technology.

7.2 Methods for the Analysis of Transgenic Loci

Most gene transfer experiments are phenotype driven, by which we mean that successfully transformed plants tend to be identified on the basis of the phenotype conferred by the transgene rather than the structure of the transgene itself. This is pertinent because the appearance of the desired phenotype is prima facie evidence that the transgene has integrated into the genome and is intact, thus allowing expression of the encoded protein. Since most transgenic plants are regenerated under selection for the product of a selectable marker gene, the fact that a transgenic plant exists at all indicates that at least one intact copy of the marker gene is present in the genome. Similarly, the phenotypes conferred by any other transgenes can be used as evidence to support successful integration and expression. This information is of limited value, however, because it divides all plants into just two categories – (a) plants transformed with at least one intact transgene and (b) plants not transformed at all or transformed with a nonfunctional transgene. It provides no quantitative information, yet every gene transfer experiment produces a population of plants with a range of phenotypes reflecting the level of transgene expression. Since the same input DNA is used in each case, the only explanation for quantitative differences in phenotype is differences in the structure and activity of the integrated transgenes.

The technique used most commonly for a definitive analysis of transgenic loci is the Southern blot, in which genomic DNA is digested with one or more restriction enzymes, fractionated by agarose gel electrophoresis, denatured, transferred to a membrane, and hybridized to a labeled probe. Many different types of information can be obtained from Southern blots depending on the restriction enzymes and probes used. One of the most common strategies is to use an enzyme that cuts once within the transgene in combination with a probe that hybridizes to the body of the transgene. This generates DNA fragments whose size depends on the distance between a fixed point in the transgene (the restriction site) and the adjacent restriction site in the genomic DNA, which of course varies according to the site of insertion. Where multiple copies of a transgene have integrated, a single cutter enzyme tends to generate a unique pattern of bands that serves as a genetic fingerprint of that plant and all its descendants, thereby helping to identify clonal relatives of the original transformant and allowing transgene segregation to be followed through generations. Because it is unlikely that any of the transgene copies will generate identical-sized bands (unless they are perfect concatemers, in which case the band size will correspond exactly to the size of the transgene), this method also provides an estimate of transgene copy number. A variant of the technique is to use a probe that hybridizes to the vector backbone instead of the transgene body, which helps to identify inserts of vector DNA.

Another handy method is to use an enzyme that cuts twice in the transgene and liberates a specific DNA cassette in combination with a probe that hybridizes to that cassette. If all copies of the transgene are intact, there should be only one hybridizing band, corresponding to the size of the cassette, and the intensity of the hybridization signal will be proportional to the number of transgene copies (since one cassette should be released from each integrated transgene copy). Copy number determination is best achieved by "spiking" genomic DNA from an untransformed plant with a known amount of transforming plasmid DNA, and then digesting this and the genomic DNA from genuine transgenic plants. It may be useful to set up a series of control DNA samples, containing for example one, five, and ten copies of the transforming plasmid per genome equivalent of DNA. In this way, a calibration curve of signal intensities relative to copy number can be generated, onto which any transgenic plant can be mapped. Band sizes greater or smaller than the diagnostic fragment indicate truncations or rearrangements. It is also useful to digest genomic DNA with an enzyme known not to cut within the transgene. If all copies of the transgene have integrated at a single locus as a concatemer, digestion with such enzymes should liberate the locus as a single, high-molecular-weight fragment. Thus, the presence of two or more bands suggests either the presence of two or more independent transgenic loci (this can be confirmed by segregation analysis as discussed below), or the presence of interspersed genomic DNA between transgene copies at a single locus (this can be confirmed, if required, by fiber-FISH as discussed below).

The polymerase chain reaction (PCR) is a rapid technique that can be used to confirm transgene integration through the use of primer combinations that generate a transgene-specific product. Although quicker than Southern blots and more amenable to multiplexing, false positives can occur through the amplification of episomal plasmid DNA so the PCR should only be used indicatively, with Southern blots used for definitive confirmation of DNA integration. Long PCR is a variation that allows larger products to be amplified and is potentially useful for analyzing larger transgenic loci (Mehlo et al. 2000). Another PCR variant, real-time quantitative PCR, is now used for the rapid estimation of transgene copy number by comparison with a control sample in which a single-copy endogenous gene is amplified. The relative signal intensities of the control and transgenic samples reveal the transgene copy numbers in the transgenic plants (Li et al. 2004; Yang et al. 2005). Although the PCR can show the presence or absence of a transgene and provide a dependable copy number estimate, it provides little in the way of information about the structure of a transgenic locus unless the genomic flanking sequences are already known. DNA sequencing is the highest-resolution transgene analysis method, and permits the precise definition of structural organization and rearrangement. It also allows the nature of transgene-genomic and transgene-transgene junctions to be investigated at the nucleotide level, and the integration site in genomic DNA to be identified. Prior to sequencing, the transgenic locus must be isolated, which can be achieved by standard molecular cloning or through a variety of methods such as inverse PCR, thermal asymmetric interlaced (TAIL)-PCR, or plasmid rescue (Twyman and Kohli 2003). The complete sequencing of a transgenic papaya variety, SUNUP, was recently reported, allowing intricate structural analysis of the transgenic locus (Ming et al. 2008).

All these methods involve the identification of discrete DNA fragments of precise length, which is useful for fine structural analysis but not for the characterization of transgenic loci at the chromosome level. In this context, fluorescence in situ hybridization (FISH) can be useful as it allows target sequences to be identified in isolated DNA fibers, interphase chromatin, and even metaphase chromosomes. FISH involves the use of fluorescently labeled nucleic acid probes to identify particular target sequences, revealing higher-order transgene organization and the distribution of integration sites. FISH to metaphase chromosomes allows the insertion sites to be mapped cytogenetically and simultaneous analysis in interphase allows the nuclear territory of transgenes to be determined (Abranches et al. 2000). Fiber-FISH on extended chromatin gives an overview of locus structure, revealing the presence of single-copy inserts, transgene concatemers, and interspersed genomic DNA (Jackson et al. 2001). The resolution of FISH lies somewhere between that of genetic segregation and Southern blot hybridization and can provide important correlative data for both techniques.

Segregation analysis involves studying the transmission of particular DNA sequences or the phenotypes thus conferred over several generations of transgenic plants. Closely linked transgene copies are unlikely to be separated by recombination, while widely separated loci are likely to segregate at meiosis in some plants. This allows the number of transgenic loci to be determined. Problems with phenotype analysis include the misleading results caused by epigenetic gene silencing, but analysis of DNA sequence segregation by Southern blot hybridization can be highly informative.

7.3 Locus Structure in Plants Transformed by *Agrobacterium tumefaciens*

7.3.1 Principles of Gene Transfer

Agrobacterium tumefaciens is a soil pathogen that colonizes wounded plant cells and induces the formation of a tumor (or crown gall) that produces special amino acid derivatives called opines, which the bacteria are able to use as a carbon and nitrogen source. The ability of virulent *Agrobacterium* strains to induce tumor growth and opine synthesis, and the capacity to utilize opines, is conferred by a resident tumor-inducing plasmid (Ti-plasmid). During the colonization process, a segment of DNA from this plasmid called the transferred DNA (T-DNA) is transferred to the plant nuclear genome. The T-DNA encodes enzymes that synthesize auxins and cytokinins, resulting in unregulated cell proliferation, and enzymes that synthesize opines from standard amino acids (reviewed by Gelvin 2003).

The Ti plasmid is a naturally occurring vector for plant transformation, but wild-type Ti-plasmids are not suitable vectors for genetic engineering in plants because they are too big to manipulate, and the oncogenes contained in the T-DNA cause uncontrolled proliferation of transformed plant cells and prevent efficient regeneration. The T-DNA must therefore be moved to a smaller, more convenient vector, and disarmed by deleting the oncogenes. A marker gene must also be included to allow transformed cells to be propagated. T-DNA transfer is controlled by about 30 genes located in a separate virulence (*vir*) region of the Ti plasmid, and these must be supplied in trans using a binary vector system if the T-DNA is placed on a smaller plasmid. Modern binary vectors contain multiple unique cloning sites within the T-DNA, a *lacZ* marker gene for blue-white selection of recombinants, and a choice of selectable markers to identify transformed plant cells (Hellens et al. 2000).

The transfer mechanism is pertinent to the resulting locus structure. The T-DNA is flanked by 25-bp imperfect direct repeats known as border sequences, which are not transferred to the plant genome intact, but they are required for the transfer process. T-DNA transfer is mediated by the *vir*A and *vir*G gene products, which transduce external signals and activate other *vir* genes resulting in the construction of a pilus for DNA transfer, and the release of the T-DNA by an endonuclease comprising the products of the *vir*D1 and *vir*D2 genes. This introduces either single-strand nicks or a double-strand break at the 25-bp borders of the T-DNA. It is thought that the intermediate formed (a double stranded T-DNA or a single T-strand) may depend on the virulence functions particular to the *Agrobacterium* strain (Steck 1997). Either the left or right border sequence can initiate T-DNA transfer, although it is more usual for initiation to occur at the right border due to the presence of an adjacent overdrive sequence, which is recognized by the VirC1 and VirC2 proteins and acts as a transfer enhancer (Shaw et al. 1984). For this reason, deletion of the right T-DNA border severely reduces the efficiency of transfer, whereas deletion of the left border has little effect (Jen and Chilton 1986).

The VirD2 protein remains covalently attached to the $5'$ end of the processed T-DNA strand and has been proposed to protect the T-DNA against nucleases, to target the DNA to the plant cell nucleus, and to help integrate it into the plant genome (Tzfira et al. 2000).

7.3.2   T-DNA Locus Structure

Most investigations of T-DNA transfer have suggested that there may be preferential integration into transcription units, with up to 90% of events occurring in genes (e.g., Lindsey et al. 1993). In petunia, FISH analysis showed that T-DNA inserts were found preferentially at the gene-dense distal chromosome sites (Wang et al. 1995; Ten Hoopen et al. 1996), and a comparative analysis in *Arabidopsis* and rice showed that T-DNA inserted randomly in the *Arabidopsis* genome (which is globally gene-rich, with little repetitive DNA) but homed in on the 10–20% of the rice genome known to be gene-dense while avoiding the more widespread heterochromatic regions (Barakat et al. 2000). It has also been suggested that T-DNA integration occurs preferentially in regions showing microhomology to the T-DNA borders (Matsumoto et al. 1990), which may also be enriched in the transcribed part of the genome. The bias in T-DNA insertion is valuable for gene-tagging experiments, ensuring a high gene hit rate in genomes with large tracts of gene-poor heterochromatin while showing little bias among different genes (Jeon et al. 2000; Weigel et al. 2000; Hsing et al. 2007; Wan et al. 2009).

The structure and complexity of transgenic loci generated by *Agrobacterium* depends on the strain, plant species, and explant type, but generally gives rise to lower transgene copy numbers than direct transformation methods. An informative experiment was performed by Cheng et al. (1997) by transforming wheat using both *Agrobacterium* and particle bombardment. Of 26 *Agrobacterium*-mediated transformants, more than one-third contained a single T-DNA insert, half contained 2–3 copies, and the remainder (about 15%) contained 4–5 copies. There were no transformants containing more than five T-DNAs. In contrast, from the population of 77 bombarded transformants, only 13 (17%) contained a single copy of the transgene. The maximum number of transgene copies in this population was not reported. Hu et al. (2003) also observed more complex transgene insertions from particle bombardment than from *Agrobacterium*-mediated techniques. More recently, similar experiments in barley showed that all the *Agrobacterium*-derived lines contained 1–3 copies of the transgene, while 60% of the transgenic lines derived by particle bombardment contained more than eight copies (Travella et al. 2005). Dai et al. (2001) found in rice that the average transgene copy numbers were 1.8 for *Agrobacterium*-derived lines and 2.7 for plants obtained by particle bombardment. However, Khanna and Raina (2002) observed multiple transgene insertions in rice transformants generated through both techniques together with the transfer of partial T-DNA fragments.

The organization of integrated T-DNA sequences differs among *Agrobacterium* strains, but a common feature of nopaline-type derivatives such as C58 is the preferential integration of T-DNA as dimers with an inverted repeat configuration, linked either at the left or right borders (Jones et al. 1987; Jorgensen et al. 1987). Where cotransformation is carried out with two T-DNAs containing different markers, the different T-DNAs were often present as heterodimer inverted repeats, preferentially around the right border (De Block and Debrouwer 1991). Similarly, cotransformation of rice with the vectors pGreen and pSoup (each containing different selectable and visible markers) resulted in 56% of plants with the two T-DNAs cointegrated, although there was also a high proportion of plants containing separate integration events (Afolabi et al. 2004). In contrast, Spielmann and Simpson (1986) carried out transformation using the octopine *Agrobacterium* strain LBA4404. They found only two integration events among the 22 characterized transformants that resulted in dimer formation, while most of the rest were single-copy integrations. When cotransformation experiments were carried out with this strain (McKnight et al. 1987), three double transformants were obtained and in all cases the two T-DNAs were genetically unlinked. These results suggest that the virulence functions carried by a particular *Agrobacterium* strain strongly influence the structure of the transgene locus.

Another important aspect of locus structure is the amount and types of transgene rearrangement. Occasionally, it has been reported that T-DNA has undergone spontaneous rearrangement prior to or during integration (e.g., Offringa et al. 1990; Puchta et al. 1992), and this has been demonstrated directly by fiber-FISH in potato (Wolters et al. 1998). In some cases, rearrangements may be induced by specific recombinogenic sequences such as the CaMV 35S promoter (Kohli et al. 1999), which may have been responsible for T-DNA rearrangements in some transgenic potato lines (Porsch et al. 1998). In many cases, however, rearrangements may reflect "collateral damage" occurring spontaneously during the transfer process. Afolabi et al. (2004) found that nonintact T-DNAs were present in >70% of transgenic rice lines, in most cases reflecting loss of the mid to right border portion of the T-DNA. Similarly, Rai et al. (2007) found that about 50% of rice plants transformed with a T-DNA containing the phytoene synthase (*psy*) and phytoene desaturase (*crt*I) genes showed evidence of T-DNA rearrangements, and in the majority of cases the rearrangements occurred in the *crt*I expression cassette, which was adjacent to the right T-DNA border.

7.3.3 T-DNA Integration Mechanism

A number of groups have investigated the structure of genomic/T-DNA and T-DNA/T-DNA junctions in plants and have concluded that integration occurs by illegitimate recombination (see Salomon and Puchta 1998; Somers and Makarevistch 2004). A strand invasion mechanism of integration has been proposed (reviewed by Tinland 1996), in which the 3' end of the T-strand initiates the integration process by hybridizing to a short region of homology in the plant genome, the second strand being completed by primer extension of the plant DNA. Other models suggest conversion of the T-strand into a double-stranded intermediate, which integrates at the site of naturally occurring chromosome breaks via double-strand DNA break repair. This is supported by experiments that show transformation efficiency increases following UV irradiation, which generates nicks and breaks in genomic DNA. However, since T-DNA integration occurs normally, if less frequently, in DNA repair mutants, it is possible both mechanisms occur simultaneously.

DNA repair models argue that proteins encoded by the host plant have a much more important role in T-DNA integration than *Agrobacterium* proteins, such as VirD2, which are imported into the plant with the T-DNA. However, since VirD2 protein remains covalently attached to the 5' end of the T-strand during transfer it is also likely to influence integration (Ward and Barnes 1988). In an in vitro assay, VirD2 can ligate together a cleaved T-DNA border sequence but cannot ligate T-DNA to other genomic targets unless plant cell extracts are also present (Pansegrau et al. 1993; Ziemienowicz et al. 2000).

Plant proteins are certainly required for integration, as a number of *Arabidopsis* mutants have been identified that are deficient for T-DNA insertion. The role of DNA strand break repair in T-DNA integration was supported by the discovery of *Arabidopsis* mutants *uvh1* and *rad5*, which are hypersensitive to UV and gamma irradiation, respectively, and show a low frequency of stable transformation by *Agrobacterium*. Since these mutants showed normal levels of transient expression, it was suggested that they caused deficiencies in the repair of radiation-induced breaks and that break repair is essential for T-DNA integration (Sonti et al. 1995). However, Nam et al. (1998) showed that *uvh1* is no less transformation proficient than wild-type plants and that *rad5* is deficient for both transient and stable transformation, indicating that the dysfunction affects a process occurring much earlier than T-DNA integration. Other mutants resistant to *Agrobacterium* transformation (*rat* mutants) have been identified, and five are thought to be blocked at the point of T-DNA integration (Nam et al. 1999). One of the corresponding genes, *rat5*, encodes a histone protein, suggesting that efficient T-DNA integration is dependent on chromatin structure at the integration site.

Much can be learned about the T-DNA integration mechanism by the inspection of borders, especially the borders between adjacent T-DNA sequences in multicopy insertions. The formation of heterodimers during cotransformation argues in favor of T-DNA concatemerization prior to integration. Although inverted repeats around the right border are often precise, those around the left border and those separating direct T-DNA repeats are often characterized by the insertion of variable-sized regions of filler DNA, which may be derived from the T-DNA sequence or from plant genomic DNA (De Buck et al. 1999; Kumar and Fladung 2000, 2002). This suggests either the simultaneous integration of multiple T-DNAs at a single locus, or a two-phase mechanism, in which a primary T-DNA integration event stimulates further secondary integrations in the same area, similar to those proposed for particle bombardment (see Sect. 7.4.3). Zhu et al. (2006) carried out a comprehensive study of T-DNA border characteristics in a population of transgenic rice plants including 156 T-DNA/genomic DNA junctions, 69 T-DNA/T-DNA junctions, and 11 T-DNA/vector backbone junctions, which included 171 left borders and 134 right borders. Conserved cleavage was observed in 6% of left and 43% of right borders, microhomology was observed in 58% of T-DNA/ genomic DNA, 43% of T-DNA/T-DNA, and 82% of T-DNA/vector junctions, mostly at left borders, and about one-third of the T-DNA/genomic DNA and T-DNA/T-DNA junctions showed evidence of filler DNA (up to 344 bp). This was derived mainly from the T-DNA region adjacent to the breakpoint and/or from the rice genomic DNA flanking the T-DNA integration site, with T-DNA/T-DNA filler DNA showing the greatest complexity. Interestingly, when two T-DNAs were integrated in the inverted repeat configuration, significant truncation was always observed in one of the two T-DNAs, whereas with direct repeat configuration, large truncations were rare. These data suggested no single integration mechanism could account for all observations, but the presence of filler DNA at many of the junctions argued that a template-driven DNA synthesis mechanism must be involved, probably reflecting abortive gap repair through a synthesis-dependent strand annealing (SDSA) process. For example, a 16-bp filler DNA that was identical to a reversed T-DNA fragment close to the right border was observed at a left/right border junction. This was most likely produced by invasion of the $3'$ end of a T-DNA into another T-DNA near the right border in reverse orientation during recombination or interaction of these two T-DNAs. When the right border is not protected by VirD2, it is subjected to $5'$ exonuclease degradation that creates a free $3'$ end in its complementary strand. This $3'$ end is able to invade another template to produce filler DNA at the right border end. Multiple template switches can be used to explain the origin of complex filler DNA structures, and longer regions of homologous DNA might reflect a single-strand annealing process in addition to SDSA.

7.3.4 Cotransfer of Vector Backbone Sequences

Agrobacterium was initially thought to be a clean transformation method because the T-DNA is more or less precisely defined (cleavage occurs at a precise position within the right border repeat and the cleavage site at the left border varies by about 100 bp). However, it is now evident that T-DNA transfer is much less precise than originally envisaged, and 25–30% of transformants may commonly contain vector sequences linked to the T-DNA insert, indicating that the cleavage reaction during T-DNA transfer can be rather inefficient (Martineau et al. 1994; Rai et al. 2007). Other studies have shown that in some systems, the frequency of vector backbone transfer can reach as high as 66% (Afolabi et al. 2004). Ramanathan and Veluthambi (1995) constructed binary vectors, in which the selectable marker was located outside the left T-DNA border. In accordance with the T-DNA transfer mechanisms discussed above, it was considered likely that this strategy would catch those transfer events in which transfer, initiated at the right border, overran the left border and terminated somewhere along the plasmid backbone. Surprisingly, these investigators found that none of the transformants contained any T-DNA sequences, indicating that, in these cases, transfer had initiated at the left border and had proceeded around the plasmid away from the T-DNA, presumably breaking off before completing the circuit and reaching the right border sequence. Further investigations have shown that vector sequence transfer is probably a very common event, occasionally involving the entire plasmid backbone with or without the T-DNA. Concatemers of the entire binary vector have also been seen, indicating that transfer does not necessarily terminate at the T-DNA border even after one or more complete circuits of the vector (Wenck et al. 1997). The exact structure of the insert and the presence or absence of T-DNA in recovered transgenic plants depend of course on the position of the selectable marker. In the strategy of Ramanathan and Veluthambi (1995), the external position of the marker allowed non-T-DNA transformants to be recovered. The experiments carried out by Kononov et al. (1997) are particularly informative because this group constructed binary vectors, in which a selectable marker was present within the T-DNA and a screenable marker gene was present outside either the left or the right borders of the T-DNA. Over 200 transformants were obtained under selection and 75% were shown to carry the external screenable marker gene *gus*A. Interestingly, both vectors appeared to transfer *gus*A to the plant genome with equal efficiency, suggesting that T-DNA transfer could be initiated nonselectively at either the left or right borders. It is also notable that Kononov and colleagues used three alternative *Agrobacterium* strains: LBA4404, GV3101, and EHA105, representing octopine, nopaline, and agropine-type virulence functions, respectively. There were no significant differences among the strains in terms of the frequency of vector sequence transfer. Finally, these investigators reported that they could also detect independent integration events involving plasmid backbone sequences alone. Since the selectable marker in these experiments was located within the T-DNA, such vector-only integrations must have occurred in addition to the T-DNA-linked integration events. This indicates that in the natural course of transformation, many vector-only integration events may occur, but will not be recovered under selection. It is also likely that vector-only integration events occur, undetected, in many plant transformation experiments.

7.4 Locus Structure in Plants Transformed by Direct DNA Transfer

7.4.1 Principles of Gene Transfer

A number of direct DNA transfer methods have been developed to transform plants recalcitrant to *Agrobacterium*-mediated transformation (reviewed by Twyman et al. 2002). Among these methods, particle bombardment has become the most successful because it is based on purely mechanical principles and is therefore not dependent on the biological factors that restrict the *Agrobacterium* "host range". Particle bombardment works with any plant species, variety, and explant, leaving the regeneration of fertile plants rather than the DNA transfer process itself as the only significant bottleneck (Altpeter et al. 2005). Particle bombardment involves the acceleration of small DNA-coated metal particles (either gold or tungsten) into plant tissue with sufficient force to break through the cell wall and membrane. Some of the particles reach the nucleus, where the DNA is released, probably by a simple diffusion mechanism (Altpeter et al. 2005). Notably, the foreign DNA entering a bombarded cell is naked, double-stranded, and competent for both transient episomal expression and integration into the genome. Transient expression also occurs in the process of *Agrobacterium*-mediated transformation, but the T-strand must first be converted into a double-stranded intermediate (Narasimhulu et al. 1996). Other direct DNA transfer methods are gentler, using chemicals (e.g., PEG, calcium phosphate) or physical methods (e.g., electroporation) to persuade plant protoplasts to take up DNA from the surroundings. However, this DNA must ultimately find its way to the nucleus, and integration occurs in the same way as described below for particle bombardment.

While *Agrobacterium*-mediated transformation involves a number of virulence gene products that must be supplied either on the same plasmid as the T-DNA or on a separate binary vector, particle bombardment has no such requirements because the introduction of DNA is governed entirely by external physical factors (Sanford et al. 1993). For convenience, therefore, vectors used for direct transfer are generally based on bacterial cloning plasmids, and incorporate a selectable marker and origin of replication functional in bacteria. In *Agrobacterium*-mediated transformation, the T-DNA is meant to be excised from the vector during the transformation process, and any vector backbone transfer results from inefficient processing. In contrast, there is no such processing in particle bombardment, although this can be achieved before transformation by excising the linear cassette, purifying it, and using just this cassette as the substrate for coating the metal particles (Fu et al. 2000). This practice has the interesting side effect of reducing the complexity of transgene loci as discussed in Sect. 7.4.3.

7.4.2 Transgenic Locus Structure

There have been few studies, in which integration sites generated by particle bombardment have been carefully mapped, so whether there is a preference for inserting in transcription units is not so clear as in the case of T-DNA integration. The variable nature of the input DNA linear cassette sequences should remove any sequence-dependent bias (compared to the preserved ends of the T-DNA), but as discussed in Sect. 7.3.4, the T-DNA cutting process can overshoot the left and/or right border, so it is likely that the substrates for integration are equally variable in T-DNA transfer. Chen et al. (1998) noted that in rice plants cotransformed with up to 13 plasmids, there was no preference for the integration of particular transgenes, indicating that the insertion mechanisms operated independent of input gene sequence. Svitashev et al. (2000) showed by FISH analysis of transgenic oat that integration occurred randomly with respect to the A/D and C genomes, and there was no preference for chromosomes from a particular genome. However, the majority of integration events occurred at telomeric and subtelomeric regions, which are typically gene-rich. It is also possible that this preferential integration may reflect some aspect of the nuclear architecture in oat rather than the distribution of genes, since FISH analysis of a limited number of transgenic wheat plants generated by particle bombardment showed no preferential integration in terms of the chromosome region. In the commercial papaya variety SUNUP, five of the six sequences flanking the three indentified transgene integration sites were genomic copies of plastid genes (Ming et al. 2008). Since the plastid genome is more AT-rich than typical genomic DNA, this both supports the possibility of preferential insertion in or near genes and matches the observation of AT-rich sequences at the insertion sites in other transgenic lines generated by *Agrobacterium* and direct DNA transfer.

Unlike the situation with *Agrobacterium*-mediated transformation, a vast literature has accumulated on the structure and complexity of transgenic loci generated by direct DNA transfer, particularly particle bombardment. As discussed in Sect. 7.3.2, T-DNA integration usually occurs with a low copy number, rarely exceeding five copies, and the T-DNA is generally intact. In contrast, direct DNA transfer often generates much larger transgenic loci. Typically, these contain from 1 to 20 transgene copies (e.g., Klein et al. 1987; Register et al. 1994; Cooley et al. 1995; Dai et al. 2001; Travella et al. 2005). The structure of such loci is highly variable, comprising single copies, tandem or inverted repeats, concatemers, intact transgenes, truncated and rearranged sequences, and interspersed genomic DNA. The analysis of transgenic cereal plants by FISH to extended DNA fibers, metaphase chromosomes, and interphase chromatin has revealed a higher-order level of organization where discrete integration events are interspersed by large fragments of genomic DNA, up to several hundred kilobase pairs in length. This organization, which generates immense (megabase) transgenic loci, appears unique to particle bombardment, and could thus reflect the nature of the transformation process itself (see Sect. 7.4.3).

A useful overview of transgene organization in wheat has been reported by Jackson et al. (2001) using the technique of fiber-FISH. This study showed that transgene loci in bombarded wheat plants can be organized in three ways. The simplest arrangement, described as a type III locus, is characterized by a single discrete fiber-FISH signal corresponding approximately to the length of the transforming plasmid. This represents an intact, single copy transgene. Type III loci may be present uniquely in a given plant, or there may be two or more unlinked inserts representing multiple genetic loci. These two possibilities can be distinguished by FISH to metaphase chromosomes and genetic segregation analysis. Other loci, described as type I loci, are longer than the single plasmid copy yet still generate a continuous signal along the extended chromatin fiber. For example, Jackson and colleagues reported a type I transgenic locus with a continuous signal of 77 kb, representing 11 contiguous plasmid copies. Such loci represent concatemers of the transforming plasmid and are characterized by the absence of intervening genomic DNA. The presence of concatemers can also be confirmed by Southern blot analysis and sequencing across plasmid/plasmid junctions. Loci thus characterized have been described by Kohli et al. (2003) as "transgene arrays" (Fig. 7.1). Until the late 1990s, both head-to-head and head-to-tail concatemers had been sporadically reported in the literature, but it was unusual for the structure of a transgenic locus to be examined in such detail. Concatemerization is probably quite a common phenomenon. Extensive concatemerization, for example, has been reported by Hadi et al. (1996) in transgenic soybean simultaneously transformed with 12 different plasmid vectors. The remaining class of locus (type II) is the most complex. It is characterized by fiber-FISH signals that extend for a significant distance (>100 kb) over the chromosome, but which are punctuated regularly by intervening segments

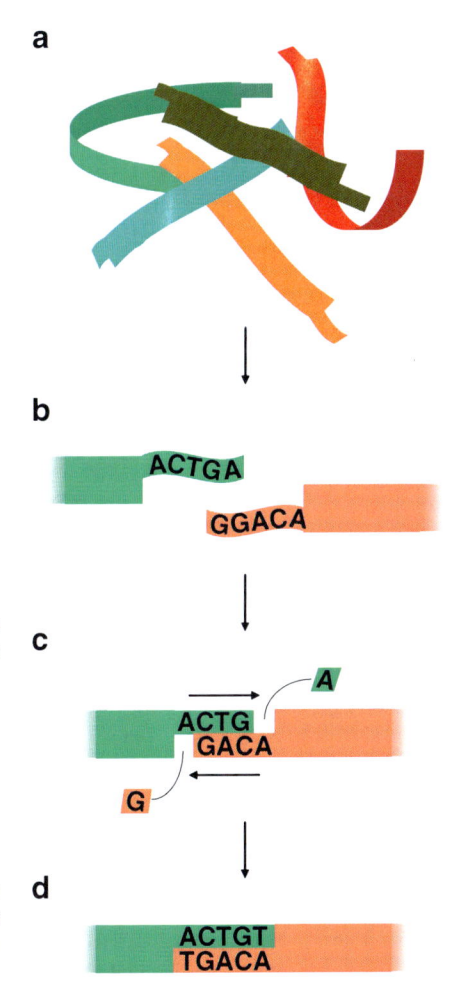

Fig. 7.1 Mechanism for transgene integration at regions of microhomology. A mixture of DNA fragments with ragged ends (**a**) interacts with a double-stranded DNA break with partially complementary ragged ends (**b**). Repair synthesis across the gap (**c**) generates a recombination junction (**d**) which may be completely conserved if the homology is precise, or may involve either the loss of terminal sequences or the insertion of filler DNA if the homology is partial

of genomic DNA (no signals). Such loci have also been identified in transgenic oat, rice, barley, and maize. In barley, for example, some transgene integration sites showed simple structures represented by one single FISH signal, whereas in others it was possible to identify up to six spots organized in a linked cluster and separated by barley DNA, making the locus several megabase pairs long (Travella et al. 2005). Kohli et al. (2003) defined such loci as "transgene clusters." Type II loci contain genomic interspersions ranging from a few tens of base pairs to approximately 10 kb (Fig. 7.2). Although dispersed over a distance of up to 100 kbp, such

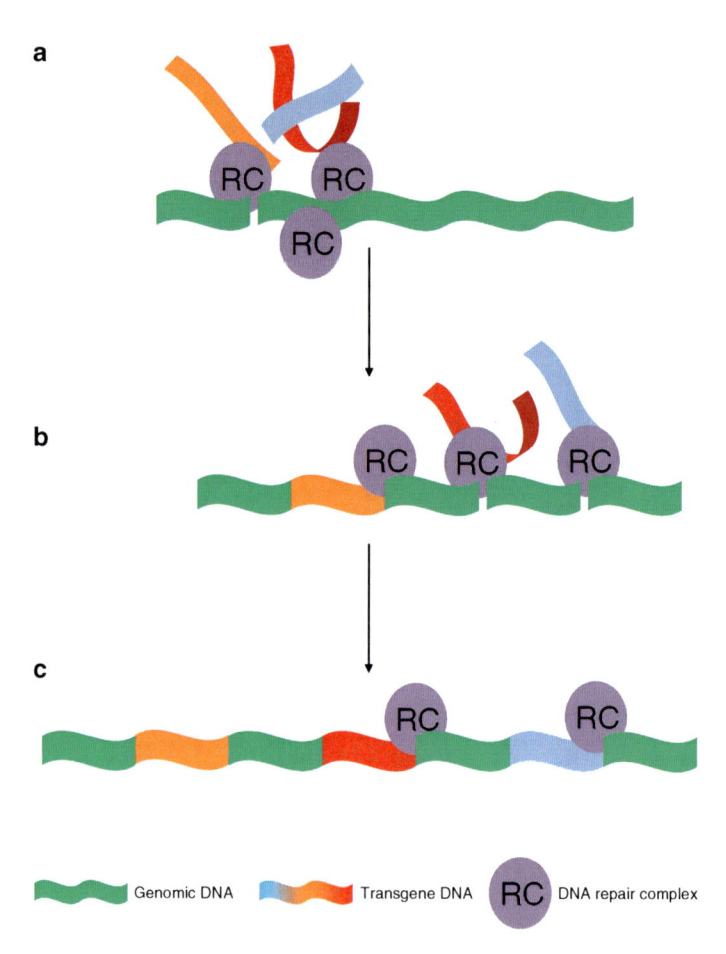

Fig. 7.2 Explanation for the formation of transgene arrays and transgene clusters interspersed with genomic DNA. A mixture of DNA fragments interacts with a double-stranded DNA break where a repair complex has already assembled (**a**). The repair complex may stitch together DNA fragments to form concatemers prior to integration, or may integrate single copies. The first integration event stimulates further repair complex activity nearby, resulting in additional nicks and breaks in the genomic DNA that act as further integration sites (**b**). This results in a cluster of transgenes (single copies and concatemers) interspersed with short regions of genomic DNA (**c**)

loci would still be expected to generate a single discrete signal if FISH analysis was applied to metaphase chromosomes due to the low resolution of this technique. However, the analysis of metaphase wheat chromosomes by FISH has revealed an unexpected third level of organization, involving the dispersion of transgene arrays and/or clusters over a larger area comprising megabase pairs of DNA (Abranches et al. 2000). Instead of discrete spots for each transgenic locus, two or more separable FISH signals were often observed, restricted to a particular chromosome region (Fig. 7.3). To be separable at the cytogenetic level, each signal must be interspersed by hundreds of kilobase pairs of genomic DNA. Similarly large genomic interspersions have been seen in transgenic oat (Svitashev and Somers 2001).

Interestingly, FISH analysis of interphase chromatin and metaphase chromosomes in the same transgenic wheat plants showed that the dispersed metaphase FISH signals could come together at interphase (Abranches et al. 2000). Occasionally, the signals clustered at a specific region of the nucleus but remained discrete.

7.4.3 Mechanisms of Transgene Integration

The analysis of plasmid/plasmid and plasmid/genomic junctions in transgenic plants generated by particle bombardment reveals features characteristic of illegitimate recombination similar to those seen for T-DNA junctions, suggesting that the same overall integration mechanisms may be involved (Svitashev et al. 2002). For example, such junctions are characterized by regions of microhomology, filler DNA, trimming of the DNA ends so sequences are lost and AT-rich elements surrounding the junction site, with similarity to topoisomerase I binding/cleavage sites (Fig. 7.1). In the analysis of multiple plasmid/plasmid junctions in 12 transgenic rice lines, Kohli et al. (1998) observed ten plants with microhomology at the junctions and two plants where junctions appeared to be generated by blunt ligation, with no overlap. A similar ratio of conserved end-joining to microhomology-mediated recombination was observed by Gorbunova and Levy (1997) and Salomon and Puchta (1998). Topoisomerase I sites were also observed adjacent to 10 out of 12 junctions characterized in transgenic *Arabidopsis* plants generated by particle bombardment (Sawasaki et al. 1998) and in four of the six junctions in the commercial SUNUP variety of papaya (Ming et al. 2008). Illegitimate recombination, therefore, appears to be responsible both for the integration of foreign DNA into the plant genome and the linking of multiple plasmid copies, which is similar to the mechanism proposed for T-DNA integration (Sect. 7.3.3).

Any model for transgene integration following particle bombardment must take into account the three-tier organization revealed in transgenic cereals: contiguous arrays, interspersed clusters, and widely dispersed FISH signals. Two-phase transgene integration mechanisms have been proposed to explain the first two levels of organization, and in such models concatemerization is proposed to occur prior to integration, while interspersion occurs during the integration process (Kohli et al.

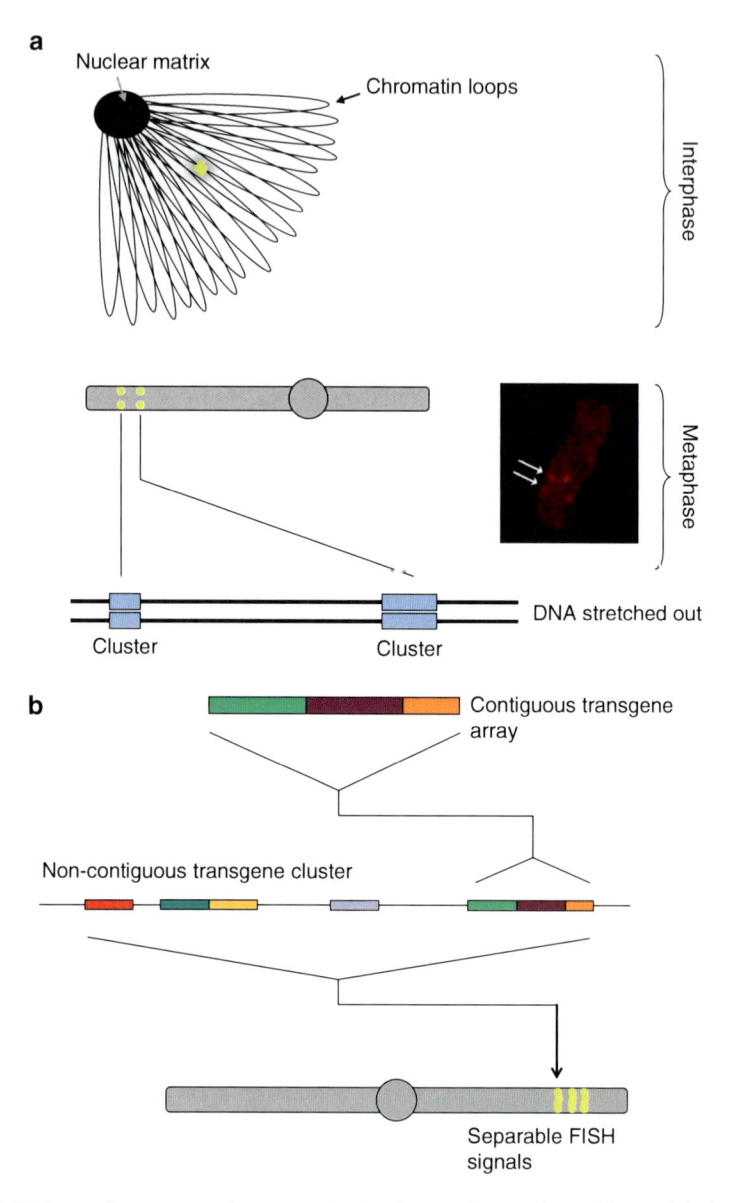

Fig. 7.3 Higher order transgene locus organization in cereals transformed by particle bombardment. Transformation occurs during interphase, when the chromatin is distributed into specific nuclear zones and territories. If a metal particle causes localized damage, DNA repair complexes will form at these sites and initiate transgene integration (**a**). During metaphase, when FISH analysis is generally carried out, loci that are brought together in interphase may be separated, resulting in multiple signals from the same transformation event (**b**). If the DNA were stretched out, this would reveal large (megabase) interspersed sequences, which have also been observed in fiber-FISH experiments

1998; Pawlowski and Somers 1998; Svitashev et al. 2002) (Fig. 7.2). In each model, penetration of the cell is proposed to elicit a wound response, which would include the induction of DNA repair enzymes, such as nucleases and ligases. The presence of these enzymes and an excess of foreign DNA would result in the linking together of several copies to form concatemers, which would be the substrates for integration. This might be stimulated by homology between individual copies of transforming plasmids, and "backbone" homology might also result in the concatemerization of plasmids carrying different transgenes in cotransformation experiments. However, cotransformation and cointegration were also shown to occur when two nonhomologous minimal cassettes were used for transformation, so homology might not be as important as the presence of free DNA ends (Fu et al. 2000). Kohli et al. (1998) suggested that transgene clusters arise in a second phase where a primary integration event occurring by illegitimate recombination at a chromosome break generates a "hot-spot" for further integration events in the same area. This might be due, for example, to the presence of local repair complexes that slide along the DNA and introduce nicks which can be exploited by more foreign DNA (Gelvin 1998). Pawlowski and Somers (1998) suggested an alternative second phase where a number of discrete transgene concatemers integrate simultaneously at a site containing multiple replication forks. Although there is no direct evidence for either mechanism, it is interesting to note that DNA integration is stimulated in rapidly dividing cells and is blocked in *Arabidopsis* mutants lacking essential components of the DNA recombination machinery.

The higher order organization of transgenic loci observed by metaphase FISH is thus far unique to particle bombardment and demands a model which takes into account the three-dimensional structure of the nucleus. In one scenario, it is possible that the transformation event affects a local region of the interphase nucleus. For example, it is possible that the metal particle causes damage to a particular area of chromatin, which is arranged in loops attached to the nuclear matrix. If the particle "skims" several loops, there will be regions of DNA damage close together in trans, but widely separated in the cis configuration were the DNA to be stretched out (Fig. 7.3). Each of these sites could act as a nucleation point where foreign DNA diffusing from the metal particle is used to patch up double-strand breaks, generating widely separated arrays and/or clusters (Abranches et al. 2000; Kohli et al. 2003). In support of this induced break and repair model, Svitashev et al. (2000) have shown that in six of 25 transgenic oat plants generated by particle bombardment, transgene integration sites were associated with rearranged chromosomes. This suggests that DNA breaks caused by incoming particles are repaired with foreign DNA and may also result in deletions, inversions, and translocations involving genomic DNA. Chromosomal rearrangements have also occasionally been seen associated with T-DNA integration (Nacry et al. 1998; Laufs et al. 1999).

The model above suggests that dispersed metaphase signals come together at interphase due to the physical position of the transgenic loci at the moment of transformation. In another scenario, the bringing together of transgene sites at interphase could represent recruitment, for example to a common transcription factory in the nucleus (Cook 1999). A further scenario involves transgenes that are brought together by virtue of their homology, perhaps as a consequence of their initial placing in the same region of the nucleus. This is an exciting prospect because the coincidence of FISH signals observed in wheat nuclei could represent a physical basis of the postulated DNA-DNA interactions that precede transcriptional transgene silencing in plants (see below).

Transgene rearrangements following particle bombardment have been widely reported in the literature, and many publications repeat the "lore" that direct DNA transfer is more likely than T-DNA transfer to generate complex rearranged loci. The number of rearrangements that can be detected depends entirely on the resolution of the method being used. Thus, careful analysis of locus structure by Southern blot hybridization, PCR, and DNA sequencing has recently shown that rearrangements may be more widespread than first envisaged in both transformation methods. The analysis of transgenic oat loci by Somers and colleagues has shown that transgene rearrangements can be extensive and extremely complex, with multiple small insertions, inversions, and deletions within any transgene, plus the presence of filler DNA (Svitashev et al. 2000). In maize, Mehlo et al. (2000) noted that every single plant among the population they analyzed showed some form of rearrangement, and they speculated that undetected "minor" rearrangements could be responsible for many instances of transgene silencing otherwise attributed to epigenetic effects (see Sect. 7.5). In particular, certain transgene rearrangements were not detectable by Southern blot hybridization because they were too subtle, but they could be picked up by PCR and sequencing. Since in most cases, Southern blot hybridization is used to determine whether a given locus is intact or rearranged, this suggests caution should be used in relying on such results, since only "major" rearrangements can be detected in this manner.

Few researchers have characterized transgene rearrangements in detail, but work by Kohli et al. (1999) has shown that rearrangements may involve palindromic sequences in the transforming plasmid, which tend for form three-dimensional structures such as hairpins and cruciforms. These investigators characterized 12 transgenic rice lines, transformed by particle bombardment, which had been shown to contain rearranged transgenes. Interestingly, they found that an imperfect palindrome in the CaMV 35S promoter was involved in one-third of all rearrangements, i.e., the sequence of this palindrome was adjacent to the rearrangement junction. Similar phenomena have been noted in T-DNA transformants containing the same promoter (Sect. 7.3.2). This sequence has the ability to adopt a cruciform secondary structure, which may stimulate recombination events. Many other promoters contain palindromic sequences of variable length within 100 bp of the transcription start site. The DNA secondary structures formed at these sites enable DNA-protein interactions for transcription under normal circumstances, but may also participate in aberrant recombination events. The fully sequenced papaya genome (Ming et al. 2008) also revealed a number of previously unidentified transgene rearrangements, i.e., a 1,533-bp fragment composed of a truncated, nonfunctional *tet*A gene and flanking vector backbone sequence, and a 290-bp nonfunctional fragment of the *npt*II gene, in addition to the intact, primary transgene conferring virus resistance.

7.5 Locus Structure and Transgene Stability

One of the most profound insights to come from the detailed analysis of transgene loci over the last decade is that many integrated transgenes contain minor rearrangements. As discussed above, these are difficult to pinpoint using low-resolution detection methods such as Southern blot hybridization and FISH, but high-resolution methods such as sequencing are rarely used as a routine analysis tool. Therefore, the impact of physical rearrangements on transgene expression is likely to be vastly underestimated, since unstable loci are often blamed on epigenetic phenomena with no further analysis to draw confirmatory evidence.

There are many factors that influence transgene stability, and these lead to highly variable expression within populations of plants generated in the same gene transfer experiment. One of the most important factors is the position effect, which reflects the influence of genomic DNA surrounding the site of transgene integration (Wilson et al. 1990). Another is the structure of the locus, including the number of transgene copies, their intactness, and their relative arrangement, which influences the likelihood of physical interactions and further recombination within the locus (physical instability) and the induction of silencing through DNA methylation and/or the production of aberrant RNA species from the locus (Heinrichs 2008).

7.5.1 Position Effects

Specific position effects result from the influence of local regulatory elements on the transgene. For example, an integrated transgene may come under the influence of a nearby enhancer, such that its expression profile is modified. The effect is transgene-specific because the enhancer interacts with regulatory elements in the transformation construct to control transcription; hence, the final expression pattern reflects the combined influence of both regulatory elements. Such effects are clearly revealed by entrapment constructs, which contain minimal control sequences linked to a visible marker gene and therefore "report" the activity of local regulatory elements (e.g., Goldsbrough and Bevan 1991).

As well as specific position effects governed by local regulatory elements, nonspecific position effects can also be generated by the surrounding chromatin architecture. Where the local environment is favorable for transgene expression, i.e., a positive position effect, it is generally taken for granted. However, nonspecific and repressive position effects reflect the integration of the transgene into a chromosomal region containing repressed chromatin (heterochromatin). The molecular features of heterochromatin, including its characteristic nucleosome structure, deacetylated histones, and hypermethylated DNA, spread into the transgene causing it to be inactivated (Pikaart et al. 1998). Analysis of the genomic context of silenced transgenes suggests that integration in the vicinity of certain repetitive DNA sequences, such as microsatellites and retrotransposon remnants, may predispose the transgene to silencing (Tanako et al. 1997). The chromosomal location is important, since in many plants, the genes are restricted to a small portion of the genome known as gene space, and the majority of the DNA is taken up by repetitive sequences. Thus, stable transgene expression has been associated with gene-rich telomeric and subtelomeric integration sites, whereas mosaic expression and silencing occurs at predominantly heterochromatic centromeric loci.

A third type of position effect reflects the tolerance of the surrounding DNA for "invasion" by foreign DNA. In this case, the effect is not automatic (as above) but is set off by the presence of the transgene. It appears that certain sequences can trigger de novo methylation, perhaps because the GC-content or sequence architecture is recognized as abnormal (reviewed by Kumpatla et al. 1998). Prokaryotic DNA may be recognized in this manner, since silencing is often associated with the presence of prokaryotic vector backbone DNA, particular binary vector sequences joining T-DNA to genomic DNA (Iglesias et al. 1997).

7.5.2   Locus Structure Effects

At least three aspects of locus structure influence transgene stability and expression: copy number, intactness, and arrangement. It is natural to assume that increasing the number of copies of a particular transgene will lead to an increase in the level of its product. However, even from the earliest plant transformation experiments, it was appreciated that multiple transgene copies could induce transgene silencing and that the phenomenon was associated with DNA methylation at the transgenic locus (e.g., Gelvin et al. 1983; Hepburn et al. 1983). A strikingly visual demonstration of this effect was provided by introducing the maize *A1* gene into mutant petunia plants with white flowers. Expression of the transgene resulted in pelargonidin production, generating a red pigment. However, it was shown that red flowers generally appeared on plants with single copy transgenes, while plants with multiple transgene copies had white or variegated flowers. Where transgene silencing had occurred, increased methylation of the transgene DNA was observed (Meyer et al. 1992). Similarly, it was thought that the amount of pigment in wild-type petunia flowers could be increased by introducing extra copies of the chalcone synthase (*chs*) gene (Napoli et al. 1990). Chalcone synthase converts coumaryol-CoA and 3-malonyl-CoA into chalcone, a precursor of anthocyanin pigments. The presence of multiple transgene copies was expected to increase the level of enzyme and hence cause stronger flower pigmentation. However, in about 50% of the plants recovered from the experiment, exactly the opposite effect was observed. The flowers were either pure white, or variegated with purple and white sectors. It appeared that integration of multiple copies of the transgene led not only to the suppression of transgene expression, but also to the cosuppression of the homologous endogenous gene.

Rooke et al. (2003) looked at the integration, inheritance, and expression of transgenes in six transgenic wheat lines generated by particle bombardment with two plasmids containing genes encoding a glutelin subunit and a selectable marker, respectively. Transgene insertion number ranged from 1 to 15, with most lines carrying multiple copies consistent with previous reports (Becker et al. 1994;

Blechl and Anderson 1996; Srivastava et al. 1996; Stoger et al. 1998; Cannell et al. 1999). Four of the transgenic loci were clusters interspersed with genomic DNA, in some cases enough to allow independent segregation which contrasts with previous reports in which cointegration and cosegregation were the norm, and independent segregation rare (Stoger et al. 1998). There was no evidence for a direct correlation between transgene copy number and expression level, and no evidence for cosuppression of endogenous glutelin genes even in multicopy lines. The presence of multiple transgene copies has been implicated in transgene silencing, but other studies in cereals have shown that multiple copies do not necessarily lead to silencing and can even enhance expression levels in proportion to copy number (Stoger et al. 1998; Gahakwa et al. 2000). In contrast, Spencer et al. (1992) failed to recover progeny expressing the marker transgenes from maize lines containing more than five or six copies of the integrated plasmid, while Cannell et al. (1999) observed silencing or a gradual reduction in marker gene expression over three generations of transgenic wheat lines. It has been suggested that the production of lines with single transgenes or low copy numbers is desirable as such lines may be more stable and less likely to exhibit transgene silencing (Finnegan and McElroy 1994), but with several studies reporting contrary results, this may indicate that chance plays a role in the impact of copy number, perhaps reflecting insertion events near to boundary elements (Sect. 7.5.3).

Variation in transgene expression levels can also result from uncontrolled differences in experimental protocols reflecting gene-environment interactions, which means that proper comparisons between transgenic lines should take place in a standardized environment. To study sources of spurious variation, transgene expression levels were quantified over five homozygous generations in two independent transgenic rice lines created by particle bombardment (James et al. 2004a, b). Both lines contained the same *gus*A expression unit which was stably inherited, and all plants were cultured and sampled using previously developed standardized protocols. Plants representative of each generation (T_2–T_6) were grown either all together or across several different growth periods. Where the plants were grown and characterized independently, the amount of extraneous variation in transgene expression levels was up to three-fold higher than in plants grown and analyzed together. This study therefore provided important evidence that the growth and analysis of all plants from all generations together, using standard operating procedures (SOP), can reduce extraneous variation associated with transgene expression and is the key to improving the reproducibility of transgenic studies conducted over multiple generations (James et al. 2004a, b).

7.5.3 Overcoming Position and Locus Structure Effects by Buffering the Transgene

As discussed earlier, analysis of the genomic context of transgene integration sites has shown that silenced transgenes are often surrounded by repetitive elements, which are sequestered into repressed chromatin. The same studies have also shown that stably expressed transgenes are often associated with matrix attachment regions (MARs) (Iglesias et al. 1997). MARS are AT-rich elements that attach chromatin to the nuclear matrix and organize it into topologically isolated loops (Holmes-Davis and Comai 1998). A number of highly expressed endogenous plant genes have also been shown to be flanked by matrix attachment regions (e.g., Chinn and Comai 1996). One strategy that has been proposed to overcome position effects is therefore to protect or buffer the transgene by flanking it with MARs prior to transformation. In this way, it is hoped that the transgene will form a discrete chromatin loop which will be isolated from surrounding chromatin.

Several experiments have been carried out in which a reporter gene such as *gus*A has been flanked by MARs. Such constructs have been introduced into transgenic plants and compared to populations containing the same reporter gene without MARs (e.g., Mlynarova et al. 1994, 1995, 1996; Van Leeuwen et al. 2001; Mlynárová et al. 2002). Generally MARs do have a positive effect on transgene expression and can significantly reduce position effects, but they cannot rescue all lines and restore full expression. It is acceptable to say that they generally reduce expression variability within a population (e.g., Breyne et al. 1992). Expression may increase as much as five-fold, but some remarkable exceptions include a 25-fold enhancement using a yeast MAR and a 140-fold enhancement using a tobacco MAR in tobacco callus (Allen et al. 1993, 1996).

7.5.4 Overcoming Position and Locus Structure Effects by Homologous Recombination

As discussed in Sects 7.3.3 and 7.4.3, transgene integration in higher plants occurs almost universally by illegitimate recombination, which may involve microhomology but is not dependent upon it. Since there is only minimal sequence relationship between the transgene and the genomic region into which it integrates, the experimenter has little control over the integration site. In other systems, notably yeast, homologous recombination is favored over illegitimate recombination if the vector carries a homology region that matches the yeast genome, allowing endogenous genes to be altered by gene targeting (Schiestl and Petes 1991). In the context of controlling transgene integration, it also allows transgenes to be inserted at specific loci, a strategy that should allow favorable sites for transgene integration to be chosen, theoretically abolishing position effects and reducing the complexity of locus structure.

Although widely used in microbial systems, homologous recombination occurs with a very low efficiency in plants (illegitimate recombination occurs about 10^5 times more frequently than homologous recombination, making genuine targeting events difficult to isolate). Only one plant species has been shown to undergo efficient nuclear homologous recombination (the moss *Physcomitrella patens*) and the results in higher plants have been much less impressive, with targeting efficiencies as low as 10^{-6} (Lee et al. 1990; Offringa et al. 1990; Miao and Lam 1995; Risseeuw et al. 1995, 1997; Kempin et al. 1997; Reiss et al. 2000; Hanin et al. 2001).

A transgene has also been repaired by homologous recombination in tobacco (Paszkowski et al. 1988). More recently, promising results have been achieved using a T-DNA-mediated gene-targeting strategy involving a long homology region in combination with a strong counterselectable marker in rice (Terada et al. 2002). Targeting frequencies of up to 1% have been achieved using this system (reviewed by Ida and Terada 2004 and Cotsaftis and Guiderdoni 2005). Gene targeting has also been reported recently in maize (D'Halluin et al. 2008).

There has also been interest in the use of zinc-finger endonucleases to make targeted double-strand breaks in the plant genome, so that homologous recombination is favored at such sites (Kumar et al. 2005). The modular nature of zinc-finger transcription factors means that recombinant DNA technology can be used to "mix and match" these DNA-binding domains to create recombinant proteins with unique sequence specificities. Zinc fingers are motifs approximately 30 amino acids in length which coordinate a Zn^{2+} ion and bind to DNA sequences three base pairs long. Combining different zinc fingers in series allows proteins to be tailor made to bind longer DNA sequences. When a nonspecific DNA endonuclease is incorporated into such a protein, it becomes a targeted DNA cutting tool (Lloyd et al. 2005; Wright et al. 2005; Zeevi et al. 2008; Cai et al. 2009).

7.5.5 Overcoming Position and Locus Structure Effects by Organelle Transformation

The plant cell contains not only a nuclear genome, but also organellar genomes in the chloroplasts and mitochondria. The chloroplast is a useful target for gene transfer because tens of thousands of chloroplasts may be present in a single plant cell, and each chloroplast may contain multiple copies of its chromosome. Genetic engineering of the plastid genome offers several advantages over nuclear transformation including that integration occurs by homologous recombination, the high copy number of transgenes in a homoplasmic cell, and the absence of gene silencing phenomena due to the lack of position and locus structure effects, and the absence of DNA methylation in the plastid genome (Daniell et al. 2005; Daniell 2006; Bock 2007). The recombination machinery is very active in chloroplasts and can induce rearrangements, as observed in some of the first tobacco transformants generated with the *aad*A selectable marker (Svab and Maliga 1993). The stability of a plastid transgene has been evaluated in soybean transformants over six generations. These transformants had integrated the *aad*A selection cassette in the intergenic region between the *rps*12/7 and *trn*V genes. Three independent homoplasmic T_0 transformation events were selected and ten plants from each event propagated to generation T_5 in the absence of selection pressure. Neither transgene rearrangement nor wild-type plastids were detected in generation T_5 by Southern blot analysis. All tested progenies were uniformly resistant to spectinomycin. Therefore, soybean transformants of generations T_0 and T_5 appear to be genetically and phenotypically identical (Dufourmantel et al. 2006).

7.5.6 Overcoming Position and Locus Structure Effects by Site-specific Recombination

Site-specific recombination is a form of recombination that occurs at short, specific recognition sites rather than DNA sequences with long regions of homology but no particular sequence specificity, as is the case for homologous recombination. Site-specific recombination is not ubiquitous – indeed different organisms encode their own very specific systems that include the cis-acting recombinogenic sites and the enzymes that recognize them and carry out the recombination event. Therefore, the target sites for site-specific recombination can be introduced easily and unobtrusively into transgenes, but recombination will only occur in a heterologous cell if a source of the specific recombinase enzyme is also supplied. As with the homologous recombination strategy discussed above, position and locus structure effects can be eliminated by introducing foreign DNA at a specific, favorable locus. A number of different site-specific recombination systems have been identified and several have been studied in detail (Sadowski 1993). The most extensively used are Cre recombinase from bacteriophage P1 (Lewandoski and Martin 1997) and FLP recombinase from the yeast *Saccharomyces cerevisiae* (Buchholz et al. 1998). These have been shown to function in many heterologous eukaryotic systems including transgenic plants (Metzger and Feil 1999). Both recombinases recognize 34-bp sites (*lox*P and *FRP*, respectively) comprising a pair of 13-bp inverted repeats surrounding an 8-bp central sequence. *FRP* possesses an additional copy of the 13-bp repeat sequence, although this is nonessential for recombination.

The Cre-*lox*P system has been used most widely in plants, often for controlled transgene excision (particularly selectable marker genes after transformation) but more recently for controlled transgene insertion (Gilbertson 2003; Lyznik et al. 2003; Puchta 2003; Marjanac et al. 2008). Marker genes are usually excised in the T_1 generation once transgene expression is verified, allowing the separately introduced *cre* gene to segregate in T_2 plants. This method has been used in many crops including wheat (Srivastava and Ow 2001, 2003), maize (Kerbach et al. 2005; Djukanovic et al. 2006; Hu et al. 2006; Vega et al. 2007, 2008), rice (Chen et al. 2004; Srivastava et al. 2004; Chawla et al. 2006; Moore and Srivastava 2006; Vega et al. 2008), potato (Kopertekh et al. 2004a,b), and tomato (Gidoni et al. 2003; Coppoolse et al. 2005). Controlled integration has been studied in transgenic plants already engineered to contain recipient *lox*P sites (Srivastava et al. 2004). In this study, three different recipient wheat lines were generated by bombarding plants with the *lox*P sequence, and these were subsequently bombarded with a *gus*A construct also containing flanking *lox*P sequences and a *cre* gene. Following transformation, about 80% of lines contained *gus*A at the recipient site, many with single-copy transgenes and others with concatemers. Both types of locus were stably inherited. There was much less variation in expression among the single copy lines (Srivastava et al. 2004).

Chawla et al. (2006) generated 18 different transgenic rice lines containing a precise single copy of *gus*A at a designated site. In seven of these lines, additional copies of the transgene integrated at random sites by illegitimate recombination, while 11 showed "clean" integration by site-specific recombination only. The single-copy lines were stable over at least four generations and showed consistent levels of expression, which doubled in homozygous plants. In contrast, the multi-copy lines showed variable expression and some fell victim to transgene silencing. Interestingly, where the site-specific and illegitimate integration loci segregated in later generations, transgene expression was reactivated in the plants carrying the site-specific integration site alone, whereas close linkage between the site-specific and random integration prevented segregation in other lines and the silencing persisted.

An exciting recent development is the GENE DELETOR system, which is a hybrid of the Cre-*lox*P and FLP-*FRT* systems. The GENE DELETOR is based on a fusion recognition site (*lox*P-*FRT*), which is inefficient when both recombinases are expressed but highly efficient when either one of the recombinases is expressed alone, giving up to 100% efficiency in populations of up to 25,000 T_1 transgenic plants (Luo et al. 2007).

Another use for Cre-*lox*P is the simplification of locus structure by resolving multicopy loci to a single transgene copy (Srivastava et al. 1999). A strategy was developed in which the transformation vector contained a transgene flanked by *lox*P sites in an inverted orientation. Regardless of the number of copies integrated between the outermost transgenes, recombination between the outermost sites resolved the integrated molecules into a single copy. The principle was proven by resolving four multicopy loci successfully into single-copy transgenes.

7.5.7 Overcoming Position and Locus Structure Effects Using Minichromosomes

In bacteria, plasmid vectors are maintained as episomal replicons to make cloning and isolating recombinant DNA a simple procedure. When it comes to expressing heterologous genes in eukaryotic cells, episomal vectors are widely used to avoid position effects, hence the development of yeast episomal vectors, yeast artificial chromosomes, mammalian plasmid vectors carrying virus origins of replication (e.g., SV40-based vectors, herpesvirus-based vectors), and plant expression vectors based on plant viruses (all of which replicate episomally). The yeast artificial chromosome system is the most relevant in this context because it allows genes of any size to be introduced into the yeast genome as an independent replicating unit that is treated by the cell as an additional chromosome. YACs comprise a yeast centromere and telomeres, the origin of replication (autonomous replicating sequence) and selectable markers. More recently, analogous systems have been developed to maintain genes as episomal minichromosomes in plants. These have many advantages for plant genetic engineering including the ability to express large transgenes or groups of transgenes, and the ability to rapidly introduce new linkage groups into diverse germplasm.

Carlson et al. (2007) created plant minichromosomes by combining the *DsRed* and *npt*II marker genes with 7–190 kb of maize genomic DNA fragments containing satellites, retroelements, and other repeat sequences commonly found in centromeres. The circular constructs were introduced into embryogenic maize tissue by particle bombardment and transformed cells were regenerated and propagated for several generations without selection. The minichromosomes were maintained as extrachromosomal replicons through mitosis and meiosis, and showed roughly Mendelian segregation ratios (93% transmission as a disome with 100% expected, 39% transmission as a monosome crossed to wild type with 50% expected, and 59% transmission in self-crosses with 75% expected). The *DsRed* reporter gene was expressed over four generations, and Southern blot analysis indicated the genes were intact.

7.6 Epigenetic Silencing Phenomena Resulting From Complex Locus Structures and High-Level Expression

As stated earlier, the earliest plant transformation experiments showed that multiple transgene copies could induce transgene silencing, in some cases associated with the cosuppression of homologous endogenes. Transgene silencing can occur through two overlapping pathways, one acting at the transcriptional level (characterized by the reduction or abolition of transcription from one or more copies of the transgene) and one acting post-transcriptionally (transcription from the silenced locus is required to initiate silencing) (Hammond et al. 2001). Transcriptional silencing is often correlated with increased methylation in the promoter regions of affected loci, and both the methylation and the silencing tend to be heritable through meiosis. Post-transcriptional silencing requires homology in the transcribed regions, which may become methylated, and the silencing effect can be reset at meiosis. Post-transcriptional silencing is also known as RNA silencing.

Transcriptional silencing occurs when transgene repeats somehow act as a trigger for de novo DNA methylation. It has been shown that inverted repeats can form secondary structures that are favored substrates for methylation, and thus it is likely that cis DNA-DNA pairing may be involved in such processes. However, transgene silencing can also occur in trans, i.e., silencing interactions may occur between unlinked loci. This has been shown, for example, in sequential transformation events with homologous transgenes, or where two plant lines carrying homologous transgenes have been crossed (Matzke and Matzke 1990, 1991). In this situation, it is likely that physical interactions between transgenes may occur to mediate silencing, and that DNA methylation may somehow be transferred from one site to another. As discussed in Sect. 7.4.3, FISH studies in transgenic wheat provide tantalizing evidence for such interactions in the interphase nucleus (Abranches et al. 2000). Since the CaMV 35S promoter is frequently used for transgene expression and can form cruciform structures that induce transgene rearrangements (Kohli et al. 1999), it may also play a role in transcriptional silencing under certain circumstances. Supporting evidence for this has been provided in studies of activation tag lines in which the CaMV 35S enhancer is used as a random insertional mutagen to hyperactivate adjacent genes and generate gain-of-function phenotypes. It has been noted that such screens using T-DNA cassettes containing the enhancer elements from the CaMV 35S promoter return a low frequency of morphological mutants (Chalfun et al. 2003). Detailed analysis revealed a correlation between the number of T-DNA insertion sites, the methylation status of the enhancer sequence and enhancer activity. All plants containing more than a single T-DNA insert were methylated on the enhancer and its activity was reduced, with the amount of methylation and the reduction of enhancer activity correlating with the number of T-DNA copies, particularly those with right border inverted repeats (Chalfun et al. 2003). Even so, methylation was still detected at a lower frequency in plants without right border inverted repeats suggesting other triggers were active in these lines.

A recurring theme in post-transcriptional silencing is the presence of double-stranded RNA. Double-stranded RNA introduced into the plant cell can trigger the catalytic degradation of homologous RNA molecules and the methylation of homologous DNA sequences in the genome (e.g., Tavernarakis et al. 2000). When carried out deliberately through the expression of hairpin RNA constructs, this process (RNA interference, RNAi) is a potent method for silencing individual genes, generating phenocopies of mutant phenotypes (e.g., see McGinnis et al. 2005; Gordon and Waterhouse 2007). It has been suggested that complex multicopy transgenic loci could also generate hairpin dsRNA, e.g., if two transgenes are present as inverted repeats, or if truncation and/or rearrangements (some perhaps undetectable by standard screening methods) generated small, aberrant dsRNA species (Jorgensen et al. 1996; Que et al. 1997; Muskens et al. 2000). Experiments designed to test this hypothesis specifically have shown that inverted repeat T-DNA configurations and arrangements of tandem repeated transgenes may not be sufficient in all cases to trigger transgene silencing (Lechtenberg et al. 2003), whereas many reports show post-transcriptional silencing in plants with intact transgenes. In such cases it has been suggested that the level of transgene expression may be an important trigger, with "runaway expression" resulting in the most potent silencing effects (Lindbo et al. 1993; Vaucheret et al. 1998; Schubert et al. 2004). Experiments comparing the frequency and potency of cosuppression by sense chalcone synthase transgenes driven by different promoters have shown that a strong promoter is required for high-frequency cosuppression of chalcone synthase genes and for the production of the full range of cosuppression phenotypes (Que et al. 1997). Indeed the correlation between transgene copy number and silencing in some systems may reflect the higher expression level in multicopy loci triggering silencing (Schubert et al. 2004) suggesting that transgenic lines escaping this effect may fall below the threshold for triggering silencing (e.g., Stoger et al. 1998).

The expression threshold model accounts for RNA silencing in intact transgenic loci but it is also possible that such loci are prone to silencing because their high expression promotes the formation of more aberrant RNA products than a poorly expressed transgene. If true, the trigger would still be aberrant dsRNA, the same as produced by complex, rearranged loci, and it should be possible to mitigate the effects and generate plants with extremely high expression levels. Several studies have shown that RNA silencing in transgenic plants is accompanied by the accumulation of incorrectly processed mRNAs (often lacking polyadenylate tails) (e.g., van Eldik et al. 1998; Metzlaff et al. 2000; Wang and Waterhouse 2000; Han and Grierson 2002) and in at least one case it has been shown specifically that tandem repeats can generate small interfering RNAs (Ma and Mitra 2002). Since dsRNA is unlikely to be generated directly from tandem repeats (as opposed to the situation with inverted repeats), the process must involve an RNA-dependent RNA polymerase. In agreement with this, Luo and Chen (2007) found that RNA silencing in transgenic *Arabidopsis* could be induced by three direct repeats of the *gus*A open reading frame, and this was dependent on the RNA-dependent RNA polymerase encoded by *RDR6*. Normal plants transformed with either three tandem copies of *gus*A or a single copy lacking a polyadenylation site were able to silence a normal *gus*A transgene cassette in trans, but there was no silencing in *rdr6* mutants, which also accumulated long RNA molecules corresponding to *gus*A read-through transcripts of various lengths. Therefore, it appears that the read-through of termination sites leading to the production of long RNA products triggers RNA silencing in an RDR6-dependent manner. A further transgenic line containing a *gus*A transgene with two polyadenylation sites produced fewer read-through transcripts, less siRNA, and therefore showed higher levels of GUS activity. Transgene silencing in tandem repeat transgenes may therefore be triggered by a defense mechanism that evolved to reduce errors caused by read-through transcription (Luo and Chen 2007).

7.7 Conclusions

Transgene integration following *Agrobacterium*-mediated transformation and direct DNA transfer occur by very similar mechanisms, involving illegitimate recombination between genomic DNA and invading transgene DNA strands, and the repair of double-stranded breaks in the host genome. There is often microhomology between the recombining partners, although direct blunt end ligation also occurs. Both transformation methods induce a wound response, resulting in the activation of nucleases, ligases, and recombinases in the host cell. The foreign DNA is simultaneously degraded and concatemerized resulting in transgene arrays containing intact and/or truncated and rearranged copies. Several integration events may occur simultaneously at a cluster of replication forks, or a primary integration event may stimulate further integrations in the local area. Regardless of the mechanism, the result is a transgene cluster interspersed with genomic DNA. In the case of particle bombardment, clusters and arrays may be widely dispersed, generating very large transgenic loci. The position of transgene integration is essentially random within the "gene space" of the plant species. The transgene is thus subject to position effects which may influence its expression, resulting in some cases in transcriptional silencing as the new DNA is sequestered into the surrounding chromatin. The structure of a transgenic locus may also induce silencing via a number of mechanisms. These include de novo DNA methylation in response to DNA-DNA interactions, the expression of aberrant RNA species (particularly small hairpin RNAs) from truncated and rearranged transgenes or partial transgenes, and the expression of aberrant RNA products from inefficiently terminated transcripts. Position effects can be reduced by buffering the transgene with matrix attachment regions or controlling the site of integration through homologous recombination or site-specific recombination. Alternatively, it may be possible to introduce the transgene into the plastid genome, which does not suffer from position effects. More recent developments such as minichromosomes may provide a method to introduce entire linkage groups and maintain them stably and episomally. Site-specific recombination can also be used to simplify locus structure, by reducing the number of repeats, which may help to reduce the likelihood of RNA silencing. Even so, many reports show that high-level transgene expression is possible in plants with multiple transgene copies, suggesting that the overall level of expression may be relevant, i.e., there may be a trigger level at which silencing is induced. This may involve the detection of high levels of transgene mRNA or may simply reflect the greater likelihood of aberrant RNA products being generated as collateral damage. The recent publication of the full draft sequence of the transgenic SUNUP papaya genome shows that the detailed characterization of the transgene sequence and its flanking regions is not an insurmountable obstacle. Perhaps such intensive analysis will, in the future, allow the accurate prediction of transgene behavior and stability in transgenic plants.

References

Abranches R, Santos AP, Wegel E, Williams S, Castilho A, Christou P, Shaw P, Stoger E (2000) Widely separated multiple transgene integration sites in wheat chromosomes are brought together at interphase. Plant J 24:713–723

Afolabi AS, Worland B, Snape JW, Vain P (2004) A large-scale study of rice plants transformed with different T-DNAs provides new insights into locus composition and T-DNA linkage configurations. Theor Appl Genet 109:815–826

Allen GC, Hall G Jr, Michalowski S, Newman W, Spiker S, Weissinger AK, Thompson WF (1996) High-level transgene expression in plant cells: effects of a strong scaffold attachment region from tobacco. Plant Cell 8:899–913

Allen GC, Hall GE Jr, Childs LC, Weissinger AK, Spiker S, Thompson WF (1993) Scaffold attachment regions increase reporter gene expression in stably transformed plant cells. Plant Cell 5:603–613

Altpeter F, Baisakh N, Beachy R, Bock R, Capell T, Christou P, Daniell H, Datta K, Datta S, Dix PJ, Fauquet C, Huang N, Kohli A, Mooibroek H, Nicholson L, Nguyen TT, Nugent G, Raemakers K, Romano A, Somers DA, Stoger E, Taylor N, Visser R (2005) Particle bombardment and the genetic enhancement of crops: myths and realities. Mol Breed 15:305–327

Barakat A, Gallois P, Raynal M, Mestre Ortega D, Sallaud C, Guiderdoni E, Delseny M, Bernardi G (2000) The distribution of T-DNA in the genomes of transgenic Arabidopsis and rice. FEBS Lett 471:161–164

Becker D, Brettschneider R, Lörz H (1994) Fertile transgenic wheat from microprojectile bombardment of scutellar tissue. Plant J 5:299–307

Blechl AE, Anderson OD (1996) Expression of a novel high-molecular-weight glutenin subunit gene in transgenic wheat. Nat Biotechnol 14:875–879

Bock R (2007) Plastid biotechnology: prospects for herbicide and insect resistance, metabolic engineering and molecular farming'. Curr Opin Biotechnol 18:100–106

Breyne P, van Montagu M, Depicker N, Gheysen G (1992) Characterization of a plant scaffold attachment region in a DNA fragment that normalizes transgene expression in tobacco. Plant Cell 4:463–471

Buchholz F, Angrand PO, Stewart AF (1998) Improved properties of FLP recombinase evolved by cycling mutagenesis. Nat Biotechnol 16:657–662

Cai CQ, Doyon Y, Ainley WM, Miller JC, Dekelver RC, Moehle EA, Rock JM, Lee YL, Garrison R, Schulenberg L, Blue R, Worden A, Baker L, Faraji F, Zhang L, Holmes MC, Rebar EJ, Collingwood TN, Rubin-Wilson B, Gregory PD, Urnov FD, Petolino JF (2009) Targeted transgene integration in plant cells using designed zinc finger nucleases. Plant Mol Biol 69(6):699–709

Cannell ME, Doherty A, Lazzeri PA, Barcelo P (1999) A population of wheat and tritordeum transformants showing a high degree of marker gene stability and heritability. Theor Appl Genet 99:772–784

Carlson SR, Rudgers GW, Zieler H, Mach JM, Luo S, Grunden E, Krol C, Copenhaver GP, Preuss D (2007) Meiotic transmission of an in-vitro-assembled autonomous maize minichromosomes. PLOS Genet 3:1965–1974

Chalfun A, Mes JJ, Mlynarova I, Aarts MGM, Angenent GC (2003) Low frequency of T-DNA based activation tagging in Arabidopsis is correlated with methylation of CaMV 35S enhancer sequences. FEBS Lett 555:459–463

Chawla R, Ariza-Nieto M, Wilson AJ, Moore SK, Srivastava V (2006) Transgene expression produced by biolistic-mediated, site-specific gene integration is consistently inherited by the subsequent generations. Plant Biotechnol J 4:209–218

Chen L, Marmey P, Taylor NJ, Brizard JP, Espinoza C, D'Cruz P, Huet H, Zhang S, de Kochko A, Beachy RN, Fauquet CM (1998) Expression and inheritance of multiple transgenes in rice plants. Nat Biotechnol 16:1060–1064

Chen SB, Liu X, Peng HY, Gong WK, Wang R, Wang F, Zhu Z (2004) Cre/lox-mediated marker gene excision in elite indica rice plants transformed with genes conferring resistance to lepidopteran insects. Acta Bot Sin 46:1416–1423

Cheng M, Fry JE, Pang S, Zhou H, Hironaka CM, Duncan DR, Conner TW, Wan Y (1997) Genetic transformation of wheat mediated by Agrobacterium tumefaciens. Plant Physiol 115:971–980

Chinn AM, Comai L (1996) The heat shock cognate 80 gene of tomato is flanked by matrix attachment regions. Plant Mol Biol 32:959–968

Cook PR (1999) The organization of replication and transcription. Science 284:1790–1795

Cooley J, Ford T, Christou P (1995) Molecular and genetic characterization of elite transgenic plants produced by electric-discharge particle acceleration. Theor Appl Genet 90:97–104

Coppoolse ER, de Vroomen MJ, van Gennip F, Hersmus BJM, van Haaren MJJ (2005) Size does matter: Cre-mediated somatic deletion efficiency depends on the distance between the target lox-sites. Plant Mol Biol 58:687–698

Cotsaftis O, Guiderdoni E (2005) Enhancing gene targeting efficiency in higher plants: rice is on the move. Transgenic Res 14:1–14

Dai S, Zheng P, Marmey P, Zhang S, Tian W, Chen S, Beachy RN, Fauquet C (2001) Comparative analysis of transgenic rice plants obtained by Agrobacterium-mediated transformation and particle bombardment. Mol Breed 7:25–33

Daniell H (2006) Production of biopharmaceuticals and vaccines in plants via the chloroplast genome. Biotechnol J 1:1071–1079

Daniell H, Kumar S, Dufourmantel N (2005) Breakthrough in chloroplast genetic engineering of agronomically important crops. Trends Biotechnol 23:238–245

De Block M, Debrouwer D (1991) Two T-DNAs co-transformed into Brassica napus by a double Agrobacterium tumefaciens infection are mainly integrated at the same locus. Theor Appl Genet 82:257–263

De Buck S, Jacobs A, Van Montagu M, Depicker A (1999) The DNA sequences of T-DNA junctions suggest that complex T-DNA loci are formed by a recombination process resembling T-DNA integration. Plant J 20:295–304

D'Halluin K, Vanderstraeten C, Stals E, Cornelissen M, Ruiter R (2008) Homologous recombination: a basis for targeted genome optimization in crop species such as maize. Plant Biotechnol J 6:93–102

Djukanovic V, Orczyk W, Gao HR, Sun XF, Garrett N, Zhen SF, Gordon-Kamm W, Barton J, Lyznik LA (2006) Gene conversion in transgenic maize plants expressing FLP/FRT and Cre/loxP site-specific recombination systems. Plant Biotechnol J 4:345–357

Dufourmantel N, Tissot G, Garcon F, Pelissier B, Dubald M (2006) Stability of soybean recombinant plastome over six generations. Transgenic Res 15:305–311

Finnegan J, McElroy D (1994) Transgene inactivation: plants fight back! Biotechnology 12:883–888

Fu X, Duc LT, Fontana S, Bong BB, Tinjuangjun P, Sudhakar D, Twyman RM, Christou P, Kohli A (2000) Linear transgene constructs lacking vector backbone sequences generate low-copy-number transgenic plants with simple integration patterns. Transgenic Res 9:11–19

Gahakwa D, Maqbool SB, Fu X, Sudhakar D, Christou P, Kohli A (2000) Transgenic rice as a system to study the stability of transgene expression: multiple heterologous transgenes show similar behaviour in diverse genetic backgrounds. Theor Appl Genet 101:388–399

Gelvin SB (1998) Multigene plant transformation: more is better! Nat Biotechnol 16:1009–1010

Gelvin SB (2003) Agrobacterium-mediated plant transformation: the biology behind the gene jockeying tool. Microbiol Mol Biol Rev 67:16–37

Gelvin SB, Karcher SJ, DiRita VJ (1983) Methylation of the T-DNA in *Agrobacterium tumefaciens* and in several crown gall tumors. Nucleic Acids Res 11:159–174

Gidoni D, Fuss E, Burbidge A, Speckmann GJ, James S, Nijkamp D, Mett A, Feiler J, Smoker M, de Vroomen MJ, Leader D, Liharska T, Groenendijk J, Coppoolse E, Smit JJ, Levin I, de Both M, Schuch W, Jones JD, Taylor IB, Theres K, van Haaren MJ (2003) Multi-functional T-DNA/Ds tomato lines designed for gene cloning and molecular and physical dissection of the tomato genome. Plant Mol Biol 51:83–98

Gilbertson L (2003) Cre-lox recombination: Cre-ative tools for plant biotechnology. Trends Biotechnol 21:550–555

Goldsbrough A, Bevan M (1991) New patterns of gene activity in plants detected using an Agrobacterium vector. Plant Mol Biol 16:263–269

Gorbunova V, Levy AA (1997) Non-homologous DNA end joining in plant cells is associated with deletions and filler DNA insertions. Nucleic Acids Res 25:4650–4657

Gordon KH, Waterhouse PM (2007) RNAi for insect-proof plants. Nat Biotechnol 25:1231–1232

Hadi MZ, McMullen MD, Finer JJ (1996) Transformation of 12 different plasmids into soybean via particle bombardment. Plant Cell Rep 15:500–505

Hammond SM, Caudy AA, Hannon GJ (2001) Post-transcriptional gene silencing by double-stranded RNA. Nat Rev Genet 2:110–119

Han Y, Grierson D (2002) Relationship between small antisense RNAs and aberrant RNAs associated with sense transgene mediated gene silencing in tomato. Plant J 29:509–519

Hanin M, Volrath S, Bogucki A, Briker M, Ward E, Paszkowski J (2001) Gene targeting in Arabidopsis. Plant J 28:671–677

Heinrichs A (2008) Small RNAs: united in silence. Nat Rev Mol Cell Biol 9:496

Hellens R, Mullineaux P, Klee H (2000) Technical focus: a guide to Agrobacterium binary Ti vectors. Trends Plant Sci 5:446–451

Hepburn AG, Clarke LE, Pearson L, White J (1983) The role of cytosine methylation in the control of nopaline synthase gene expression in a plant tumor. J Mol Appl Genet 2:315–329

Holmes-Davis R, Comai L (1998) Nuclear matrix attachment regions and plant gene expression. Trends Plant Sci 3:91–97

Hsing YI, Chern CG, Fan MJ, Lu PC, Chen KT, Lo SF, Sun PK, Ho SL, Lee KW, Wang YC, Huang WL, Ko SS, Chen S, Chen JL, Chung CI, Lin YC, Hour AL, Wang YW, Chang YC, Tsai MW, Lin YS, Chen YC, Yen HM, Li CP, Wey CK, Tseng CS, Lai MH, Huang SC,

Chen LJ, Yu SM (2007) A rice gene activation/knockout mutant resource for high throughput functional genomics. Plant Mol Biol 63:351–364

Hu Q, Nelson K, Luo H (2006) FLP-mediated site-specific recombination for genome modification in turfgrass. Biotechnol Lett 28:1793–1804

Hu T, Metz S, Chay C, Zhou HP, Biest N, Chen G, Cheng M, Feng X, Radionenko M, Lu F, Fry J (2003) Agrobacterium-mediated large-scale transformation of wheat (*Triticum aestivum* L.) using glyphosate selection. Plant Cell Rep 21:1010–1019

Ida S, Terada R (2004) A tale of two integrations, transgene and T-DNA: gene targeting by homologous recombination in rice. Curr Opin Biotechnol 15:132–138

Iglesias VA, Moscone EA, Papp I, Neuhuber F, Michalowski S, Phelan T, Spiker S, Matzke M, Matzke AJM (1997) Molecular and cytogenetic analyses of stably and unstably expressed transgene loci in tobacco. Plant Cell 9:1251–1264

Jackson SA, Zhang P, Chen WP, Phillips RL, Friebe B, Muthukrishnan S, Gill BS (2001) High resolution structural analysis of biolistic transgene integration into the genome of wheat. Theor Appl Genet 103:56–62

James C (2007) Global status of commercialized biotech/GM crops: 2007. ISAAA Brief No 37. ISAAA, Ithaca, NY

James VA, Worland B, Snape JW, Vain P (2004a) Development of a standard operating procedure (SOP) for the precise quantification of transgene expression levels in rice plants. Physiol Plant 120:650–656

James VA, Worland B, Snape JW, Vain P (2004b) Strategies for precise quantification of transgene expression levels over several generations in rice. J Exp Bot 55:1307–1313

Jen GC, Chilton MD (1986) The right border region of pTiT37 T-DNA is intrinsically more active than the left border region in promoting T-DNA transformation. Proc Natl Acad Sci USA 83:3895–3899

Jeon JS, Lee S, Jung KH, Jun SH, Jeong DH, Lee J, Kim C, Jang S, Yang K, Nam J, An K, Han MJ, Sung RJ, Choi HS, Yu JH, Choi JH, Cho SY, Cha SS, Kim SI, An G (2000) T-DNA insertional mutagenesis for functional genomics in rice. Plant J 22:561–570

Jones JDG, Gilbert DE, Grady KL, Jorgensen RA (1987) T-DNA structure and gene expression in petunia plants transformed by *Agrobacterium tumefaciens* C58 derivates. Mol Gen Genet 207:478–485

Jorgensen RA, Cluster PD, English J, Que Q, Napoli CA (1996) Chalcone synthase cosuppression phenotypes in petunia flowers: Comparison of sense vs. antisense constructs and single-copy vs. complex T-DNA sequences. Plant Mol Biol 31:957–973

Jorgensen RA, Snyder C, Jones JDG (1987) T-DNA is organized predominantly in inverted repeat structures in plants transformed with *Agrobacterium tumefaciens* C58 derivatives. Mol Gen Genet 207:471–477

Kempin SA, Liljegren SJ, Block LM, Rounsley SD, Yanofsky MF, Lam E (1997) Targeted disruption in Arabidopsis. Nature 389:802–803

Kerbach S, Lorz H, Becker D (2005) Site-specific recombination in Zea mays. Theor Appl Genet 111:1608–1616

Khanna HK, Raina SK (2002) Elite indica transgenic rice plants expressing modified Cry1Ac endotoxin of *Bacillus thuringiensis* show enhanced resistance to yellow stem borer (*Scirpophaga incertulas*). Transgenic Res 11:411–423

Klein TM, Wolf ED, Wu R, Sanford JC (1987) High-velocity microprojectiles for delivering nucleic acids into living cells. Nature 327:70–73

Kohli A, Griffiths S, Palacios N, Twyman RM, Vain P, Laurie DA, Christou P (1999) Molecular characterization of transforming plasmid rearrangements in transgenic rice reveals a recombination hotspot in the CaMV 35S promoter and confirms the predominance of microhomology-mediated recombination. Plant J 17:591–601

Kohli A, Leech M, Vain P, Laurie DA, Christou P (1998) Transgene organization in rice engineered through direct DNA transfer supports a two-phase integration mechanism mediated by the establishment of integration hot spots. Proc Natl Acad Sci USA 95:7203–7208

Kohli A, Twyman RM, Abranches A, Wegel E, Shaw P, Christou P, Stoger E (2003) Transgene integration, organization and interaction in plants. Plant Mol Biol 52:247–258

Kononov ME, Bassuner B, Gelvin SB (1997) Integration of T-DNA binary vector "backbone" sequences into the tobacco genome: Evidence for multiple complex patterns of integration. Plant J 11:945–957

Kopertekh L, Juttner G, Schiemann J (2004a) PVX-Cre-mediated marker gene elimination from transgenic plants. Plant Mol Biol 55:491–500

Kopertekh L, Juttner G, Schiemann J (2004b) Site-specific recombination induced in transgenic plants by PVX virus vector expressing bacteriophage P1 recombinase. Plant Sci 166:485–492

Kumar S, Fladung M (2000) Transgene repeats in aspen: molecular characterisation suggests simultaneous integration of independent T-DNAs into receptive hotspots in the host genome. Mol Gen Genet 264:20–28

Kumar S, Fladung M (2002) Transgene integration in aspen: structures of integration sites and mechanism of T-DNA integration. Plant J 31:543–551

Kumar S, Allen GC, Thompson WF (2005) Gene targeting in plants: fingers on the move. Trends Plant Sci 11:159–161

Kumpatla SP, Chandrasekharan MB, Iyer LM, Li G, Hall TC (1998) Genome intruder scanning and modulation systems and transgene silencing. Trends Plant Sci 3:97–104

Laufs P, Autran D, Traas J (1999) A chromosomal paracentric inversion associated with T-DNA integration in Arabidopsis. Plant J 18:131–139

Lechtenberg B, Schubert D, Forsbach A, Gils M, Schmidt R (2003) Neither inverted repeat T-DNA configurations nor arrangements of tandemly repeated transgenes are sufficient to trigger transgene silencing. Plant J 34:507–517

Lee KY, Lund P, Lowe K, Dunsmuir P (1990) Homologous recombination in plant cells after Agrobacterium-mediated transformation. Plant Cell 2:415–425

Lewandoski M, Martin GR (1997) Cre-mediated chromosome loss in mice. Nat Genet 17:223–225

Li Z, Hansen JL, Liu Y, Zemetra RS, Berger PH (2004) Using real-time PCR to determine transgene copy number in wheat. Plant Mol Biol Rep 22:179–188

Lindbo JA, Silva-Rosales L, Proebsting WM, Dougherty WG (1993) Induction of a highly specific antiviral state in transgenic plants: implications for regulation of gene expression and virus resistance. Plant Cell 5:1749–1759

Lindsey K, Wei W, Clarke MC, McArdle HF, Rooke LM, Topping JF (1993) Tagging genomic sequences that direct transgene expression by activation of a promoter trap in plants. Transgenic Res 2:33–47

Lloyd A, Plaisier CL, Carroll D, Drews GN (2005) Targeted mutagenesis using zinc-finger nucleases in Arabidopsis. Proc Natl Acad Sci USA 102:2232–2237

Luo K, Duan H, Zhao D, Zheng X, Deng W, Chen Y, Stewart CN Jr, McAvoy R, Jiang X, Wu Y, He A, Pei Y, Li Y (2007) 'GM-gene-deletor': fused loxP-FRT recognition sequences dramatically improve the efficiency of FLP or CRE recombinase on transgene excision from pollen and seed of tobacco plants. Plant Biotechnol J 5:263–274

Luo Z, Chen Z (2007) Improperly terminated, unpolyadenylated mRNA of sense transgenes is targeted by RDR6-mediated RNA silencing in Arabidopsis. Plant Cell 19:943–958

Lyznik LA, Gordon-Kamm WJ, Tao Y (2003) Site-specific recombination for genetic engineering in plants. Plant Cell Rep 21:925–932

Ma C, Mitra A (2002) Intrinsic direct repeats generate consistent post-transcriptional gene silencing in tobacco. Plant J 31:37–49

Marjanac G, De Paepe A, Peck I, Jacobs A, De Buck S, Depicker A (2008) Evaluation of Cre-mediated excision approaches in Arabidopsis thaliana. Transgenic Res 17:239–250

Martineau B, Voelker TA, Sanders RA (1994) On defining T-DNA. Plant Cell 6:1032–1033

Matsumoto S, Ito Y, Hosoi T, Takahashi Y, Machida Y (1990) Integration of Agrobacterium T-DNA into a tobacco chromosome: possible involvement of DNA homology between T-DNA and plant DNA. Mol Gen Genet 224:309–316

Matzke MA, Matzke AJM (1990) Gene interactions and epigenetic variation in transgenic. plants. Dev Genet 11:214–223

Matzke MA, Matzke AJM (1991) Differential inactivation and methylation of a transgene in plants by two suppressor loci containing homologous sequences. Plant Mol Biol 16:821–830

McGinnis K, Chandler V, Cone K, Kaeppler H, Kaeppler S, Kerschen A, Pikaard C, Richards E, Sidorenko L, Smith T, Springer N, Wulan T (2005) Transgene-induced RNA interference as a tool for plant functional genomics. RNA Interf 392:1–24

McKnight TD, Lillis MT, Simpson RB (1987) Segregation of genes transferred to one plant cell from two separate Agrobacterium strains. Plant Mol Biol 8:439–445

Mehlo L, Mazithulela G, Twyman RM, Boulton MI, Davies JW, Christou P (2000) Structural analysis of transgene rearrangements and effects on expression in transgenic maize plants generated by particle bombardment. Maydica 45:277–287

Metzger D, Feil R (1999) Engineering the mouse genome by site-specific recombination. Curr Opin Biotechnol 10:470–476

Metzlaff M, O'Dell M, Hellens R, Flavell RB (2000) Developmentally and transgene regulated nuclear processing of primary transcripts of chalcone synthase A in petunia. Plant J 23:63–72

Meyer P, Linn F, Heidmann I, Meyer H, Niedenhof I, Saedler H (1992) Endogenous and environmental factors influence 35S promoter methylation of a maize A1 gene construct in transgenic petunia and its colour phenotype. Mol Gen Genet 231:345–352

Miao ZH, Lam E (1995) Targeted disruption of the *TGA3* locus in *Arabidopsis thaliana* Plant J 7:359–365

Ming R, Hou S, Feng Y, Yu Q, Dionne-Laporte A, Saw JH, Senin P, Wang W, Ly BV, Lewis KL, Salzberg SL, Feng L, Jones MR, Skelton RL, Murray JE, Chen C, Qian W, Shen J, Du P, Eustice M, Tong E, Tang H, Lyons E, Paull RE, Michael TP, Wall K, Rice DW, Albert H, Wang ML, Zhu YJ, Schatz M, Nagarajan N, Acob RA, Guan P, Blas A, Wai CM, Ackerman CM, Ren Y, Liu C, Wang J, Wang J, Na JK, Shakirov EV, Haas B, Thimmapuram J, Nelson D, Wang X, Bowers JE, Gschwend AR, Delcher AL, Singh R, Suzuki JY, Tripathi S, Neupane K, Wei H, Irikura B, Paidi M, Jiang N, Zhang W, Presting G, Windsor A, Navajas-Pérez R, Torres MJ, Feltus FA, Porter B, Li Y, Burroughs AM, Luo MC, Liu L, Christopher DA, Mount SM, Moore PH, Sugimura T, Jiang J, Schuler MA, Friedman V, Mitchell-Olds T, Shippen DE, dePamphilis CW, Palmer JD, Freeling M, Paterson AH, Gonsalves D, Wang L, Alam M (2008) The draft genome of the transgenic tropical fruit tree papaya (Carica papaya Linnaeus). Nature 452:991–997

Mlynarova L, Jansen RC, Conner AJ, Stiekema WJ, Nap JP (1995) The MAR-mediated reduction in position effect can be uncoupled from copy number-dependent expression in transgenic plants. Plant Cell 7:599–609

Mlynarova L, Keizer L, Stiekema WJ, Nap JP (1996) Approaching the lower limits of transgene variability. Plant Cell 8:1589–1599

Mlynarova L, Loonen A, Heldens J, Jansen RC, Keizer P, Stiekema WJ, Nap JP (1994) Reduced position effect in mature transgenic plants conferred by the chicken lysozyme matrix-associated region. Plant Cell 6:417–426

Mlynárová L, Loonen A, Mietkiewska E, Jansen RC, Nap JP (2002) Assembly of two transgenes in an artificial chromatin domain gives highly coordinated expression in tobacco. Genetics 160:727–740

Moore SK, Srivastava V (2006) Efficient deletion of transgenic DNA from complex integration locus of rice mediated by Cre/lox recombination system. Crop Sci 46:700–705

Muskens MW, Vissers AP, Mol JN, Kooter JM (2000) Role of inverted DNA repeats in transcriptional and post-transcriptional gene silencing. Plant Mol Biol 43:243–260

Nacry P, Camilleri C, Courtial B, Caboche M, Bouchez D (1998) Major chromosomal rearrangements induced by T-DNA transformation in Arabidopsis. Genetics 149:641–650

Nam J, Mysore KS, Gelvin SB (1998) *Agrobacterium tumefaciens* transformation of the radiation hypersensitive *Arabidopsis thaliana* mutants *uvh1* and *rad5*. Mol Plant Microbe Interact 11:1136–1141

Nam J, Mysore KS, Zheng C, Knue MK, Matthysse AG, Gelvin SB (1999) Identification of T-DNA tagged Arabidopsis mutants that are resistant to transformation by Agrobacterium. Mol Gen Genet 261:429–438

Napoli C, Lemieux C, Jorgensen R (1990) Introduction of a chimeric chalcone synthase gene into petunia results in reversible co-suppression of homologous genes in trans. Plant Cell 2:279–289

Narasimhulu SB, Deng XB, Sarria R, Gelvin SB (1996) Early transcription of Agrobacterium T-DNA genes in tobacco and maize. Plant Cell 8:873–886

Offringa R, de Groot MJ, Haagsman HJ, Does MP, van den Elzen PJ, Hooykaas PJ (1990) Extrachromosomal homologous recombination and gene targeting in plant cells after Agrobacterium mediated transformation. EMBO J 9:3077–3084

Pansegrau W, Schoumacher F, Hohn B, Lanka E (1993) Site-specific cleavage and joining of single-stranded DNA by VirD2 protein of *Agrobacterium tumefaciens* Ti plasmids: analogy to bacterial conjugation. Proc Natl Acad Sci USA 90:11538–11542

Paszkowski J, Baur M, Bogucki A, Potrykus I (1988) Gene targeting in plants. EMBO J 7:4021–4026

Pawlowski WP, Somers DA (1998) Transgenic DNA integrated into the oat genome is frequently interspersed by host DNA. Proc Natl Acad Sci USA 95:12106–12110

Pikaart M, Feng J, Villeponteau B (1998) The polyomavirus enhancer activates chromatin accessibility on integration into the HPRT gene. Mol Cell Biol 12:5785–5792

Porsch P, Jahnke A, During K (1998) A plant transformation vector with a minimal T-DNA. II. Irregular integration patterns of the T-DNA in the plant genome. Plant Mol Biol 37:581–585

Puchta H (2003) Marker-free transgenic plants. Plant Cell Tissue Organ Cult 74:123–134

Puchta H, Kocher S, Hohn B (1992) Extrachromosomal homologous DNA recombination in plant cells is fast and is not affected by CpG methylation. Mol Cell Biol 12:3372–3379

Que Q, Wang HY, English JJ, Jorgensen RA (1997) The frequency and degree of cosuppression by sense chalcone synthase transgenes are dependent on transgene promoter strength and are reduced by premature nonsense codons in the transgene coding sequence. Plant Cell 9:1357–1368

Rai M, Datta K, Parkhi V, Tan J, Oliva N, Chawla HS, Datta SK (2007) Variable T-DNA linkage configuration affects inheritance of carotenogenic transgenes and carotenoid accumulation in transgenic indica rice. Plant Cell Rep 26:1221–1231

Ramanathan V, Veluthambi K (1995) Transfer of non-T-DNA portions of the *Agrobacterium tumefaciens* Ti plasmid pTiA6 from the left terminus of TL-DNA. Plant Mol Biol 28:1149–1154

Ramessar K, Capell T, Twyman RM, Quemada H, Christou P (2008) Trace and traceability – a call for regulatory harmony. Nat Biotechnol 26:975–978

Register JC 3rd, Peterson DJ, Bell PJ, Bullock WP, Evans IJ, Frame B, Greenland AJ, Higgs NS, Jepson I, Jiao S, Lewnau CJ, Sillick JM, Wilson HM (1994) Structure and function of selectable and non-selectable transgenes in maize after introduction by particle bombardment. Plant Mol Biol 25:951–961

Reiss B, Schubert I, Kopchen K, Wendeler E, Schell J, Puchta H (2000) RecA stimulates sister chromatid exchange and the fidelity of double-strand break repair, but not gene targeting, in plants transformed by Agrobacterium. Proc Nat Acad Sci USA 97:3358–3363

Risseeuw E, Franke-van Dijk ME, Hooykaas PJ (1997) Gene targeting and instability of Agrobacterium T-DNA loci in the plant genome. Plant J 11:717–728

Risseeuw E, Offringa R, Franke-van Dijk ME, Hooykaas PJ (1995) Targeted recombination in plants using Agrobacterium coincides with additional rearrangements at the target locus. Plant J 7:109–119

Rooke L, Steele SH, Barcelo P, Shewry PR, Lazzeri PA (2003) Transgene inheritance, segregation and expression in bread wheat. Euphytica 129:301–309

Sadowski PD (1993) Site-specific genetic recombination: hops, flips, and flops. FASEB J 7:760–767

Salomon S, Puchta H (1998) Capture of genomic and T-DNA sequences during double-strand break repair in somatic plant cells. EMBO J 17:6086–6095

Sanford JC, Smith FD, Russell JA (1993) Optimizing the biolistic process for different biological applications. Methods Enzymol 217:483–509

Sawasaki T, Takahashi M, Goshima N, Morikawa H (1998) Structures of transgene loci in transgenic Arabidopsis plants obtained by particle bombardment: junction regions can bind to nuclear matrices. Gene 218:27–35

Schiestl RH, Petes TD (1991) Integration of DNA fragments by illegitimate recombination in *Saccharomyces cerevisiae*. Proc Natl Acad Sci USA 88:7585–7589

Schubert D, Lechtenberg B, Forsbach A, Gils M, Bahadur S, Schmidt R (2004) Silencing in Arabidopsis T-DNA transformants: the predominant role of a gene-specific RNA sensing mechanism versus position effects. Plant Cell 16:2561–2572

Shaw CH, Watson MD, Carter GH, Shaw CH (1984) The right hand copy of the nopaline Ti-plasmid 25 bp repeat is required for tumour formation. Nucleic Acids Res 12:6031–6041

Somers DA, Makarevistch I (2004) Transgene integration in plants: poking or patching holes in promiscuous genomes? Curr Opin Biotechnol 15:126–131

Sonti RV, Chiurazzi M, Wong D, Davies CS, Harlow GR, Mount DW, Signer ER (1995) Arabidopsis mutants deficient in T-DNA integration. Proc Natl Acad Sci USA 92:11786–11790

Spencer TM, O'Brien JV, Start WG, Adams TR, Gordon-Kamm WJ, Lemaux PG (1992) Segregation of transgenes in maize. Plant Mol Biol 18:201–210

Spielmann A, Simpson RB (1986) T-DNA structure in transgenic tobacco plants with multiple independent integration sites. Mol Gen Genet 205:34–41

Srivastava V, Anderson OD, Ow DW (1999) Single-copy transgenic wheat generated through the resolution of complex integration patterns. Proc Natl Acad Sci USA 96:11117–11121

Srivastava V, Ariza-Nieto M, Wilson AJ (2004) Cre-mediated site-specific gene integration for consistent transgene expression in rice. Plant Biotechnol J 2:169–179

Srivastava V, Ow DW (2001) Single-copy primary transformants of maize obtained through the co-introduction of a recombinase-expressing construct. Plant Mol Biol 46:561–566

Srivastava V, Ow DW (2003) Rare instances of Cre-mediated deletion product maintained in transgenic wheat. Plant Mol Biol 52:661–668

Srivastava V, Vasil V, Vasil IK (1996) Molecular characterization of the fate of transgenes in transformed wheat. Theor Appl Genet 92:1031–1037

Steck TR (1997) Ti plasmid type affects T-DNA processing in Agrobacterium tumefaciens. FEMS Microbiol Lett 147:121–125

Stoger E, Williams S, Keen D, Christou P (1998) Molecular characteristics of transgenic wheat and the effect on transgene expression. Transgenic Res 7:463–471

Svab Z, Maliga P (1993) High-frequency plastid transformation in tobacco by selection for a chimeric *aadA* gene. Proc Natl Acad Sci USA 90:913–917

Svitashev SK, Ananiev E, Pawlonski WP, Somers DA (2000) Association of transgene integration sites with chromosome rearrangements in hexaploid oat. Theor Appl Genet 100:872–880

Svitashev SK, Pawlowski WF, Makarevitch I, Plank DW, Somers DA (2002) Complex transgene locus structures implicate multiple mechanisms for plant transgene rearrangement. Plant J 32:433–445

Svitashev SK, Somers DA (2001) Genomic interspersions determine the size and complexity of transgene loci in transgenic plants produced by microprojectile bombardment. Genome 44:691–697

Tanako M, Egawa H, Ikeda J, Wakasa K (1997) The structures of integration sites in transgenic rice. Plant J 11:353–361

Tavernarakis N, Wang SL, Dorovkov M, Ryazanov A, Driscoll M (2000) Heritable and inducible genetic interference by double-stranded RNA encoded by transgenes. Nat Genet 24:180–183

Ten Hoopen R, Robbins TP, Fransz PF, Montijn BM, Oud O, Gerats A, Nanninga N (1996) Localization of T-DNA insertions in petunia by fluorescence in situ hybridization: physical evidence for suppression of recombination. Plant Cell 8:823–830

Terada R, Urawa H, Inagaki V, Tsugane K, Lida S (2002) Efficient gene targeting by homologous recombination in rice. Nat Biotechnol 20:1030–1034

Tinland B (1996) The integration of T-DNA into plant genomes. Trends Plant Sci 1:178–184

Travella S, Ross SM, Harden J, Everett C, Snape JW, Harwood WA (2005) A comparison of transgenic barley lines produced by particle bombardment and Agrobacterium-mediated techniques. Plant Cell Rep 23:780–789

Twyman RM, Christou P, Stoger E (2002) Genetic transformation of plants and their cells. In: Oksman-Caldentey KM, Barz W (eds) Plant biotechnology and transgenic plants. Marcel, New York, pp 111–141

Twyman RM, Kohli A (2003) Genetic modification: Insertional and transposon mutagenesis. In: Thomas B, Murphy DJ, Murray B (eds) Encyclopedia of applied plant sciences. Elsevier Science, London, UK, pp 369–377

Tzfira T, Rhee Y, Chen MH, Kunik T, Citovsky V (2000) Nucleic acid transport in plant-microbe interactions: the molecules that walk through the walls. Annu Rev Microbiol 54:187–219

van Eldik GJ, Litiere K, Jacobs JJ, Van Montagu M, Cornelissen M (1998) Silencing of beta-1, 3-glucanase genes in tobacco correlates with an increased abundance of RNA degradation intermediates. Nucleic Acids Res 26:5176–5181

Van Leeuwen W, Mlynárová L, Nap JP, van der Plas LH, van der Krol AR (2001) The effect of MAR elements on variation in spatial and temporal regulation of transgene expression. Plant Mol Biol 47:543–554

Vaucheret H, Beclin C, Elmayan T, Feuerbach F, Godon C, Morel JB, Mourrain P, Palauqui JC, Vernhettes S (1998) Transgene-induced gene silencing in plants. Plant J 16:651–659

Vega JM, Yu WC, Han FP, Kato A, Peters EM, Birchler JA (2007) Transfer of a Cre/lox recombination system to maize for gene targeting and chromosome engineering. Chromosome Res 15:1–11

Vega JM, Yu WC, Han FP, Kato A, Peters EM, Zhang ZJ, Birchler JA (2008) Agrobacterium-mediated transformation of maize (*Zea mays*) with Cre-lox site specific recombination cassettes in BIBAC vectors. Plant Mol Biol 66:587–598

Wan S, Wu J, Zhang Z, Sun X, Lv Y, Gao C, Ning Y, Ma J, Guo Y, Zhang Q, Zheng X, Zhang C, Ma Z, Lu T (2009) Activation tagging, an efficient tool for functional analysis of the rice genome. Plant Mol Biol 69:69–80

Wang J, Lewis ME, Whallon JH, Sink KC (1995) Chromosome. mapping of T-DNA inserts in transgene. Petunia. by in situ hybridization. Transgenic Res 4:241–246

Wang MB, Waterhouse PM (2000) High-efficiency silencing of a beta-glucuronidase gene in rice is correlated with repetitive transgene structure but is independent of DNA methylation. Plant Mol Biol 43:67–68

Ward ER, Barnes WM (1988) VirD2 protein of *Agrobacterium tumefaciens* very tightly linked to 5' end of T-strand DNA. Science 242:927–930

Weigel D, Ahn JH, Blázquez MA, Borevitz JO, Christensen SK, Fankhauser C, Ferrándiz C, Kardailsky I, Malancharuvil EJ, Neff MM, Nguyen JT, Sato S, Wang ZY, Xia Y, Dixon RA, Harrison MJ, Lamb CJ, Yanofsky MF, Chory J (2000) Activation tagging in Arabidopsis. Plant Physiol 122:1003–1013

Wenck A, Czakó M, Kanevski I, Márton L (1997) Frequent collinear long transfer of DNA inclusive of the whole binary vector during Agrobacterium-mediated transformation. Plant Mol Biol 34:913–922

Wilson C, Bellen HJ, Gehring WJ (1990) Position effects on eukaryotic gene expression. Ann Rev Cell Biol 6:679–714

Wolters AMA, Trindade LM, Jacobsen E, Visser RGF (1998) Fluorescence in situ hybridization on extended DNA fibres as a tool to analyse complex T-DNA loci in potato. Plant J 13:837–847

Wright DA, Townsend JA, Winfrey RJ Jr, Irwin PA, Rajagopal J, Lonosky PM, Hall BD, Jondle MD, Voytas DF (2005) High-frequency homologous recombination in plants mediated by zinc-finger nucleases. Plant J 44:693–705

Yang L, Ding J, Zhang C, Jia J, Weng H, Liu W, Zhang D (2005) Estimating the copy number of transgenes in transformed rice by real-time quantitative PCR. Plant Cell Rep 23:759–763

Zhu QH, Ramm K, Eamens AL, Dennis ES, Upadhyaya NM (2006) Transgene structures suggest that multiple mechanisms are involved in T-DNA integration in plants. Plant Sci 171:308–322

Zeevi V, Tovkach A, Tzfira T (2008) Increasing cloning possibilities using artificial zinc finger nucleases. Proc Natl Acad Sci USA 105:12785–12790

Ziemienowicz A, Tinland B, Bryant J, Gloeckler V, Hohn B (2000) Plant enzymes but not Agrobacterium VirD2 mediate T-DNA ligation in vitro. Mol Cell Biol 20:6317–6322

Chapter 8
Organelle Transformations

Anjanabha Bhattacharya

8.1 Introduction

The world population is increasing at a rapid pace. Agricultural production must match population growth in the near seeable future amid fears of climate change. However, complex traits like crop productivity are difficult to manipulate and take time using conventional breeding (Zuker et al. 1998; Mishra and Srivastava 2004). The availability of a limited gene pool and the failure of wide-crosses among crop varieties in conventional breeding have led to the exploitation of genetic transformation in generating high-yielding crop varieties.

Nucleus transformation is the target of choice for the development of transgenic varieties using one of the several techniques of transformation available today. However, the nuclear genome is large and contains several copies of the same gene, presence of introns, cis-elements, and as such. Therefore, unpredictable results are obtained when transgene(s) are integrated in different parts of the nuclear genome because of positional effect, including, but not limited to, gene silencing or lower levels of transgene expression (Kumar et al. 2004) and off-target influence, thus presenting unusual challenges in their commercialization and restricting consumer acceptability. Removing this analogy required researchers to start looking for other organelles in the plant cell, such as plastids and mitochondria, that could be targeted for genetic transformation. The plastids may differentiate to become chloroplast (green pigment storage plastids), chromoplast (pigment storage organs in fruits), elaioplast (lipid storage), or amyloplast (starch storage). Immature plastids are called pro-plastids. The evolutionary lineage suggests that these organelles were primitively free-living prokaryotes (cyanobacteria), and with the evolution of land plants, they began to form a permanent symbiotic relationship with the host cell and

A. Bhattacharya
National Environmental Sound Production Agriculture Laboratory, University of Georgia, Tifton, GA 31794, USA
e-mail: anjanabha.bhattacharya@gmail.com

C. Kole et al. (eds.), *Transgenic Crop Plants*,
DOI 10.1007/978-3-642-04809-8_8, © Springer-Verlag Berlin Heidelberg 2010

became a part and parcel of the plant cell (Lopez-Juez and Pyke 2005). Over time, plastids began to lose the vital genes associated with their independent existence and became dependent on the plant cell machinery to evolve and replicate (Pyke 2007; Provorov 2005). Thus, a sort of endosymbiosis developed between chloroplast and the nucleus, which involved import of many proteins from the cytoplasm (Lopez-Juez and Pyke 2005; Bhattacharya et al. 2007). Organelles like chloroplast and mitochondria have low copy number of genes, several origins of replication, and divide independent of the nucleus; thus, addition of multitransgene in the open reading frame (ORF) results in high levels of gene expression. The first reported case of *Chlamydomonas* chloroplast transformation was by Boynton et al. (1988) almost two decades ago, and subsequent reports of tobacco plastid transformation were by Khan and Maliga (1999).

There is also the absence of epigenetic interference with the inserted transgenes (Bock 2007). The plastid transformation increases the precision of genetic engineering by targeted homologous recombination at specific sites. Specific plastid sequences act as flanking regions for the gene of interest and selectable marker, and are targeted to locate specifically homologous region in the plastid genome thus making precise integration at specific location in plastids. The plastids have several origins of replication (Scharff and Koop 2007) of prokaryotic evolution, and this is essential because they do not depend on nuclear division for transcription. This in turn eliminates the problem of gene silencing; plastids were, therefore, identified as the target of choice for crop transformation. Organelle transformation can also be used to harvest any specific trans proteins as they tend to accumulate at very high levels, thus becoming important in pharmaceutical industries (Buhot et al. 2006). Bohne et al. (2007) concluded that mitochondrial genes and substantial part of plasteome are transcribed by related RNA polymerases. There are also reports of using cross species plastids for effective transformation of the most recalcitrant species (Kuchuk et al. 2006).

8.2 Overview of Organelle Transformation

8.2.1 Plastids

Plastids are of prokaryotic evolution and their genome size is 120–180 kb (Wakasugi et al. 2001). They are related to cyanobacteria and now have become an inseparable part of the plant cell. They have small genome size and their specific sequences are targeted (sequence identified from their sequenced genome; Ruhlman et al. 2006) for construction of suitable vectors for genetic transformation aiming at crop improvement and for therapeutic proteins. Moreover, data mining of the plastid genome could help to identify suitable endogenous flanking sequences for construction of plastid vectors as depicted in Fig. 8.1. A generalized organelle transformation strategy commonly in use is explained in Fig. 8.2.

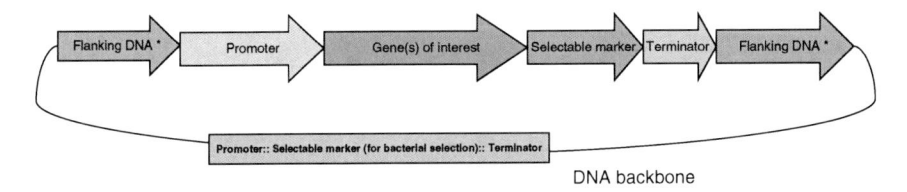

DNA backbone

Fig. 8.1 Typical construction of an organelle transformation vector. *Endogenous flanking border sequence homologous to organelle DNA sequence for precise integration and recombination. Such sequences can be indentified by data mining organelle genome sequences

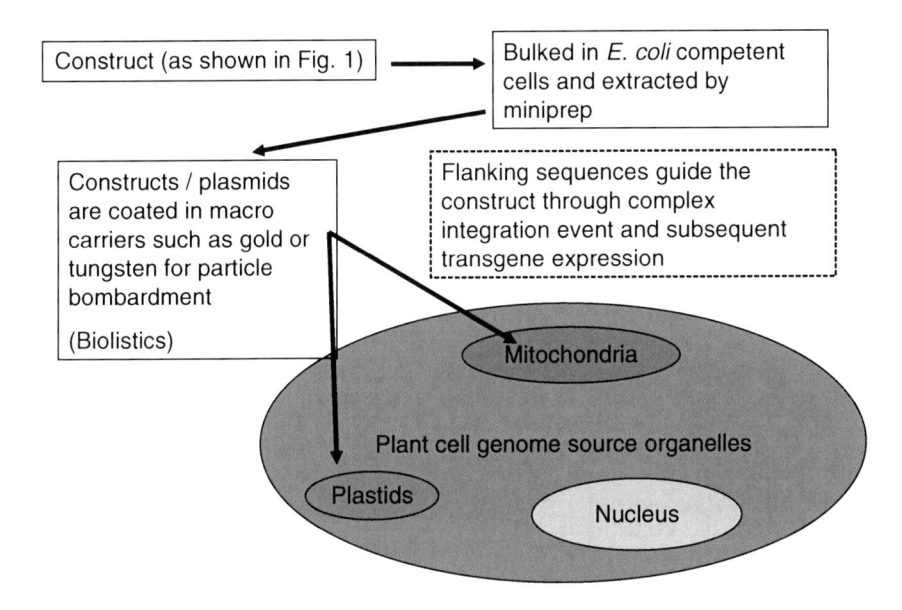

Fig. 8.2 Organelle genetic transformation

The additional advantage of molecular pharming is that the complex enzymes present within the chloroplast forms complex proteins with correct folding thus reducing price incurred for any other additional processing (Koya and Daniell 2005). The plastid acts as a giant factory that produces 100 times more protein than the conventional nuclear transformation (Ishida 2005; Nugent and Joyce 2005). Thus, the pharmaceutical industries are more interested in utilizing these natural bioreactors in plants and reduce cost associated with drug production. Further, the processed protein is free from contaminant, which is a major concern with proteins purified from animal systems. Some successful examples of plastid transformation are listed in Table 8.1. Further, Lutz and Maliga (2007) explained the feasibility of obtaining marker-free transgenic plants with homology based excision, excision by phage recombinase, cointegration of marker gene with the gene of interest.

Table 8.1 Some examples of successful organelle transformation

Plant	Gene introduced	Specific border regions specific for organelle targeting	Organelle involved	Reference
Tobacco	Aminoglycoside 3′-adenyltransferase (*aadA*) [Spectinomycin resistance] and green fluorescent protein (*gfp*)	*trnI* and *trnA*	Chloroplast	Davarpanah et al. (2009)
Tobacco	*HIV Gag (Pr 55)*	chloroplast targeting sequence	Plastids	Scotti et al. (2009)
Sugarbeet	Aminoglycoside 3′-adenyltransferase (*aadA*) and green fluorescent protein (*gfp*)	*rrn16* and *rps12* (intergenic region of plastome)	Chloroplast	De Marchis et al. (2009)
Tobacco and tomato	*HIV antigen p24*	p24-Nef fusion gene cassettes	Plastids	Zhou et al. (2008)
Tobacco	*HPV-16 L1* antigen	5′-UTR and *N*-terminal coding segment of a plastid gene, *L1P*, *L1V*	Plastids	Lenzi et al. (2008)
Tomato	Beta-cyclase gene	Intergenic region of plastid gene	Plastids	Wurbs et al. (2007)
Tobacco	Human serum albumin (*HAS*)	Intergenic region of plastid gene	Plastids	Fernáandez-San Millan et al. (2007)
Arabidopsis	YFP (Yellow Flurocescent Protein) fused with plastoglobulin 34 (*PGL34*)	Chloroplast targeting sequence	Plastids	Vidi et al. (2007)
Tobacco	Transformation protocol	Plastid targeting sequences and *cre- loxP*	Plastids	Lutz et al. (2006)
Poplar	*Gfp*	Poplar plastid genes, *accD* and *rbcL*	Plastids	Okumura et al. (2006)
Tobacco	*EBV VCA* antigen	Transplastomic plants	Plastids	Lee et al. (2006b)
Lettuce	*gfp*:: *addA* gene	Plastid genome sequences allowing its targeted insertion between the *rbcL* and *accD* genes	Plastids	Kanamoto et al. (2006)
Rice	*gfp*:: addA	Plastid targeting sequence	Plastids	Lee et al. (2006b)
Petunia	*aadA* :: *gusA* genes	Chloroplast targeting sequence	Plastids	Zubko et al. (2004)
Cotton	*aphA-6* and *nptII*	Plastid transformation vector	Via tobacco chloroplast	Kumar et al. (2004)
Soybean	*aadA*	Plastome sequences targeted between the *trnV* gene and the *rps12/7* operon	Plastids	Dufourmantel et al. (2004)

8.2.2 Mitochondria

The mitochondrial genome is much larger than the plastid genome but considerably smaller than the nuclear genome. The mitochondrial genome size varies from 1,500 to 2,300 kbp in cucumber to about 150 kbp in yeast (Havey et al. 2002). The similarity between plastids and mitochondria raised the possibility that mitochondrial gene may function in plastids and could be included in strategies to transform plastids (Maliga 2004) by designing suitable constructs. Further, Weber-Lotfi et al. (2009) reported that mitochondrial DNA can accept linear DNA molecules. However, the formulation of effective techniques of transforming mitochondria unlike plastids is still underway. There are only a few reports of genetic transformation of mitochondria, for example, in *Saccharomyces cerevisiae* and *Chlamydomonas reinhardtii* (Bonnefoy et al. 2007).

8.3 Achievements and Technique Used in Organelle Transformation

8.3.1 Plastids

8.3.1.1 Biolistics Transformation

Development of Transplasteomic Plants Expressing Florescent Genes or Selectable Antibiotic Markers

Davarpanah et al. (2009) transformed plastids of *Nicotina tobacum* following biolistics (standard gene gun) protocol originally adopted by Jeong et al. (2004) by subjecting leaf blades (~3 cm long) previously grown in tissue culture medium with *gfp* and *addA* marker gene (spectinomycin resistance). Similarly, De Marchis et al. (2009) used particle bombardment of sugar beet (*Beta vulgaris*) petioles (of 0.5-cm pieces' size) placed on Petri dishes to obtain transplasteomic plants expressing Gfp (Green Fluorescent Protein). Lee et al. (2006b) obtained transgenic plants expressing *gfp::addA* genes thus devising a proper system for rice plastid transformation. Okumura et al. (2006) used biolistic transformation with gold particles (0.6-μm diameter) and 900-psi rupture disk (Bio-Rad Laboratories) to introduce *gfp* genes in poplar. Skarjinskaia et al. (2003) reported plastid transformation of *Lesquerella fendleri* using biolistics for *gfp* and *aadA* transgene. This was the first reported plastid transformable species beyond *Arabidopsis thaliana* of the Brassicaceae family.

Development of Transplasteomic Plants Producing Proteins for Pharmaceutical and Nutraceutical Industries

Scotti et al. (2009) bombarded [Bio-Rad (Hercules, CA, USA) PDS-1000/He] *Nicotiana tabacum* leaves with 0.6-μm diameter gold particles coated with pFA1 or pNS40 plasmid DNA to obtain plants expressing HIV Gag (Pr 55) proteins [a Gag Pr55 protein is a derivative protein from human immunodeficiency virus (HIV), andmyristoyl coenzyme A]. Zhou et al. (2008) obtained transplasteomic tobacco and tomato plants expressing HIV antigen 24 by particle bombardment. Lenzi et al. (2008) obtained transplastomic plants after biolistic transformation using tungsten (1 μm) or gold particles (0.6 μm) as macrocarrier, and DuPont PDS1000He biolistic device. The plasmid was carrying a chimeric gene encoding the L1 protein, in the native viral [L1(v) gene] form or in the form of a synthetic sequence suitable for plant plastids [L1(p) gene] under the influence of a plastid gene. Lutz et al. (2006) used biolistic transformation to transform plastid genome by homologous recombination. The transgene excision was achieved by plastid targeted Cre gene expression from nucleus. Thus, they showed the feasibility of using Cre-Lox system in plants. Lee et al. (2006a) showed the feasibility of production of vaccine EBV (Epstein-Barr Virus antigen, which is a source of major cause of malignancies originating from lymphoid and epithelial cells) from transplasteomic tobacco cv. SR1. Moreover, transplastomic tobacco has been cited in several research papers for the production of vaccine antigen against plague, tetanus, anthrax, insulin like growth factors by using biolistic transformation (Daniell et al. 2001; Watson et al. 2004).

Development of Transplasteomic Plants for Crop Improvement

Wurbs et al. (2007) developed a system to facilitate the feasibility of genetically modifying nutritionally important biochemical pathways in nongreen plastids by introducing β-cyclase transgene in the chloroplast genome of tomato. Leelavathi et al. (2003) expressed xylanase enzyme in the chloroplast of tobacco plants. Overproduction of xylanase did not affect plant growth unlike nuclear expressed trans-xylanase enzymes. This provided an example of superior performance of organellar transformation over nuclear transformation. Salt tolerance is another trait, which needs immediate attention. Plant cells tend to accumulate osmoprotectants like glycine betaine to overcome salt-induced stress. Overexpression of such genes in plastids or mitochondria might help a plant to perform well in saline conditions. Kumar et al. (2004) showed cotton plastids can be transformed by plasteomic vectors and this technology can be used for cotton improvement. Dufourmantel et al. (2004) reported the feasibility of transforming plastids of soybean crop, which is an important leguminous and oilseed crop, grown extensively in many parts of the world. There have been reports of development of insect and disease resistance traits, salt- and drought-tolerance traits among important agronomic traits (reviewed by Daniell et al. 2005). Hou et al. (2003) reported chloroplast transformation of oilseed rape and introduction of a specific gene (insect resistance gene cry1Aa10) between rps7 and ndhB genes of the plastid genome, thus pioneering a new method for oilseed rape genetic improvement by chloroplast bioengineering.

8.3.1.2 Electroporation

There are only a few reports of using electroporation for transforming plastids in plants. Rangasamy et al. (1997) established plastid targeted transformation of pea using electroporation technique for the expression of ATP:citrate lyase in proto-plasts. To et al. (1996) were able to introduce and express *cat* and *gus* genes in isolated spinach plastids by electroporation with pHD203-GUS. Carrot cell cultures have been effectively transformed by electroporation, carrying transplasteomic cells. The rate of growth is very high in culture medium and thus can be used for oral vaccine production programs (Daniell et al. 2005).

8.3.1.3 Other Systems

There are other numerous novel transformation systems, which include microabla-tion (Kajiyama et al. 2008), electrophoresis of embryos, the pollen-tube pathway, microinjection (Neuhaus and Spangenberg 1990; Holm et al. 2000), microbeads' transformation, laser cell perforation (Weber et al. 1988), liposome-mediated gene transfer (Ciboche 1990), and ultrasonification (Joersbo and Brunstedt 1992). These techniques might play an important role in organelle transformation.

8.3.2 Mitochondria

A robust transformation strategy targeting mitochondria instead of the nucleus will allow scientists to use reverse genetics strategy to study trans-mitochondrial gene expression (Havey et al. 2002). This will further establish the efficacy of using target mitochondrial gene to develop transgenic crops. Villarreal et al. (2009) indi-cated the possibility of overexpressing mitochondria gamma carbon anhydrase-2 gene that causes male sterility in *Arabidopsis*. Possibility of using such genes specifically targeted in mitochondria can help to achieve self-incompatible lines for crop improvement programs. Almost two decades earlier, Kemble et al. (1988) showed the possibility of transforming mitochondria of *Brassica napus* hybrid plants via protoplast fusion mediated by polyethylene glycol (PEG) or electropora-tion with recombinant vectors. Traits like cytoplasmic male sterility and nonchro-mosomal stripe mutations of maize have their genomic basis in mitochondria (reviewed by Havey et al. 2002).

8.4 Conclusions

Organelle transformation has opened many vista for crop improvement, production of industrial grade enzymes, biomolecules, pharmaceuticals, and nutraceuticals. More than 40 plastids from different species have been sequenced; thus, a vast plethora of information is available for data mining and development of new vectors for plastid transformation (Maier and Schmitz-Linneweber 2004). It overcomes the limitations imposed by the nuclear transformation techniques by introducing site-specific genes into plastid genome. Organelle transformation has the advantage as genetic information are maternally inherited against biparental inheritance of the nuclear genome (Daniell 2007). Besides the *cre-lox* system (for preventing transgene escape) can be used as an additional safeguard (Corneille et al. 2003) that can eliminate the risk of transgene escape by pollen release (pollen do not carry plastids or mitochondria). In addition, increasing biosafety of transplasteomic crops (Ruf et al. 2007) should result in more acceptable technology for developing transgenic crops. Organelle transformation is more predictable, can insert many (trans)genes at the same time, is not constrained by gene silencing, and facilitates accumulation of transgene-induced metabolites at high concentration. However, only a limited number of crop species have been exploited for plastid transformation, viz., tobacco (Scotti et al. 2009), petunia (Zubko et al. 2004), tomato (Wurbs et al. 2007) , poplar (Okumura et al. 2006), lettuce, rice, soybean (Dufourmantel et al. 2004), cotton (Kumar et al. 2004). Among them, tobacco has been by far the most successful as compared to other plant species. The flanking sequences used for designing vector for plastid transformation should have 100% homology between plastid genome of different species. Less homology between flanking regions has shown to result in lower transformation efficiency (Zubko et al. 2004). Therefore, emphasis should be paid on species-specific flanking sequences for plastid transformation particularly for field crop species to avoid any complications (Daniell et al. 2005). Thus, the need of the hour is to look at various techniques of transformation as these still remain a bottle neck in the development of transgenic variety using organelle transformation particularly for cultivated crop species.

References

Bhattacharya D, Archibald JM, Weber APM, Reyes-Prieto A (2007) How do endosymbionts become organelles? Understanding early events in plastid evolution. Bioassays 29:1239–1246

Bock R (2007) Plastid biotechnology: prospects for herbicide and insect resistance, metabolic engineering and molecular farming. Curr Opin Biotechnol 18:100–106

Bohne AV, Ruf S, Boerner T, Bock R (2007) Faithful transcription initiation from a mitochondrial promoter in transgenic plastids. Nucleic Acid Res 35:7256–7266

Boynton JE, Gillham NW, Harris EH, Hosler JP, Johnson AM, Jones AR et al (1988) Chloroplast transformation in Chlamydomonas with high velocity microprojectiles. Science 240: 1534–1538

Buhot L, Horvath E, Medgyesy P, Lerbs-Mache S (2006) Hybrid transcription system for controlled plastid transgene expression. Plant J 46:700–707

Bonnefoy N, Remacle C, Fox TD (2007) Genetic transformation of *Saccharomyces cerevisiae* and *Chlamydomonas reinhardtii* mitochondria. In: Pon LA, Schon EE (eds) Methods in Cell Biology 80(26): 525–547, Academic Press, Mitochondria

Ciboche M (1990) Liposome-mediated transfer of nucleic acids in plant protoplasts. Physiol Plant 79:173–176

Corneille S, Lutz KA, Azhagiri AK, Maliga P (2003) Identification of functional lox sites in the plastid genome. Plant J 35:753–762

Daniell H, Lee SB, Panchal T, Wiebe PO (2001) Expression of cholera toxin B subunit gene and assembly as functional oligomers in transgenic tobacco chloroplasts. J Mol Biol 311: 1001–1009

Davarpanah SJ, Jung SH, Kim YJ, Park YI, Min SR, Liu JR, Jeong WJ (2009) Stable plastid transformation in *Nicotiana benthamiana*. J Plant Biol 52:244–250

Daniell H, Kumara S, Dufourmantel N (2005) Breakthrough in chloroplast genetic engineering of agronomically important crops. Trends Biotechnol 23:238–245

Daniell H (2007) Transgene containment by maternal inheritance: effective or elusive? Proc Natl Acad Sci USA 17:6879–6880

De Marchis F, Wang YX, Stevanato P, Arcioni S, Bellucci M (2009) Genetic transformation of the sugar beet plastome. Transgenic Res 18:17–30

Dufourmantel N, Pelissier B, Garcon F, Peltier G, Ferullo JM, Tissot G (2004) Generation of fertile transplastomic soybean. Plant Mol Biol 55:479–489

Fernández-San Millán A, Farran I, Molina A, Mingo-Castel AM, Veramendi J (2007) Expression of recombinant proteins lacking methionine as N-terminal amino acid in plastids: human serum albumin as a case study J Biotechnol 127(4):593–604

Havey MJ, Lilly JW, Bohanec B, Bartoszewski G, Malepszy S (2002) Cucumber: a model angiosperm for mitochondrial transformation? J Appl Genet 43:1–17

Holm PB, Olsen O, Schnorf M, Brinch-Pedersen H, Knudsen S (2000) Transformation of barley by microinjection into isolated zygote protoplasts. Transgenic Res 9:21–32

Hou BK, Zhou YH, Wan LH, Zhang ZL, Shen GF, Chen ZH, Hu ZM (2003) Chloroplast transformation in oilseed rape. Transgenic Res 12:111–114

Ishida K (2005) Protein targeting into plastids: a key to understanding the symbiogenetic acquisitions of plastids. J Plant Res 118:237–245

Jeong SW, Jeong WJ, Woo JW, Choi DW, Liu JR (2004) Dicistronic expression of the green fluorescent protein and antibiotic resistance genes in the plastid for tracking and selecting plastid transformed cells in tobacco. Plant Cell Rep 22:747–751

Joersbo M, Brunstedt J (1992) Sonication: a new method for gene transfer to plants. Plant Physiol 85:230–234

Kanamoto H, Yamashita A, Asao H, Okumura S, Takase H, Hattori M, Yokota A, Tomizawa K (2006) Efficient and stable transformation of *Lactuca sativa* L. cv. Cisco (lettuce) plastids. Transgenic Res 15:205–217

Kajiyama S, Joseph B, Inoue F, Shimamura M, Fukusaki E, Tomizawa K, Kobayashi A (2008) Transient gene expression in guard cell chloroplasts of tobacco using ArF excimer laser microablation. J Biosci Bioeng 106:194–198

Kemble RJ, Barsby TL, Yarrow SA (1988) Ectopic expression of mitochondrial gamma carbonic anhydrase 2 causes male sterility by anther indehiscence. Plant Mol Biol 70:471–485

Khan MS, Maliga P (1999) Fluorescent antibiotic resistance marker for tracking plastid transformation in higher plants. Nat Biotechnol 17:910–915

Koya V, Daniell H (2005) OBPC Symposium: Maize 2004 and Beyond – Recent advances in chloroplast genetic engineering. In Vitro Cell Dev Biol-Plant 41:388–404

Kuchuk N, Sytnyk K, Vasylenko M, Shakhovsky A, Komarnytsky I, Kushnir S, Gleba Y (2006) Genetic transformation of plastids of different Solanaceae species using tobacco cells as organelle hosts. Theor Appl Genet 113:519–527

Kumar S, Dhingra A, Daniell H (2004) Stable transformation of the cotton plastid genome and maternal inheritance of transgenes. Plant Mol Biol 56:203–216

Lee MYT, Zhou YX, Lung RWM, Chye ML, Yip WK, Zee SY, Lam E (2006a) Expression of viral capsid protein antigen against Epstein-Barr virus in plastids of *Nicotiana tabacum* cv. SR1. Biotechnol Bioeng 94:1129–1137

Lee SM, Kang KS, ChungH Y, SH XuXM, Lee SB, Cheong JJ, Daniell H, Kim M (2006b) Plastid transformation in the monocotyledonous cereal crop, rice (*Oryza sativa*) and transmission of transgenes to their progeny. Mol Cells 21:401–410

Leelavathi S, Gupta N, Maiti S, Ghosh A, Reddy VS (2003) Overproduction of an alkali- and thermostable xylanase in tobacco chloroplasts and efficient recovery of the enzyme. Mol Breed 11:59–67

Lenzi P, Scotti N, Alagna F, Tornesello ML, Pompa A, Vitale A, De Stradis A, Monti L, Grillo S, Buonaguro FM, Maliga P, Cardi T (2008) Translational fusion of chloroplast-expressed human papillomavirus type 16 L1 capsid protein enhances antigen accumulation in transplastomic tobacco. Transgenic Res 17:1091–1102

Lopez-Juez E, Pyke KA (2005) Plastids unleashed: their development and their integration in plant development. Int J Dev Biol 49:557–577

Lutz KA, Svab Z, Maliga P (2006) Construction of marker-free transplastomic tobacco using the Cre-loxP site-specific recombination system. Nat Protocol 2:900–910

Lutz KA, Maliga P (2007) Construction of marker-free transplastomic plants. Curr Opin Biotechnol 18:107–114

Maier R, Schmitz-Linneweber C (2004) Plastid genomes. In: Daniell H, Chase CD (eds) Molecular biology and biotechnology of plant organelles. Springer, Dordrecht, pp 1–12

Maliga P (2004) Plastid transformation in higher plants. Annu Rev Plant Biol 55:289–313

Millan AFS, Farran I, Molina A, Mingo-Castel AM, Veramendi J (2007) Expression of recombinant proteins lacking methionine as N-terminal amino acid in plastids: Human serum albumin as a case study. Biotechnol J 127:593–604

Mishra P, Srivastava AK (2004) Molecular investigations in ornamental floricultural plants. J Plant Biotechnol 6:131 143

Nugent JM, Joyce SM (2005) Producing human therapeutic proteins in plastids. Curr Pharm Des 11:2459–2470

Neuhaus G, Spangenberg G (1990) Plant transformation by microinjection techniques. Physiol Planta 79:213–217

Okumura S, Sawada M, Park YW, Hayashi T, Shimamura M, Takase H, Tomizawa KI (2006) Transformation of poplar (Populus alba) plastids and expression of foreign proteins in tree chloroplasts. Transgenic Res 15:637–646

Provorov NA (2005) Molecular basis of symbiogenic evolution: from free-living bacteria towards organelles. Zh Obshch Biol 66:371–388

Pyke KA (2007) Plastid development and differentiation. In: Bock R (ed) Topics in current genetics: plastid development. Springer, Heidelberg, p 512

Rangasamy D, Ratledge C, Woolston CJ (1997) Plastid targeting and transient expression of rat liver ATP: Citrate lyase in pea protoplasts. Plant Cell Rep 16:700–704

Ruhlman T, Lee SB, Jansen RK, Hostetler JB, Tallon LB, Town CD, Daniell H (2006) Complete plastid genome sequence of Daucus carota: implications for biotechnology and phylogeny of angiosperms. BMC Genomics 7:222

Ruf S, Karcher D, Bock R (2007) Determining the transgene containment level provided by chloroplast transformation. Proc Natl Acad Sci USA 17:6998–7002

Scharff LB, Koop HU (2007) Targeted inactivation of the tobacco plastome origins of replication A and B. Plant J 50:782–794

Scotti N, Alagna F, Ferraiolo E, Formisano G, Sannino L, Buonaguro L, De Stradis A, Vitale A, Monti L, Grillo S, Buonaguro FM, Cardi T (2009) High-level expression of the HIV-1 Pr55 (gag) polyprotein in transgenic tobacco chloroplasts. Planta 229:1109–1122

To KY, Cheng MC, Chen LFO, Chen SCG (1996) Introduction and expression of foreign DNA in isolated spinach chloroplasts by electroporation The Plant Journal 10:737–743

Vidi PA, Kessler F, Brehelin C (2007) Plastoglobules: a new address for targeting recombinant proteins in the chloroplast. BMC Biotechnol 7:4

Villarreal F, Martán V, Colaneri A, González-Schain N, Perales M, Martán M, Lombardo C, Braun HP, Bartoli C and Zabaleta E (2009) Ectopic expression of mitochondrial gamma carbonic anhydrase 2 causes male sterility by anther indehiscence Plant Mol Biol 70:471–485

Wakasugi T, Tsudzuki T, Sugiura M (2001) The genomics of land plant chloroplasts: gene content and alteration of genomic information by RNA editing. Photosynth Res 70:107–118

Watson J, Koya V, Leppla SH, Daniell H (2004) Expression of *Bacillus anthracis* protective antigen in transgenic chloroplasts of tobacco, a non-food/feed crop. Vaccine 22:4374–4384

Weber G, Monajembashi S, Greulich KO, Wolfrum J (1988) Genetic manipulation of plant-cells and organelles with a laser microbeam. Plant Cell Tissue Organ Cult 12:219–222

Weber-Lotfi F, Ibrahim N, Boesch P, Cosset A, Konstantinov Y, Lightowlers RN, Dietrich A (2009) Developing a genetic approach to investigate the mechanism of mitochondrial competence for DNA import. Biochim Biophys Acta 1787(5):320–327

Wurbs D, Ruf S, Bock R (2007) Contained metabolic engineering in tomatoes by expression of carotenoid biosynthesis genes from the plastid genome. Plant J 49:276–288

Zhou F, Badillo-Corona JA, Karcher D, Gonzalez-Rabade N, Piepenburg K, Borchers AMI, Maloney AP, Kavanagh TA, Gray JC, Bock R (2008) High-level expression of human immunodeficiency virus antigens from the tobacco and tomato plastid genomes. Plant Biotechnol J 6:897–917

Zubko MK, Zubko EI, van Zuilen K, Meyer P, Day A (2004) Stable transformation of petunia plastids. Transgenic Res 13:523–530

Zuker A, Tzfira T, Vainstein A (1998) Genetic engineering for cut-flower improvement. Biotechnol Adv 16:33–79

Chapter 9
Biosynthesis and Biotransformation

Hajiem Mizukami and Hiroaki Hayashi

9.1 Introduction

Plant secondary metabolites, the so-called natural products, are used as flavors, food additives, and pharmaceuticals. Since complex natural products are not economically produced by total chemical synthesis except for some small molecules, most of the useful metabolites are still obtained from wild or cultivated plant resources. Alternatively, plant cell culture technique would be a potential method to produce these useful secondary metabolites, and commercial production of some useful secondary metabolites has been achieved (Kolewe et al. 2008; Smetanska 2008). Furthermore, biotransformation by plant cell cultures could be a useful method to convert natural products or unnatural synthetic products into chemically different products of economical importance. This chapter will focus on biosynthesis and biotransformation of useful chemical compounds using plant cell culture technology.

9.2 Biosynthesis of Useful Secondary Metabolites by Plant Cell Cultures

Higher plants produce small organic molecules of diverse structures, such as alkaloids, terpenoids, and flavonoids, which are localized in the specific organs of intact plants. Extensive studies showed that some of the useful metabolites are produced by plant cell suspension cultures, while others are not produced by the undifferentiated cells. Some of them are produced by organ culture, especially root cultures and hairy root cultures. However, some useful metabolites are produced

H. Mizukami (✉)

Department of Pharmacognosy, Graduate School of Pharmaceutical Sciences, Nagoya City University, 3-1 Tanabe-dori, Mizuho-ku Nagoya, 467-8603, Japan

e-mail: hajimem@phar.nagoya-cu.ac.jp

C. Kole et al. (eds.), *Transgenic Crop Plants*,
DOI 10.1007/978-3-642-04809-8_9, © Springer-Verlag Berlin Heidelberg 2010

neither by undifferentiated cells nor by hairy root cultures. In this section, useful secondary metabolites will be discussed under three groups, based on the strategies of production by plant cell cultures: (1) metabolites produced by cell suspension cultures, (2) metabolites produced by hairy root cultures, (3) metabolites that are hardly produced by plant cell/organ cultures. These three classes of metabolites will be discussed using seven selected metabolites as examples in the following subsections, with a focus on the secondary metabolites used for pharmaceuticals.

9.2.1 Useful Secondary Metabolites Produced by Cell Suspension Cultures

Table 9.1 shows some examples of useful secondary metabolites produced by cell suspension cultures. Two outstanding examples of the metabolites are shikonin and paclitaxel, both of which are commercially produced by undifferentiated plant cells. Thus, in this subsection, production of shikonin by *Lithospermum erythrorhizon* cells and paclitaxel by *Taxus* cells will be discussed.

9.2.1.1 Shikonin

Shikonin derivatives, fatty acid esters of shikonin, are red pigments isolated from the roots of *L. erythrorhizon* (Family Boraginaceae) that have been used as a traditional medicine in Japan and China in the form of an ointment to treat wound, burn, and anal hemorrhage (Tabata 1996). Shikonin derivatives have also been used as a dyestuff from ancient times. These red pigments exhibit various pharmacological activities including wound healing, anti-inflammatory, antitumor, and antimicrobial activities (Papageorgiou et al. 1999).

Callus cultures established from seedlings of *L. erythrorhizon* successfully produced shikonin derivatives when the callus was incubated in the dark (Tabata et al. 1974), and repeated cell selection resulted in the establishment of a high shikonin producing culture strain, M18, whose shikonin content was higher than that of the intact roots (Mizukami et al. 1978). Production of shikonin derivatives

Table 9.1 Examples of useful secondary metabolites produced by cell suspension cultures

Metabolite	Plant species	Application	Reference
Shikonin (quinone)	*Lithospermum erythrorhizon*	Dyes, cosmetic	Tabata et al. (1974)
Paclitaxel (diterpenoid)	*Taxus* spp.	Antitumor	Yukimune et al (1996)
Berberine (alkaloid)	*Coptis japonica* *Thalictrum minus*	Antibacterial	Sato and Yamada (1984) Nakagawa et al. (1984)
Sanguinarin (alkaloid)	*Papaver somniferum*	Antibacterial	Eilert et al. (1985)
Ginseonoside (saponin)	*Panax ginseng*	Tonic	Furuya et al. (1983)
Soyasaponin (saponin)	*Glycyrrhiza glabra*	Hepatoprotective	Hayashi et al. (1990)

was completely inhibited by light and a synthetic plant hormone, 2,4-dichlorophe-
noxyacetic acid (2,4-D). These dramatic regulations of shikonin biosynthesis
prompted further experiments to identify many positive-regulating factors such as
agar, copper ion, oligogalacturonide, and methyl jasmonate, as well as many
negative-regulating factors including light, 2,4-D, ammonium ion, and glutamine
(Tabata 1996; Yazaki et al. 1997). Furthermore, establishment of the production
medium M9 containing no ammonium ion but nitrate ion for pigment formation
(Fujita et al. 1981) led to the commercial production of shikonin by the Mitsui
Petrochemical Company (Tabata and Fujita 1985). Thus *L. erythrorhizon* cell
culture system provides us with a model system suitable to elucidate the regulatory
mechanism of secondary metabolism in higher plants.

   Biosynthesis pathway of shikonin and its related compounds has been extensively studied, whereas the subsequent steps of shikonin biosynthesis are still
unknown (Tabata 1996). Figure 9.1 depicts the biosynthetic pathways of shikonin
and the related constituents in *L. erythrorhizon* cultures. Shikonin is biosynthesized

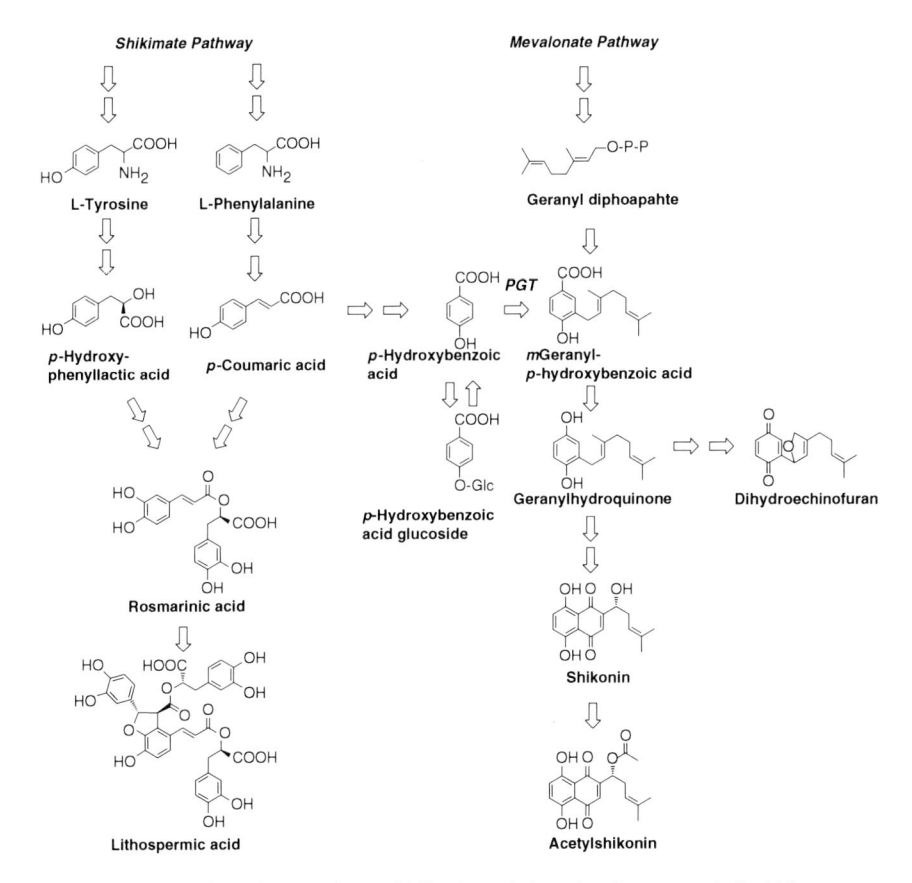

Fig. 9.1 Biosynthetic pathways of acetylshikonin and the related compounds in *Lithospermum
erythrorhizon*

from *m*-geranyl-*p*-hydroxybenzoic acid, a coupling product of *p*-hydroxybenzoic acid derived from the shikmate pathway with geranyl diphosphate derived from isoprene units. This coupling step is catalyzed by *p*-hydroxybenzoate geranyltransferase (PGT), which plays a crucial role in the regulation of shikonin biosynthesis (Heide et al. 1989). Two PGT cDNAs have been isolated and characterized from *L. erythrorhizon*, and the regulation of their expression by various physical and biochemical factors coincides with that of shikonin biosynthesis (Yazaki et al. 2002). It is also noteworthy that geranyl moiety of shikonin was shown to be formed via the mevalonate pathway (Li et al. 1998), whereas most of monoterpenes are produced via the MEP pathway, the so-called non-mevalonate pathway (Rohmer 1999).

Not only shikonin derivatives but also dihydroechinofuran, *p*-hydroxybenzoate (PHB) glucoside, and caffeic acid oligomers are accumulated in the cultured *L. erythrorhizon* cells (Tabata 1996; Yamamoto et al. 2002). Production of both shikonin and dihydroechinofuran, an unusual metabolite derived from *m*-geranyl-*p*-hydroxybenzoic acid, is induced by addition of oligogalacturonides or methyl jasmonate to the cultures (Tani et al. 1993; Yazaki et al. 1997), although the accumulation of dihydroechinofuran precedes that of shikonin during the induction by the elicitors. PHB glucoside is regarded as a storage form of PHB, a biosynthetic intermediate of the shikonin biosynthesis, and the accumulation of PHB glucoside is induced by light irradiation, which inhibits the shikonin biosynthesis (Tabata 1996). In addition, the cultured *Lithospermum* cells produce large amounts of caffeic acid oligomers, such as rosmarinic acid, lithospermic acid B, and (+)-rabdosiin derived from phenylpropanoid pathway, whose biosynthesis is regulated differently from that of shikonin derivatives (Yamamoto et al. 2002).

9.2.1.2 Paclitaxel

Paclitaxel isolated from the bark of pacific yew, *Taxus brevifolia* (Family Taxaceae), is an antimitotic drug used in chemotherapy of breast, ovarian, and lung cancers. Since the supply of paclitaxel by isolation from the bark of the slow-growing yew tree is limited, alternative source of paclitaxel was indispensable for its clinical application (Frense 2007). However, the economic production of paclitaxel by total chemical synthesis has not yet been achieved. Alternatively, paclitaxel was produced by semi-synthesis from 10-deacetylbaccatin III, which is a biosynthetic intermediate of paclitaxel and could be obtained from the needle of European yew, *Taxus baccata*, in relatively large amounts. As a more economical source to supply paclitaxel, *Taxus* cell suspension cultures is promising (Frense 2007). Production of paclitaxel by plant cell suspension cultures has been intensively studied, and production of paclitaxel in cell suspension cultures was shown to be significantly up-regulated by methyl jasmonate (Yukimune et al. 1996). Now, paclitaxel is commercially produced by plant cell culture technique using the undifferentiated cells of *Taxus* plants (Frense 2007; Kolewe et al. 2008).

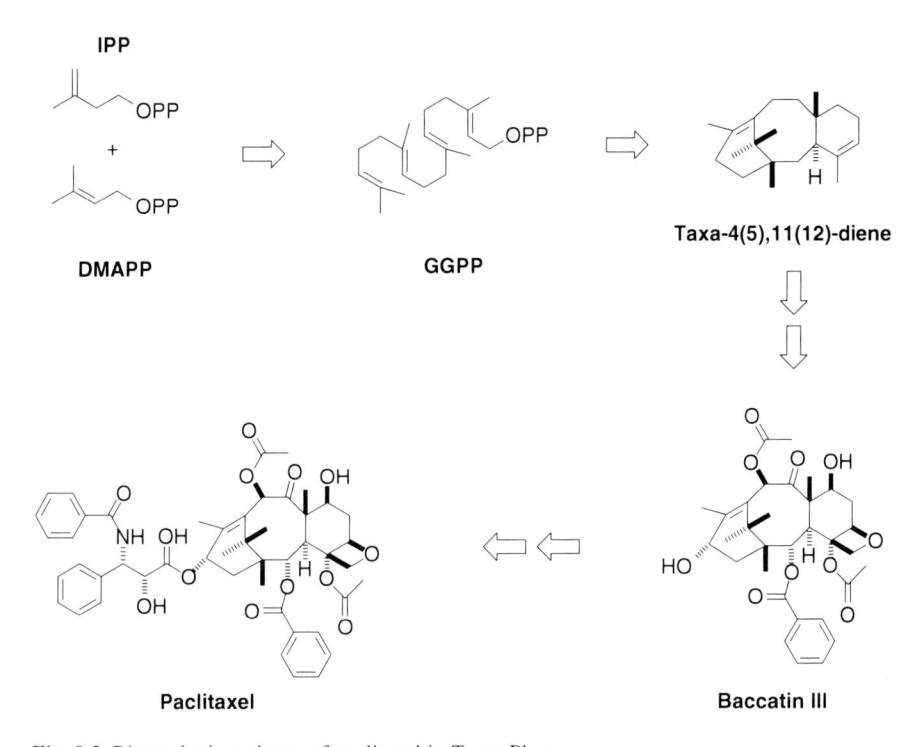

Fig. 9.2 Biosynthetic pathway of paclitaxel in *Taxus* Plants

Since paclitaxel and its analogs are important anticancer drugs, biosynthesis of paclitaxel has been extensively studied to characterize various genes encoding the enzymes involved in the paclitaxel biosynthesis (Jennewein and Croteau 2001; Frense 2007). Figure 9.2 describes the biosynthetic pathway of paclitaxel. Geranylgeranyl diphosphate derived from the MEP pathway was cyclized by a terpene synthase, taxa-4(5),11(12)-diene synthase, into a cyclic diterpene, taxa-4(5),11 (12)-diene. The cDNA encoding taxa-4(5),11(12)-diene synthase has been cloned and characterized (Wildung and Croteau 1996). Taxa-4(5),11(12)-diene is further hydroxylated and acylated by several complex steps to produce paclitaxel. Many of genes involved in the later steps have also been characterized (Jennewein and Croteau 2001; Frense 2007).

9.2.1.3   Other Useful Metabolites Produced in the Undifferentiated Cells

Besides shikonin and paclitaxel, many useful secondary metabolites are produced by plant cell suspension cultures as shown in Fig. 9.3. Regarding alkaloids, an antimicrobial isoquinoline alkaloid berberine is produced in the undifferentiated cells of *Coptis japonica* (Sato and Yamada 1984) and *Thalictrum minus* (Nakagawa et al. 1984). Berberine is one of the few secondary metabolites whose biosynthetic

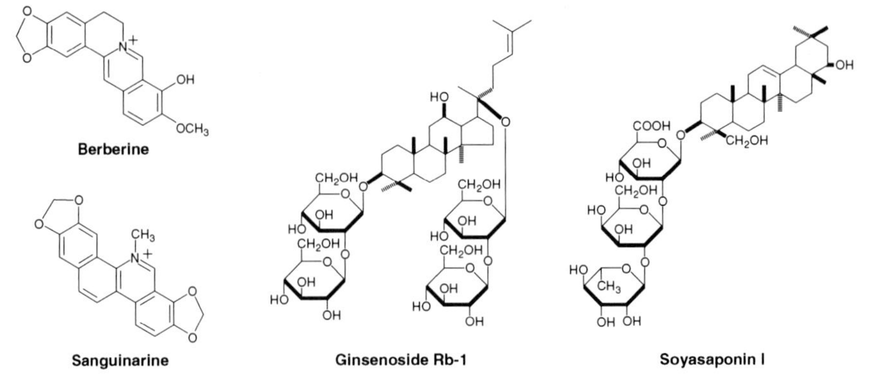

Fig. 9.3 Structures of useful metabolites produced by cell suspension cultures

pathway has been completely characterized (Hashimoto and Yamada 1994). Another antimicrobial alkaloid sanguinarin is produced in the fungal elicitor treated cell suspension cultures of the opium poppy, *Papaver somniferum*, whereas the narcotic alkaloid morphine is not produced in the undifferentiated opium cells (Eilert et al. 1985). Biosynthesis of morphine will be discussed under Sect. 9.2.2.3.

Certain triterpene saponins are also produced by undifferentiated cells of plant cell cultures. Cell suspension cultures of *Panax ginseng* produce ginsenosides, which are active constituents of ginseng, a famous tonic used in the Far East Asia (Furuya et al. 1983). The content and the composition of ginsenosides in cell suspension cultures are almost the same as those in the roots of the cultivated ginseng. Cell suspension cultures of *Glycyrrhiza glabra* (licorice) produce soyasaponins, common triterpene saponins in legumes. However, they do not produce the sweet tritepene saponin, glycyrrhizin, which is localized in the thickened roots of the intact plants (Hayashi et al. 1990). Triterpenoid biosynthesis in licorice will be further discussed under Sect. 9.2.3.2.

9.2.2 Useful Secondary Metabolites Produced by Hairy Root Cultures

Many plant cell suspension cultures have failed to produce useful secondary metabolites that are produced in the respective intact plants. Alternatively, some of these metabolites are produced by organ cultures, such as root cultures and shoot cultures (Kolewe et al. 2008; Smetanska 2008). In particular, hairy root cultures obtained by transforming plant cells with *Agrobacterium rhizogenes* have been used to produce useful secondary metabolites that are not produced in cell suspension cultures (Georgiev et al. 2007). Table 9.2 includes some examples of useful secondary metabolites that are not produced by the undifferentiated cells

Table 9.2 Examples of useful secondary products produced by hairy root cultures

Metabolite	Plant species	Application	Reference
Camptothecin (alkaloid)	*Camptotheca acuminata*	Antitumor	Lorence et al. (2004)
	Ophiorrhiza pumila		Saito et al. (2001)
Hypscyamine (alkaloid)	*Atropa belladonna*	Anticholinergic	Kamada et al. (1986)
	Datura stramonium		Payne et al. (1987)
Scopolamine (alkaloid)	*Scopolia japonica*	Anticholinergic	Mano et al. (1986)
	Duboisia leichhardtii		Mano et al. (1989)
Morphine (alkaloid)	*Papaver somniferum*	Narcotic analgesic	Park and Facchini (2000)
Artemisinin (sesquiterpene)	*Artemisia annua*	Antimalarial	Weathers et al. (1994)
Saikosaponin (saponin)	*Bupleurum falcatum*	Anti-inflammatory	Kim et al. (2006)

but produced by the hairy root cultures. In this subsection, production of tropane alkaloids and camptothecin will be discussed.

9.2.2.1 Tropane Alkaloids

Tropane alkaloids, hyoscyamine and scopolamine, are isolated from certain Solanaceaus plants including *Atropa belladonna, Datura innoxia, Duboisia leichhardtii, Hyoscyamus niger*, and *Scopolia japonica*. They are muscarinic antagonists used for treatment of various gastrointestinal disorders. Callus cultures of *Scopolia parviflora* produce only trace amounts of tropane alkaloids, whereas the root-differentiated tissue accumulates higher content of tropane alkaloids (Tabata et al. 1972). Similar results were observed in other tropane alkaloid producing plants, such as *A. belladonna, Datura stramonium*, and *D. leichhardtii*. In contrast, undifferentiated cell suspension cultures of *H. niger* produced small amount of tropane alkaloids (Yamada and Hashimoto 1982; Yamada and Endo 1984). In the intact plants, tropane alkaloids are produced mainly in the root and are translocated into the aerial parts, which have the storage organs of these alkaloids (Waller and Nowacki 1978). Thus, hairy root cultures of these tropane alkaloid producing plants were established to produce alkaloids. As expected, hairy root cultures of *S. japonica* (Mano et al. 1986), *A. belladonna* (Kamada et al. 1986), *Datura stramonium* (Payne et al. 1987), and *D. leichhardtii* (Mano et al. 1989) were shown to produce high levels of tropane alkaloids.

Biosynthetic pathway of tropane alkaloids is extensively studied using the root cultures as shown in Fig. 9.4 (Hashimoto and Yamada 1994). Tropane alkaloids are esters of tropine and tropic acid, which are derived from arginine and phenylalanine, respectively. The most characterized enzyme in the biosynthetic pathway of tropane alkaloids is hyoscyamine-6-hydroxylase (H6H), a 2-oxoglutarate-dependent dioxygenase, involved in the conversion of hyoscyamine into scopolamine. This enzyme is localized in the pericycle of the root (Hashimoto et al. 1991). This

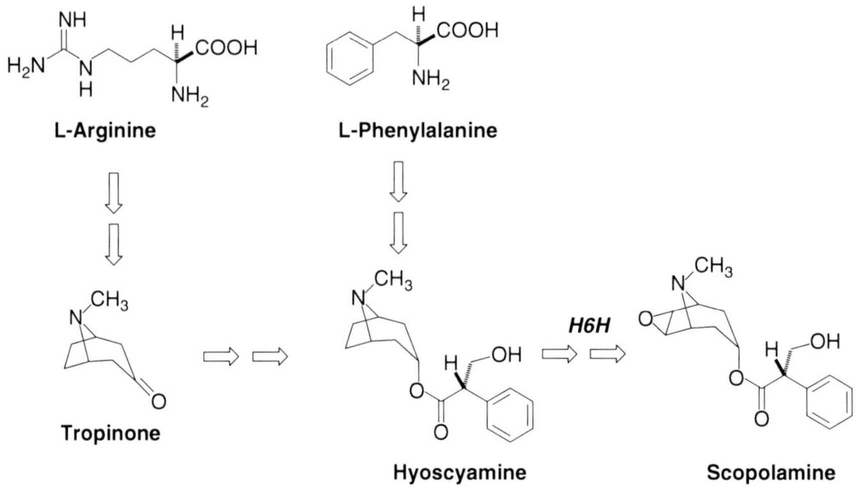

Fig. 9.4 Biosynthetic pathway of tropane alkaloids

enzyme is involved not only in the hydroxylation of hyoscyamine to 6-hydroxy-hyoscyamine but also in the epoxidation of 6-hydroxyhyoscyamine to scopolamine. H6H cDNA has been isolated from *H. niger* (Matsuda et al. 1991), and metabolic engineering using the H6H gene resulted in transgenic *A. belladonna* with the preferential accumulation of scopolamine, which has a higher commercial demand than hyoscyamine (Yun et al. 1992).

9.2.2.2 Camptothecin

Camptothecin is an anticancer compound isolated from *Camptotheca acuminata* (Family Nyssaceae). Although camptothecin itself is not used for cancer chemotherapy due to its toxicity, the semisynthetic compounds derived from camptothecin, such as irinotecan and topotecan, are clinically important antitumor drugs in the world (Fig. 9.5). Since economical total synthesis of camptothecin has not been achieved, plant cell cultures can be an alternative method to produce camptothecin (Lorence and Nessler 2004).

Cell suspension cultures of *C. acuminata* were established, but they produced only trace amounts of camptothecin (Sakato et al. 1974). Later, hairy root cultures of *C. acuminata* were found to produce significant amount of camptothecin (Lorence et al. 2004). Camptothecin is produced not only by *C. acuminata* but also by many other species of taxonomically unrelated families including *Nothapodytes foetida* (Family Icacinaceae) and *Ophiorrhiza pumila* (Family Rubiaceae) (Lorence and Nessler 2004). Although undifferentiated callus cultures of *O. pumila* produced no camptothecin (Kitajima et al. 1998), hairy root cultures of *O. pumila* produced high levels of camptothecin (Saito et al 2001). Furthermore, a part of camptothecin produced by *O. pumila* hairy roots was excreted into culture media, and the excreted

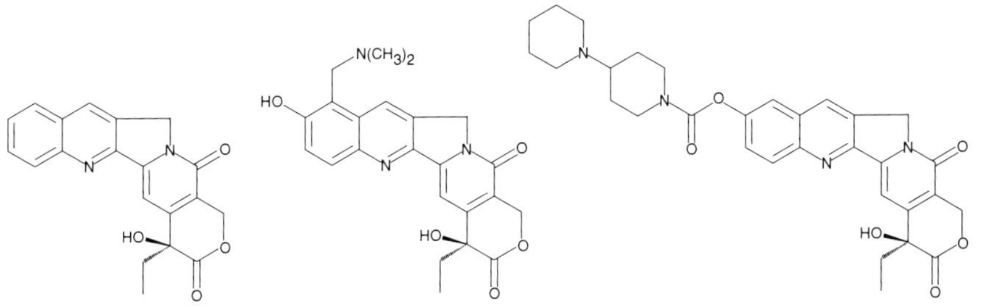

Fig. 9.5 Camptothecin and anticancer drugs (topothecan and irinotecan) derived from camptothecin

camptothecin in the medium was efficiently recovered by resin. This system is feasible to supply camptothecin from plant cell cultures.

Camptothecin is a monoterpene indole alkaloid biosynthesized from a key intermediate strictosidine. Strictosidine is the condensation product of tryptamine, a monoamine alkaloid derived from shikimate pathway, with secologanin, an iridiod glycoside (for the biosynthetic pathway see Fig. 9.8). It has been shown that the secologanin moiety in the structure of comptothecin is derived from monoterpene formed via MEP pathway (Yamazaki et al. 2004). cDNAs encoding tryptamine decarboxylase and strictosidine synthase, two key enzymes in the monoterpene indole alkaloid biosynthesis, have already been isolated and characterized (Yamazaki et al. 2003). However, later steps of camptothecin biosynthesis are so far unknown.

9.2.2.3   Morphine

Morphine, an important narcotic analgesic, is a benzylisoquinoline alkaloid produced by opium poppy, *Papaver sominiferum* (Family Papaveraceae), which is one of the most thoroughly investigated model plants to elucidate the regulation of alkaloid biosynthesis in higher plants (Facchini and De Luca 2008). Although cell suspension cultures of *P. somniferum* produce no detectable amount of morphine and codeine, the undifferentiated cells treated with fungal elicitors accumulate sanguinarine, another antimicrobial alkaloid distributed in some plants of the family Papaveraceae (Eilert et al. 1985). In contrast, hairy root culture of *P. somniferum* was shown to produce small amount of morphine (Park and Facchini 2000).

Biosynthetic pathways of morphine and sanguinarine have been almost completely characterized (Facchini et al. 2007). Figure 9.6 illustrates the biosynthetic pathways of morphine and related alkaloids. These benzylisoquinoline alkaloids share the central intermediate (S)-norcoclaurine, which is produced by norcoclaurine synthase from dopamine and 4-hydroxyphenylacetaldehyde. (S)-Norcoclaurine is converted into the branching intermediate (S)-reticuline. The first committed step in the morphine biosynthesis involves the conversion of (S)-reticuline into its (R)-epimer. On the other hand, the conversion of (S)-reticuline to

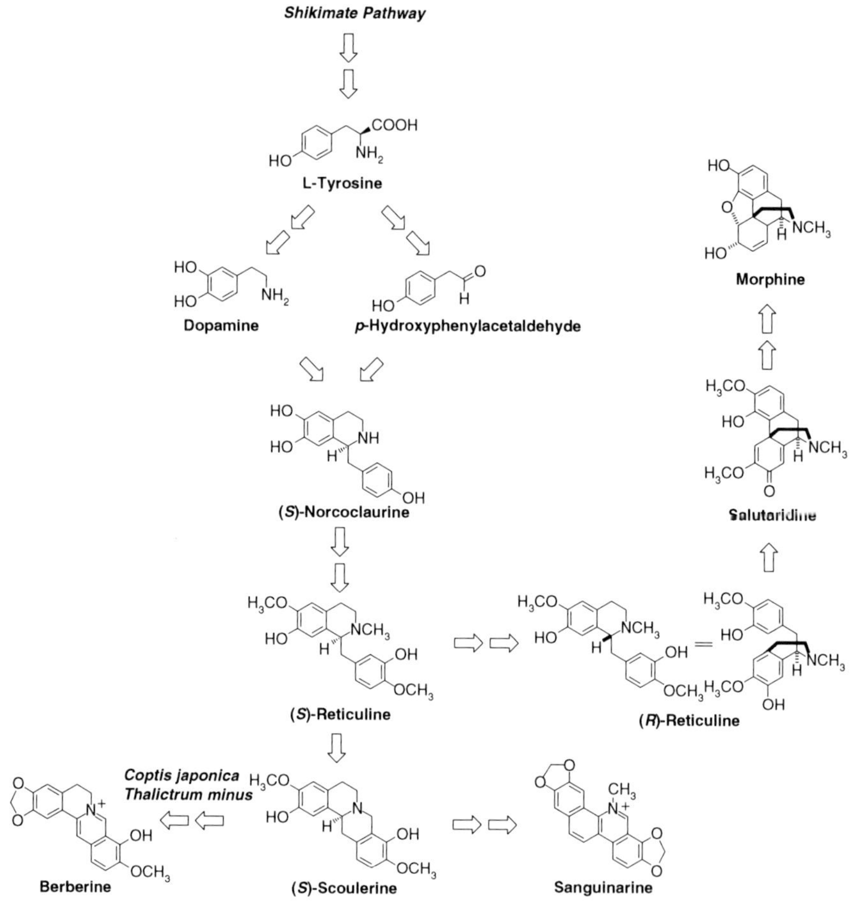

Fig. 9.6 Biosynthetic pathway of morphine and related alkaloids in *Papaver somniferum*

(S)-scoulerine by berberine bridge enzyme leads to sanguinarine in *P. somniferum* as well as to berberine in *C. japonica* (Hashimoto and Yamada 1994; Facchini et al. 2007). Many genes involved in the alkaloid biosynthesis have been so far identified from opium poppy. Extensive studies including the biosynthesis pathway, gene regulation, and metabolic engineering of morphine and related alkaloids are reviewed elsewhere (Facchini et al. 2007; Facchini and De Luca 2008).

9.2.2.4 Other Useful Metabolites Produced by Hairy Root Cultures

In addition to the metabolites mentioned earlier, many other useful secondary products are produced by hairy root cultures, whereas they are not produced by cell suspension cultures. Saikosaponins (Fig. 9.7) are oleanane-type triterpene saponins with antiallergic and anti-inflammatory activities, and are isolated from

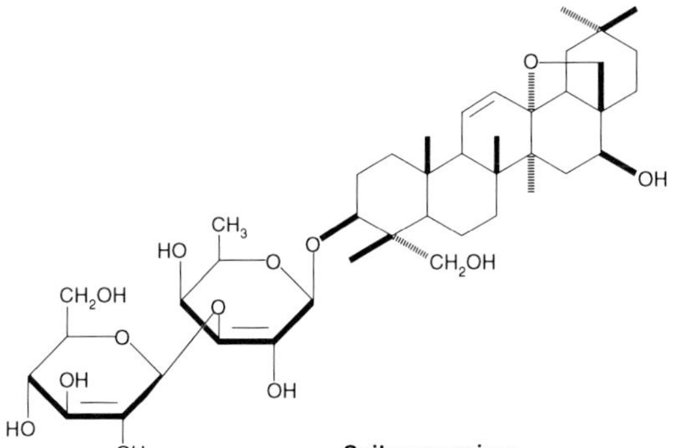

Fig. 9.7 Structures of useful metabolites produced by hairy root cultures

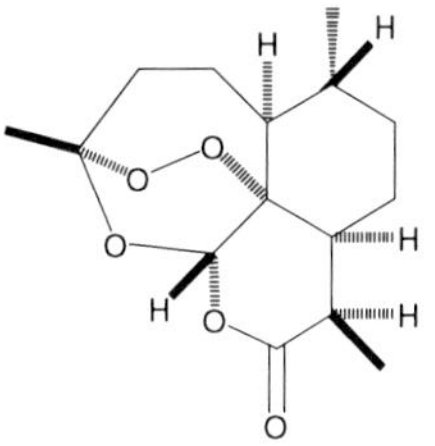

the root of *Bupleurum falcatum*. Although callus cultures of *B. falcatum* produce no detectable amounts of saikosaponins, adventitious roots differentiated from the callus (Uomori et al. 1974) as well as hairy root cultures (Kim et al. 2006) of *B. falcatum* are capable of producing large amount of saikosaponins. Artemisinin (Fig. 9.7), an endoperoxide sesquiterpene lactone isolated from *Artemisia annua*, has a potent antimalarial activity against the chloroquin-resistant malarial parasite *Plasmodium falciparum*; thus, this compound is one of the targets for plant cell cultures (Liu et al. 2006). No artemisinin was produced in the cell suspension cultures of *A. annua,* whereas only trace amount of artemisinin was found in multiple shoot cultures (Paniego and Giuletti 1994). Artemisinin was produced by hairy root cultures of *A. annua*, suggesting that the commercial production of artemisinin by hairy root culture is feasible (Weathers et al. 1994).

9.2.3   Useful Secondary Products That Are Hardly Produced by Plant Cell Cultures

Despite the extensive efforts to produce useful secondary products by the plant cell cultures, some target metabolites are hardly produced by the cell suspension cultures as well as by the hairy root cultures. Examples of these useful products are listed in Table 9.3. In this subsection, dimeric monoterpene indole alkaloids, vincristine and vinblastine in *Catharanthus roseus* and glycyrhrizin in *G. glabra*, are described.

9.2.3.1   Vinca Alkaloids

Madagascar periwinkle, *C. roseus* (syn. Vinca; Family Apocynaceae), produces the Vinca alkaloids vincristine and vinblastine, which are mitotic inhibitors used for clinical treatment of cancer. Both vincristine and vinblastine are dimeric

Table 9.3 Examples of useful secondary metabolites that are hardly produced by plant cell cultures

Metabolite	Plant species	Application	Culture method	Reference
Vinblastine (alkaloid)	*Catharanthus roseus*	Antitumor	Cell suspension	Eilert et al. (1987)
	Hairy root			Parr et al. (1988)
Glycyrrhizin (saponin)	*Glycyrrhiza glabra*	Sweetener	Cell suspension	Hayashi et al. (1988)
	Hairy root			Toivonen and Rosenqvit (1995)

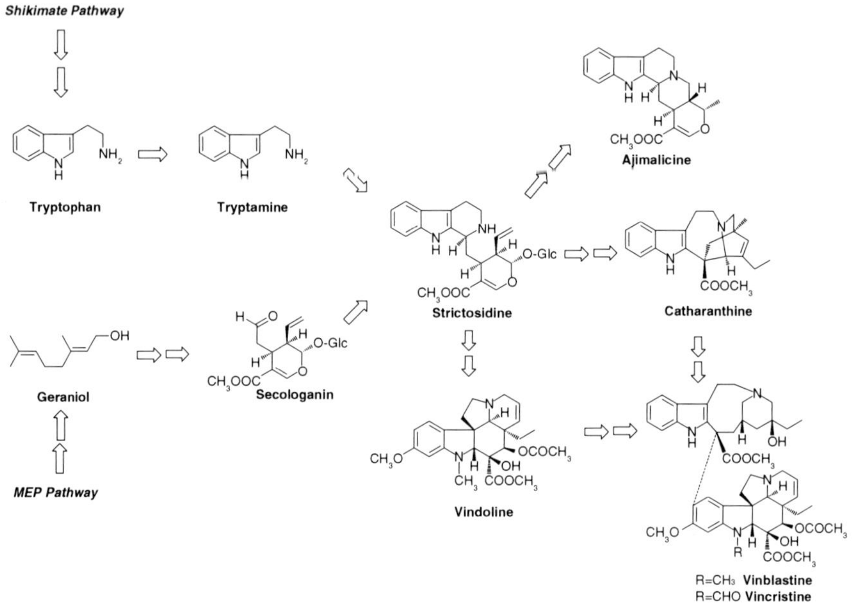

Fig. 9.8 Biosynthetic pathways of vinblastine and vincristine in *Catharanthus roseus*

monoterpene indole alkaloids with complex chemical structures (Facchini and De Luca 2008). The content of Vinca alkaloids is low in the field-grown plants, which are the actual sources to supply these alkaloids at the present. Thus, an alternative method to produce these alkaloids is necessary, and plant cell culture technique would be an attractive method. However, despite the enormous efforts, the dimeric alkaloids are not produced sufficiently by cell suspension culture (Eilert et al. 1987) or by hairy root cultures of *C. roseus* (Parr et al. 1988; Toivonen et al. 1989). To overcome this difficulty, extensive studies including chemistry, biochemistry, and molecular biology have been carried out. In fact, *C. roseus* is one of the most thoroughly investigated medicinal plants. This subject has been well reviewed elsewhere (Facchini and De Luca 2008).

Figure 9.8 depicts the biosynthetic pathways of Vinca alkaloids. Monoterpene indole alkaloids are biosynthesized from a key intermediate, strictosidine, which is the condensation product of tryptamine, a monoamine derived from shikimate pathway, and secologanin, an iridiod glycoside derived from nonmevalonate pathway. Although hairy root cultures of *C. roseus* produce only trace amounts of dimeric monoterpene indole alkaloids, they produce monomeric alkaloids, such as ajmalicine and catharanthine. However, the hairy roots produce no detectable amount of vindoline, which is one of the building blocks of the dimeric indole alkaloids (Toivonen et al. 1989).

9.2.3.2   Glycyrrhizin

Glycyrrhizin is a sweet oleanane type triterpene saponin isolated from the roots and stolons of *G. glabra* (licorice) of the family Fabaceae. Dried roots and stolons of licorice have been used as an important crude drug from ancient times (Shibata 2000). Glycyrrhizin is also used as an anti-inflammatory drug for treatment of hepatitis. Cell suspension cultures of *G. glabra* produced no detectable amount of glycyrrhizin (Hayashi et al. 1988). Glycyrrhizin also was not produced by hairy root cultures of *G. glabra* (Toivonen and Rosenqvit 1995). In the intact plant of *G. glabra,* the accumulation of glycyrrhizin is localized in the thickened roots and stolons, and glycyrrhizin is not contained in the rootlets, leaves, stems, and seeds (Hayashi et al. 1993). Instead of glycyrrhizin, cultured licorice cells produced two triterpenoid constituents, viz. soyasaponins and betulinic acid. Soyasaponins are also oleanane-type triterpene saponin, and are localized mainly in the seed and rootlet of *G. glabra*. Betulinic acid, a lupane-type triterpene distributed widely in higher plants, is mainly localized in the cork layer of the thickened licorice roots.

Figure 9.9 shows the biosynthetic pathways of glycyrrhizin and the related triterpenoids. These triterpenoids and sterols share a common key intermediate, 2,3-oxidosqualene, which is formed via the mevalonate pathway (Rohmer 1999). 2,3-Oxidosqualene is further converted by three oxidosqualene cyclases, β-amyrin synthase (bAS), lupeol synthase (LUS), and cycloartenol synthase (CAS) into the three cyclization products, respectively, leading to the end-products. cDNAs of these three oxidosqualene cyclases are characterized from *G. glabra,* and mRNA levels of the oxidosqualene cyclases were differently regulated in the intact plants and cultured cells of *G. glabra*. The levels of their mRNAs correlate with the accumulation of respective end products indicating that the transcription of oxidosqualene cyclase genes is an important regulatory step for triterpenoid biosynthesis (Hayashi et al. 2003, 2004). The following steps of the saponin biosynthesis pathway are oxidations and glycosylations of triterpenes. Recently, β-amyrin 11-oxidase, a cytochrome P450 involved in glycyrrhizin biosynthesis, has been characterized (Seki et al. 2008). Metabolic engineering of a saponin-producing plant or microorganism is an attractive approach to produce unique and useful saponins in the future.

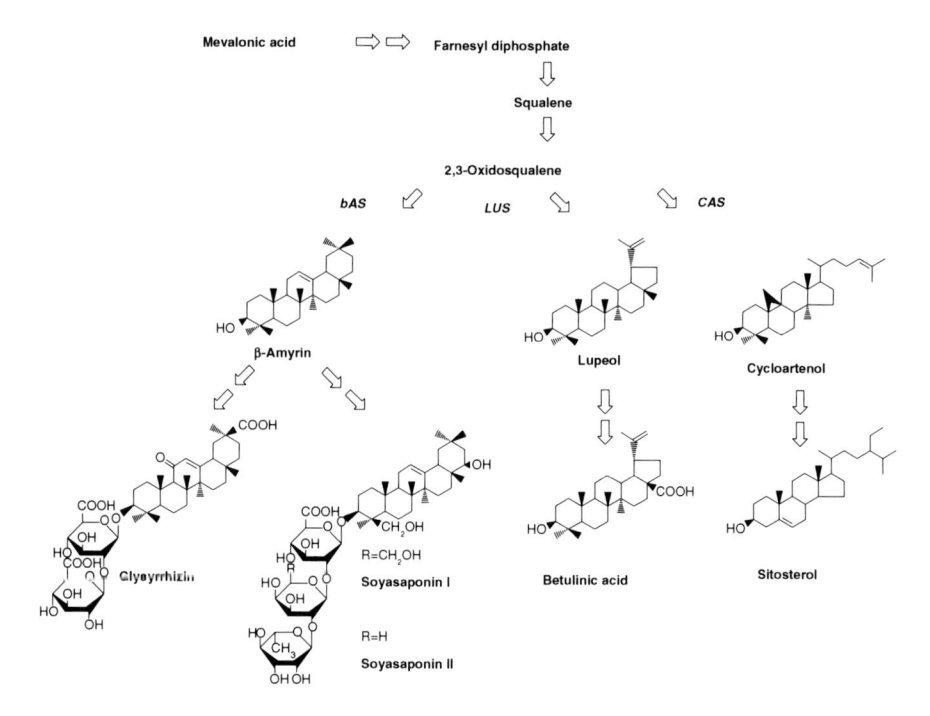

Fig. 9.9 Biosynthetic pathways of glycyrrhizin and related triterpenoids in *Glycyrrhiza glabra*

9.3   Biotransformation

In addition to synthesizing secondary metabolites de novo from carbon sources supplemented in the culture medium, plant cells are able to carry out biotransformation reactions on substrates exogenously added to the culture medium. Not only secondary metabolites but also xenobiotic compounds can be used as substrates of biotransformation. Such biotransformation reactions have been investigated as one of the major targets for biotechnological application of plant cell culture systems because plant cells can catalyze the stereospecific- and/or region-specific modification of organic compounds that are not easily carried out by chemical synthesis or by microorganisms (Rao and Ravishankar 1999). Biotransformation reactions include oxido-reduction, hydroxylation, glycosyl conjugation, acylation, methylation, etc. (Suga and Hirata 1990). This chapter describes biotransformations catalyzed by cultured plant cells, focusing on examples with potential importance for industrial application.

9.3.1   Glycosylation

Higher plants synthesize a wide range of glycosides as secondary metabolites, and are capable of conjugating sugar residues not only to endogenous metabolic intermediates but also xenobiotics. Glycosyl conjugation of lipophilic low molecular weight compounds is an efficient tool to enhance water solubility, to improve stability, and thereby to increase bioavailability and to modify biological activity. Chemical synthesis of glycosides is usually difficult because it involves multiple blocking/deblocking steps before any product can be obtained. Thus, glycosylation of various organic compounds has attracted attention as one of the targets in biotechnological application of plant cell culture systems. Here we describe some examples of practical interests.

9.3.1.1   Arbutin

Arbutin is a monoglucoside of hydroquinone and a main bioactive compound contained in *Arctostaphylos uva-ursi* that has been traditionally used as a urethral disinfect. Arbutin also inhibits the melanin formation in human (Akiu et al. 1988) and is used as an ingredient of cosmetics. Glucosylation of hydroquinone to arbutin was first shown using *Datura innoxia* cell suspension cultures (Tabata et al. 1976). Later, it was shown that *C. roseus* cells in culture can efficiently convert hydroquinone to arbutin, particularly when medium concentration of sucrose is increased (Yokoyama et al. 1990). A large-scale production system up to 20-L jar fermenter was established by culturing *C. roseus* cells at high density followed by continuous supply of hydroquinone to the medium (Fig. 9.10). The arbutin yield could be increased as high as 9.2 g/L (corresponding to 45% of cell dry weight) and the conversion rate from hydroquinone was 98% (Inomata et al. 1991).

9.3.1.2   Curcumin

Curcumin (diferuloylmethane) is a yellow pigment of turmeric (dried rhizome of *Curcuma longa*). It has been primarily used as a food colorant but has attracted increased attention because of its potent pharmacological activities (Maheshwari et al. 2006). However, its low water solubility limits further pharmacological exploration and practical application. Screening of cell cultures from ten different

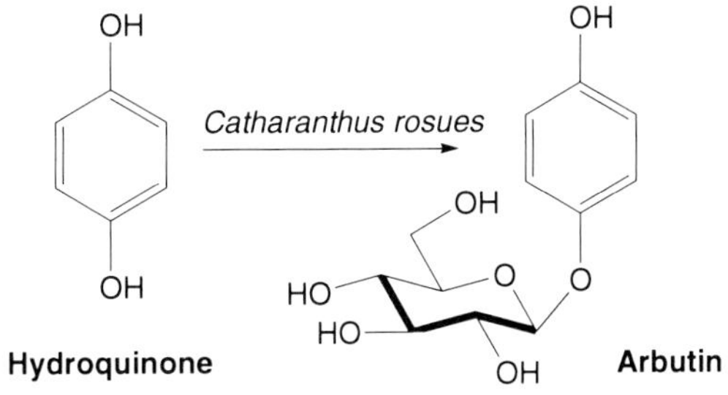

Fig. 9.10 Glucosylation of hydroquinone in *Catharanthus roseus* cell suspension cultures

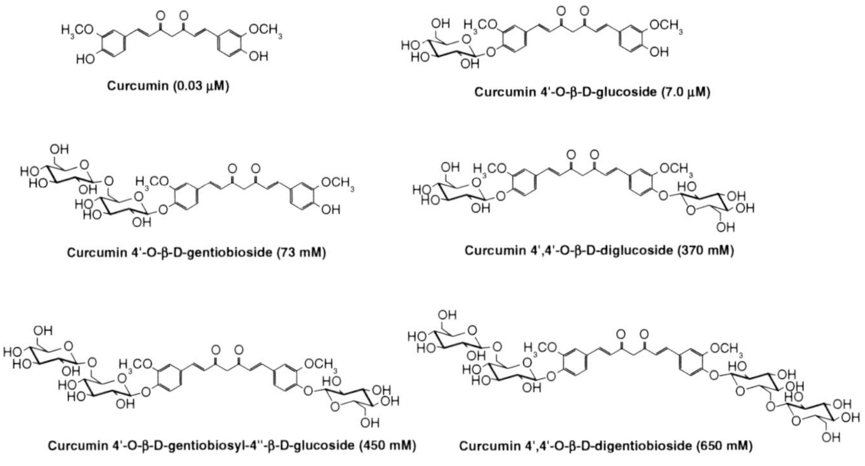

Fig. 9.11 Curcumin glucosides produced from curcumin by *Catharanthus roseus* cell suspension cultures Water solubility of the each compound is shown in a parenthesis

plant species revealed that *C. roseus* cells converted exogenously supplied curcumin to a series of curcumin glucosides as shown in Fig. 9.11 (Kaminaga et al. 2003). Although the water solubility of curcumin monoglucoside was increased by only 230-fold, the solubility was increased about two million-fold and about 20 million-fold in the case of curcumin monogentiobioside and curcumin digentiobioside, respectively, compared to the solubility of curcumin. The result indicated that conjugation of at least two glucose residues to curcumin enhanced dramatically the water solubility. An effective chemoenzymatic system for glucosylation of curcumin was established using a recombinant glucosyltransferase from *C. roseus* (Masada et al. 2007). However, high cost of UDP-glucose, a donor substrate for the enzymatic glucosylation, still makes the biotransformation using cultured cells a method of choice although the product yield should be improved.

9.3.1.3　Capsaicin

Capsaicin is a major pungent compound in hot red peppers (fruits of *Capsicum annum*). It exhibits analgesic activity through stimulation of vanilloid 1 receptor and has been used as a topical analgesic as well as a therapeutic drug against allergic rhinitis (Bley 2004). However, incomplete solubility of capsaicin sometimes leads to variable results of capsaicin activity (Kopec et al. 2008) and limits pharmacological exploitation. Capsaicin was first shown to be glucosylated to the monoglucoside by *Coffea arabica* cultured cells (Kometani et al. 1993). Recently, *C. roseus* suspension cultures was shown to convert capsaicin and 8-nordihydro-capsaicin to their monoglucosides (main products) together with β-primeveroside

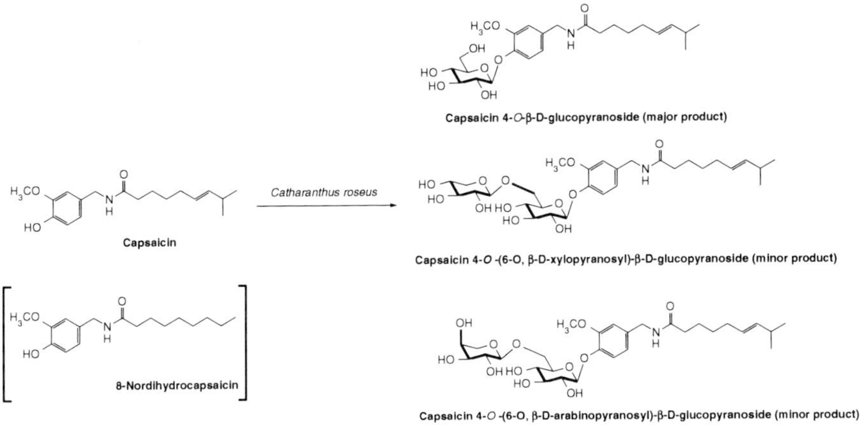

Fig. 9.12 Glycosylation products of capsaicin by *Catharanthus roseus* cell suspension cultures

(xylopyronosylglucopyranoside) and β-vicianoside (arabinopyranosylglucopyranoside), as shown in Fig. 9.12. The glycosylation rate was slightly higher for 8-nordihydrocapsaicin than for capsaicin (Shimoda et al. 2007). Since capsaicin monoglucoside was reported to be 1/100 times pungent as compared to capsaicin (Mihara et al. 1992), while exhibiting higher activity for reducing serum and liver lipid levels in vivo (Tani et al. 2003), it may be interesting to clarify the physiochemical and pharmacological characteristics of these capsaicinoid glycosides.

9.3.1.4 Monoterpene Alcohols

Higher plants produce a variety of flavors. These compounds are likely to sublime and have usually low solubility in water. Glucosylation of these volatile compounds has been one of the targets for biotransformation in plant cell cultures.

Menthol is a monoterpene compound contained in the essential oils from *Mentha* plants and has been used as a refreshing flavor for foods, medicines, and cosmetics. However, practical use of menthol is limited because of its low water solubility. *Eucalyptus perririana* cells glucosylated either (+)- or (−)-menthol mainly to their gentiobisosides (Orihara et al. 1991). Although glucosylation yield was much lower for (+)-menthol than that for (−)-menthol, a triglucoside, 2,6-di-*O*-(β-D-glucopyranosyl)-β-D-glucopyranoside, was produced only from (+)-menthol as a minor product. Monoterpene alcohols such as thymol, carvacrol, and eugenol were also efficiently converted to their genitiobiosides presumably via the corresponding monoglucosides by *E. perriniana* cell suspension cultures, and the conversion rate reached as high as about 90% in case of thymol (Shimoda et al. 2006), as shown in Fig. 9.13.

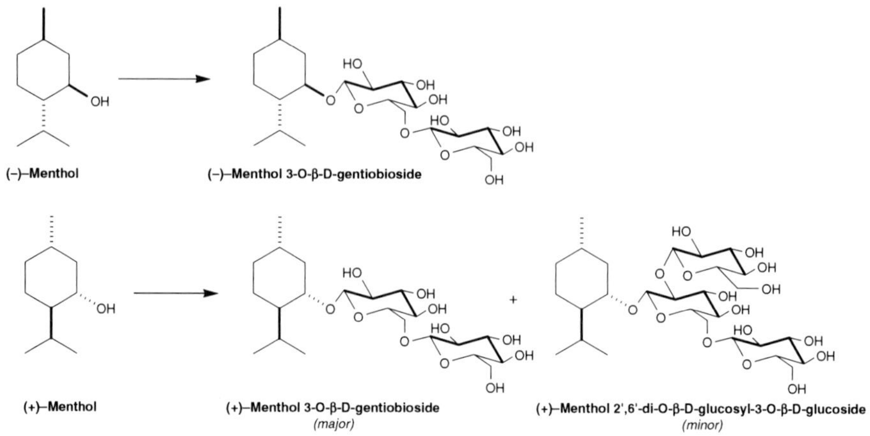

Fig. 9.13 Glycosylation products of (+)-menthol and (-)-menthol by *Eucalyptus perririana* cell suspension cultures

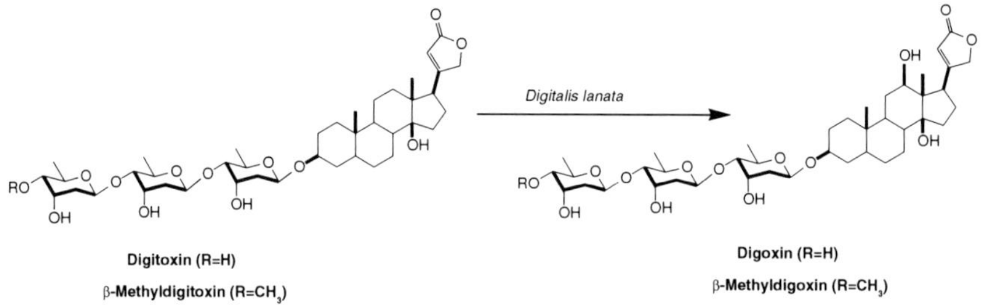

Fig. 9.14 Hydroxylation of digitoxin and β-methyldigitoxin by *Digitalis lanata* cell suspension cultures

9.3.2 Hydroxylation

Regio- and stereo-selective introduction of oxygenated functions at various positions of exogenously supplied molecules by plant cell cultures has been widely investigated because this may lead to modification in biochemical and pharmacological activities of the particular compound.

9.3.2.1 Digitoxin

Digitoxin and digoxin (12β-hydroxydigitoxin) are cardenolides accumulated in *Digitalis purpurea* and *D. lanata*. Although both compounds have been used for treatment of chronic heart diseases, digoxin has superior pharmacological and pharmacodynamic properties compared to digitoxin. In contrast, the amount of each compound recovered is by far in excess in favor of digitoxin. This led to the investigation of regio- and stereo-specific hydroxylation of digitoxin to produce digoxin by plant cell cultures (Alfermann et al. 1980). *D. lanata* cells were found to hydroxylate digitoxin at 12-position to produce 12β-hydroxydigitoxin (digoxin) (Fig. 9.14). β-Methyldigitoxin was a more suitable substrate than digitoxin because the former compound was efficiently converted to β-methyldigoxin with few by-products. Methyldigoxin yield reached 800 mg/L when *D. lanata* cells were cultured in a 20-L stirred tank reactor (Spieler et al. 1985).

The biocatalytic ability of 12β-hydroxylation seems to be specific to *D. lanata* cells because digitoxigenin was hydroxylated at 1β- or 5β-position by *Straophanthus gratus, S. intermedius,* and *Daucus carota* cells (Furuya et al. 1988; Kawaguchi et al. 1989) and at 16β-position by *D. purpurea* cells (Hirotani and Furuya 1980). It may be interesting to investigate the pharmacological activities of these hydroxylated digitoxins at 1β-, 5β-, or 16β-position because they have not been found in nature.

9.3.3 Miscellaneous

9.3.3.1 Podophyllotoxin

Podophyllotoxin in an antineoplasmic lignan isolated from *Podophyllum peltatum.* Etoposide is an important anticancer drug chemically derived from podophyllotoxin and clinically used for treatment of small cell lung carcinoma. Etoposide is synthesized through chemical conversion of podophylltoxin to 4'-demethylepipodophyllotoxin, which is then attached with the carbohydrate unit leading to etoposide. Because the present route requires isolation of podophyllotoxin from *P. peltatum,* an alternative route to 4'-demethylepipodophyllotoxin was exploited using synthetic dibenzylbutanolide as a substrate. *P. peltatum* cells in culture converted exogenously added dibenzylbutanolide to a podophyllotoxin analog with a 50% conversion rate (Kutney et al. 1993). Later, it was found that the cyclization of dibenzylbutanolide was performed with peroxidase excreted into culture broth from cells of different species such as *Nicotiana sylvestris* (Botta et al. 2001) and shoot cultures of *Halplophyllum patavinum* (Puricelli et al. 2003). Although a chemical process in removing the isopropyl group in the biotransformation product and regeneration of the methylenedioxy function was completed, the final product is still a C1-isomer of 4'-demethylepipodophyllotoxin (Fig. 9.15).

9.3.3.2 Scopolamine

Hyoscyamine-6-hydroxylase (H6H) catalyzes conversion of hyoscyamine to its epoxide scopolamine which has higher pharmacological values. Transgenic tobacco cell cultures expressing an H6H transgene derived from *Hyoscyamus*

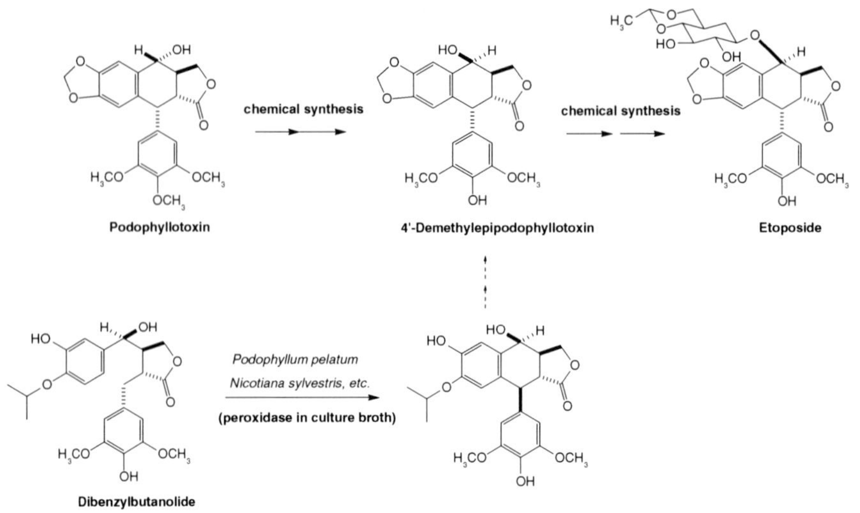

Fig. 9.15 Biotransformation of a dibenzylbutanolide to a podophyllotoxin analog by plant cell cultures

muticus efficiently converted exogenously supplied hyoscyamine to scopolamine. The productivity of scopolamine reached to 36 mg/L when 200 mg/L hyoscyamine was added to the cells cultured in a 5-L turbine stirred bioreactor, corresponding to 18% conversion (Moyano et al. 2007).

References

Akiu S, Suzuki Y, Fujinuma Y, Asahara T, Fukuda M (1988) Inhibitory effect of arbutin in melanogenesis: biochemical study in cultured B16 melanoma cells and effect on the UV-induced pigmentation in human skin. Proc Jpn Soc Investig Dermatol 12:138–139

Alfermann AW, Schuller I, Reinhard E (1980) Biotransformation of cardiac glycosides by immobilized cells of digitalis lanata. Planta Med 40:218–223

Bley KR (2004) Recent developments in transient receptor potential vanilloid receptor agonist-based therapies. Expert Opin Investig Drugs 13:1445–1456

Botta B, Monache GD, Misiti D, Vitali A, Zappia G (2001) Aryltetralin ligrans: chemistry, pharmacology and biotransformations. Curr Med Chem 8:1363–1381

Eilert U, DeLuca V, Kurz WGW, Constabel F (1987) Alkaloid formation by habituated and tumorous cell suspension cultures of *Catharanthus roseus*. Plant Cell Rep 6:271–274

Eilert U, Kurz WGW, Constabel F (1985) Stimulation of sanguinarine accumulation in *Papaver somniferum* cell cultures by fungal elicitors. J Plant Physiol 119:65–76

Facchini PJ, Hagel JM, Liscombe DK, Loukania N, MacLeod BP, Samanani N, Zulak KG (2007) Opium poppy: blueprint for an alkaloid factory. Phytochem Rev 6:97–124

Facchini PJ, De Luca V (2008) Opium poppy and madagascar periwinkle: model non-model systems to investigate alkaloid biosynthesis in plants. Plant J 54:763–784

Frense D (2007) Taxanes: perspectives for biotechnological production. Appl Microbiol Biotechnol 73:1233–1240

Fujita Y, Hara Y, Suga C, Morimoto T (1981) Production of shikonin derivatives by cell suspension cultures of *Lithospermum erythrorhizon*. II. A new medium for the production of shikonin derivatives. Plant Cell Rep 1:61–63

Furuya T, Kawaguchi K, Hirotani M (1988) Biotransformation of digitoxigenin by cell suspension cultures of *Strophanthus gratus*. Phytochemistry 27:2129–2133

Furuya T, Yoshikawa T, Orihara Y, Oda H (1983) Saponin production in cell suspension cultures of *Panax ginseng*. Planta Med 48:83–87

Georgiev MI, Pavlov AI, Bley T (2007) Hairy root type plant in vitro systems as sources of bioactive substances. Appl Microbiol Biotechnol 74:1175–1185

Hashimoto T, Hayashi A, Amano Y, Kohno J, Iwanari H, Usuda S, Yamada Y (1991) Hyoscyamine 6β-hydroxylase, an enzyme involved in tropane alkaloid biosynthesis, is localized at the pericycle of the root. J Biol Chem 266:4648–4653

Hashimoto T, Yamada Y (1994) Alkaloid biogenesis: molecular aspects. Annu Rev Plant Physiol Plant Mol Biol 45:257–285

Hayashi H, Fukui H, Tabata M (1988) Examination of triterpenoids produced by callus and cell suspension cultures of *Glycyrrhiza glabra*. Plant Cell Rep 7:508–511

Hayashi H, Sakai T, Fukui H, Tabata M (1990) Formation of soyasaponins in licorice cell suspension cultures. Phytochemistry 29:3127–3129

Hayashi H, Fukui H, Tabata M (1993) Distribution pattern of saponins in different organs of *Glycyrrhiza glabra*. Planta Med 59:351–353

Hayashi H, Huang P, Inoue K (2003) Up-regulation of soyasaponin biosynthesis by methyl jasmonate in cultured cells of *Glycyrrhiza glabra*. Plant Cell Physiol 44:404–411

Hayashi H, Huang P, Takada S, Obinata M, Inoue K, Shibuya M, Ebizuka Y (2004) Differential expression of three oxidosqualene cyclase mRNAs in *Glycyrrhiza glabra*. Biol Pharm Bull 27:1086–1092

Heide L, Nishioka N, Fukui H, Tabata M (1989) Enzymatic regulation of shikonin biosynthesis in *Lithospermum erythrorhizon* cell cultures. Phytochemistry 28:1873–1877

Hirotani M, Furuya T (1980) Biotransformation of digitoxigenin by cell suspension cultures of *Digitalis purpurea*. Phytochemistry 19:531–534

Inomata S, Yokoyama M, Seto S, Yanagi M (1991) High level production of arbutin from hydroquinone in suspension cultures of Catharanthus roseus. Plant Cell Physiol 36:315–319

Jennewein S, Croteau R (2001) Taxol: biosynthesis, molecular genetics, and biotechnological applications. Appl Microbiol Biotechnol 57:13–19

Kamada H, Okamura N, Satake M, Harada H, Shimomura K (1986) Alkaloid production by hairy root cultures in *Atropa belladonna*. Plant Cell Rep 5:239–242

Kaminaga Y, Nagatsu A, Akiyama T, Sugimoto N, Yamazaki T, Maitani T, Mizukami H (2003) Production of unnatural glucosides of curcumin with drastically enhanced water solubility by cell suspension cultures of *Catharanthus roseus*. FEBS Lett 555:311–316

Kawaguchi K, Hirotani M, Furuya T (1989) Biotransformation of digitoxigenin by cell suspension cultures of *Strophanthus intermedius*. Phytochemistry 28:1093–1097

Kim YS, Cho JH, Ahn J, Hwang B (2006) Upregulation of isoprenoid pathway genes during enhanced saikosaponin biosynthesis in the hairy roots of *Bupleurum falcatum*. Mol Cell 22:269–274

Kitajima M, Fischer U, Nakamura M, Ohsawa M, Ueno M, Takayama H, Unger M, Stoeckigt J, Aimi N (1998) Anthraquinones from *Ophiorrhiza pumila* tissue and cell cultures. Phytochemistry 48:107–111

Kolewe ME, Gaurav V, Roberts SC (2008) Pharmaceutically active natural product synthesis and supply via plant cell culture technology. Mol Pharm 5:243–256

Kometani T, Tanimoto H, Nishijima H, Kanabara I, Okada S (1993) Glucosylation of capsaicin by cell suspension cultures of *Coffea arabica*. Biosci Biotechnol Biochem 57:2192–2193

Kopec SE, Irwin RS, DeBellis RJ, Bohlke MB, Maher TJ (2008) The effects of Tween-80 on the integrity of solutions of capsaicin: useful information for performing tussigenic challenges. Cough 4:3

Kutney JP, Chen YP, Gao S, Hewitt GM, Kuri-Brena F, Milanova RK, Stoynov N (1993) Studies with plant cell cultures of *Podophyllum peltatum* L. II. Biotransformation of dibenzylbutanolides to lignans. Development of a "biological factory" for lignan synthesis. Heterocycles 36:13–20

Li SM, Hennig S, Heide L (1998) Shikonin: a geranyl diphosphate-derived plant hemiterpenoid formed via the mevalonate pathway. Tetrahedron Lett 39:2721–2724

Liu C, Zhao Y, Wang Y (2006) Artemisinin: current state and perspectives for biotechnological production of an antimalarial drug. Appl Microbiol Biotechnol 72:11–20

Lorence A, Medina-Boliver F, Nessler CL (2004) Camptothecin and 10-hydroxycamptothecin from *Camptotheca acuminata* hairy root. Plant Cell Rep 22:437–441

Lorence A, Nessler CL (2004) Camptothecin, over four decades of surprising findings. Phytochemistry 65:2735–2749

Maheshwari RK, Singh AK, Gaddipati J, Srimal RC (2006) Multiple biological activities of curcumin: a short review. Life Sci 27:2081–2087

Mano Y, Nabeshima S, Matsui C, Ohkawa H (1986) Production of tropane alkaloids by hairy root cultures. Agric Biol Chem 50:2715–2722

Mano Y, Ohkawa H, Yamada Y (1989) Production of tropane alkaloids by hairy root cultures of Duboisia leichhardtii transformed by *Agrobacterium rhizogenes*. Plant Sci 59:191–201

Masada S, Kawase Y, Nagatoshi M, Oguchi Y, Terasaka K, Mizukami H (2007) An efficient chemoenzymatic production of small molecule glucosides with *in situ* UDP-glucose recycling. FEBS Lett 581:2562–2567

Matsuda J, Okabe S, Hashimoto T, Yamada Y (1991) Molecular cloning of hyoscyamine 6beta-hydroxylase, a 2-oxoglutarate-dependent dioxygenase, from cultured roots of Hyoscyamus niger. J Biol Chem 266:9460–9464

Mihara S, Hiraoka K, Kameda W (1992) Synthesis of N-[4-(R-D-glucopyranosyloxy)-3-methoxybenzyl] nonamide and its β-anomer. J Agric Food Chem 40:2057–2059

Mizukami H, Konoshima M, Tabata M (1978) Variation in pigment production in *Lithospermum erythrorhizon* callus cultures. Phytochemistry 17:95–97

Moyano E, Palazón J, Bonfill M, Osuna L, Cusinó RM, Oksman-Caldentey KM, Piñol MT (2007) Biotransformation of hyoscyamine into scopolamine in transgenic tobacco cell cultures. J Plant Physiol 164:521–524

Nakagawa K, Konagai A, Fukui H, Tabata M (1984) Release and crystallization of berberine in the liquid medium of *Thalictrum minus* cell suspension cultures. Plant Cell Rep 3:254–257

Orihara Y, Miyatake H, Furuya T (1991) Triglucosylation on the biotransformation of (+)-menthol by cultured cells of *Eucaryptus perriniana*. Phytochemistry 30:1843–1845

Paniego NB, Giuletti AM (1994) *Artemisia annua* L. dedifferentiated and differentiated cultures. Artemisinin production. Plant Cell Tissue Organ Cult 36:163–168

Papageorgiou VP, Assimopoulou AN, Couladouros EA, Hepworth D, Nicolaou KC (1999) The chemistry and biology of alkannin, shikonin, and related naphthazarin natural products. Angew Chem Int Ed 38:270–300

Park SU, Facchini PJ (2000) *Agrobacterium rhizogenes*–mediated transformation of opium poppy, *Papaver somniferum* L., and California poppy, *Eschscholzia californica* Cham., root cultures. J Exp Bot 347:1005–1016

Parr AJ, Peerless ACJ, Hamill JD, Walton NJ, Robins RJ, Rhodes MJC (1988) Alkaloid production by transformed root cultures of *Catharanthus roseus*. Plant Cell Rep 7:309–312

Payne J, Hamill JD, Robins RJ, Rhodes MJC (1987) Production of hyoscyamine by hairy root cultures of *Datura stramonium*. Planta Med 53:474–478

Puricelli L, Caniato R, Monache GD (2003) Biotransformation of a dibenzylbutanolide to podophyllotoxin analogues by shoot cultures of *Haplophyllum patavinium*. Chem Pharm Bull 51:848–850

Rohmer M (1999) The discovery of a mevalonate-independent pathway for isoprenoid biosynthesis in bacteria, algae, and higher plants. Nat Prod Rep 16:565–574

Rao RS, Ravishankar GA (1999) Biotransformation of isoeugenol to vanilla flavour metabolites and capsaicin in freely suspended and immobilized cell cultures of *Capsicum frutescens*: study of the influence of β-cyclodextrin and fungal elicitor. Process Biochem 35:341–348

Saito K, Sudo H, Yamazaki M, Koseki-Nakamura M, Kitajima M, Takayama H, Aimi N (2001) Feasible production of camptothecin by hairy root culture of *Ophiorrhiza pumila*. Plant Cell Rep 20:267–271

Sakato K, Tanaka H, Mukai N, Misawa M (1974) Isolation and identification of camptothecin from cells of *Camptotheca acuminata* suspension cultures. Agric Biol Chem 38:217–218

Sato F, Yamada Y (1984) High berberine producing cultured *Coptis japonica* cells. Phytochemistry 23:281–285

Seki H, Ohyama K, Sawai S, Mizutani M, Ohnishi T, Sudo H, Akashi T, Aoki T, Saito K, Muranaka T (2008) Licorice β-amyrin 11-oxidase, a cytochrome P450 with a key role in the biosynthesis of the triterpene sweetener glycyrrhizin. Proc Natl Acad Sci USA 105:14204–14209

Shibata S (2000) A drug over the millennia: pharmacognosy, chemistry, and pharmacology of licorice. Yakugaku Zasshi 120:849–862

Shimoda K, Kondo Y, Nishida T, Hamada H, Nakajima N, Hamada H (2006) Biotransformation of thymol, carvacrol, and eugenol by cultured cells of *Eucalyptus perriniana*. Phytochemistry 67:2256–2261

Shimoda K, Kwon S, Utsuki A, Ohiwa S, Katsuragi H, Yonemoto N, Hamada H, Hamada H (2007) Glycosylation of capsaicin and 8-nordihydrocapsaicin by cultured cells of *Catharanthus roseus*. Phytochemistry 68:1391–1396

Smetanska I (2008) Production of secondary metabolites using plant cell cultures. Adv Biochem Eng Biotechnol 111:187–228

Spieler H, Alfermann AW, Reinhard E (1985) Biotransformation of b-methyldigitoxin by cell cultures of *Digitalis lanata* in airlift and stirred tank reactors. Appl Microbiol Miotechnol 23:1–4

Suga T, Hirata T (1990) Biotransformation of exogeneous substrates by plant cell cultures. Phytochemistry 29:2393–2406

Tabata M, Yamamoto H, Hiraoka N, Konoshima M (1972) Organization and alkaloid production in tissue cultures of *Scopolia parviflora*. Phytochemistry 11:949–955

Tabata M, Mizukami H, Hiraoka N, Konoshima M (1974) Pigment formation of callus cultures of in *Lithospermum erythrorhizon*. Phytochemistry 13:927–932

Tabata M, Ikeda F, Hiraoka N, Konoshima M (1976) Glucosylation of phenolic compounds by *Datura innoxia* cell suspension cultures. Phytochemistry 15:1225–1229

Tabata M, Fujita Y (1985) Production of shikonin by plant cell cultures. In: Zaitlin M, Day P, Hollander A (eds) Biotechnology in Plant Science. Acad Press, Orland, pp 207–218

Tabata M (1996) The mechanism of shikonin biosynthesis in *Lithospermum* cell cultures. Plant Tissue Cult Lett 13:117–125

Tani M, Takeda K, Yazaki K, Tabata M (1993) Effects of oligogalacturonides on biosynthesis of shikonin in *Lithospermum* cell cultures. Phytochemistry 34:1285–1290

Tani Y, Fujioka T, Hamada H, Kunimatsu M, Furuichi Y (2003) Favorable effects of vanillylnonamide-b-D-glucoside on serum lipids in hyperlipidemic rats. J Jpn Soc Nutr Food Sci 56:181–187

Toivonen L, Balsevich J, Kurz WGW (1989) Indole alkaloid production by hairy root cultures of *Catharanthus roseus*. Plant Cell Tissue Organ Cult 18:79–93

Toivonen L, Rosenqvit H (1995) Establishment and growth characteristics of *Glycyrrhiza glabra* hairy root cultures. Plant Cell Tissue Organ Cult 41:249–258

Uomori A, Seo S, Tomita Y (1974) Studies on the constituents in tissue cultures of Bupleurum falcatum L. Syoyakugaku Zasshi 28:152–160

Waller GR, Nowacki EK (1978) Alkaloid Biology and Metabolism in Plant. Plenum Publ, New York, USA

Weathers PJ, Cheetham RD, Follansbee E, Teoh K (1994) Artemisinin production by transformed roots of *Artemisia annua*. Biotechnol Lett 16:1281–1286

Wildung MR, Croteau R (1996) A cDNA clone for taxadiene synthase, the diterpene cyclase that catalyzes the committed step of taxol biosynthesis. J Biol Chem 271:9201–9204

Yamada Y, Endo T (1984) Tropane alkaloid production in cultured cells of *Duboisia leichhardtii*. Plant Cell Rep 3:186–188

Yamada Y, Hashimoto T (1982) Production of tropane alkaloids in cultured cells of *Hyoscyamus niger*. Plant Cell Rep 1:101–103

Yamamoto H, Zhao P, Yazaki K, Inoue K (2002) Regulation of lithospermic acid B and shikonin production in *Lithospermum erythrorhizon* cell suspension cultures. Chem Pharm Bull 50:1086–1090

Yamazaki Y, Sudo H, Yamazaki M, Aimi N, Saito K (2003) Camptothecin biosynthetic genes in hairy roots of *Ophiorrhiza pumila*: Cloning, characterization and differential expression in tissues and by stress compounds. Plant Cell Physiol 44:395–403

Yazaki K, Takeda K, Tabata M (1997) Effects of methyl jasmonate on shikonin and dihydroechinofuran production in *Lithospermum* cell cultures. Plant Cell Physiol 38:776–782

Yazaki K, Kunihisa M, Fujisaki T, Sato F (2002) Geranyl diphosphate: 4-hydroxybenzoate geranyltransferase from *Lithospermum erythrorhizon*. Cloning and characterization of a key enzyme in shikonin biosynthesis. J Biol Chem 277:6240–6246

Yamazaki Y, Kitajima M, Arita M, Takayama H, Sudo H, Yamazaki M, Aimi N, Saito K (2004) Biosynthesis of camptothecin. In silico and in vitro tracer study from [1-13C] glucose. Plant Physiol 134:161–170

Yokoyama M, Inomata S, Seto S, Yanagi M (1990) Effect of sugars on the glucosylation of exogenous hydroquinone by *Catharanthus roseus* cells in suspension culture. Plant Cell Physiol 31:551–555

Yukimune Y, Tabata H, Higashi Y, Hara Y (1996) Methyl jasmonate induced overproduction of paclitaxel and baccatin III in Taxus cell suspension cultures. Nat Biotechnol 14:1129–1132

Yun DJ, Hashimoto T, Yamada Y (1992) Metabolic engineering of medicinal plants: transgenic *Atropa belladonna* with an improved alkaloid composition. Proc Natl Acad Sci USA 89:11799–11803

Chapter 10
Metabolic Engineering of Pathways and Gene Discovery

Miloslav Juříček, Chandrakanth Emani, Sunee Kertbundit, and Timothy C. Hall

10.1 Introduction

Humans have been manipulating the genetic information of plants throughout the history of agriculture. In this respect, every new plant variety or animal race is a result of the introduction of novel metabolic changes. This process has been slowly advancing for millennia. However, with the discovery of biochemical pathways and later with the introduction of gene manipulation techniques in 1970s, the pace greatly speeded up. Already in the mid-1980s, many of the compounds and enzymes participating in metabolic pathways were linked to their cloned genes, which can then be used for engineering the plant metabolism. Soon, novel products from plants appeared including, vaccines and other pharmaceuticals, plastics, and proteins that may render certain plants as effective tools for environmental decontamination. These products were a result of the manipulation of plant endogenous biochemical pathways and thus the novel field of science-metabolic engineering was born. Metabolic engineering can be defined as the targeted and purposeful modification of metabolic pathways in an organism for the improved use of cellular pathways for chemical transformation, energy transduction, and macromolecular synthesis or breakdown, potentially benefiting the society by producing biological substitutes for toxic chemicals, increasing agricultural production, improving industrial fermentation processes, producing completely new compounds, or by understanding the molecular mechanism underlying medical conditions in order to develop new cures (Kurnaz 2005).

Among several organisms, plants have rapidly become the main object of metabolic engineering. This may be attributed to the higher interest in plants over

T.C. Hall (✉)
Institute of Developmental and Molecular Biology, Department of Biology, Texas A&M University, College Station, TX 77843-3155, USA
e-mail: tim@idmb.tamu.edu

C. Kole et al. (eds.), *Transgemic Crop Plants*, DOI 10.1007/978-3-642-04809-8_10, © Springer-Verlag Berlin Heidelberg 2010

bacteria or other organisms considered which was stimulated by potential commercial applications resulting from engineering the resistance against pests and diseases and later also improving the contents of metabolites already used in medicine or developing novel medicines. Engineering of secondary metabolites seemed the easiest way to obtain this goal. However, more than hundred thousand metabolites have already been identified and this may be only a fraction of the total amount in plant kingdom. Clearly, detailed mapping of metabolic pathways and their engineering will be an enormous task. Moreover, even detailed understanding of the biochemical processes may not be sufficient because an interaction between various pathways in the total metabolic network, enzyme complexes, compartmentation, feedback inhibition, and/or gene expression regulation may completely change the story. Despite these obstacles, a number of successful cases were reported during the last 30 years, some of them are summarized later.

10.2   The Beginnings and Early Years of Metabolic Engineering

Historical archives of plant sciences related to exploiting plants as natural chemical factories are replete with examples of utilizing plants as sources of medicinal compounds. One of the earliest examples is the medicinally valuable St. John's wort discovered by the Greek physician, Hippocrates, in the fifth century BC. In the present times, St. John's wort is part of the medical research and clinical trials aimed at determining its efficacy for a wide variety of ailments such as depression, cancer, inflammation, and viral infections. The ancient Indian medical discipline of *Ayurveda* effectively illustrates the use of plants as derivatives of medicines in combating various ailments. In modern medicine, one-quarter of prescription drugs are of plant origin (Fischer and Emans 2000).

The technology of extracting useful compounds from plants evolved into what we now know as "plant molecular pharming." In 1983, Murai et al. (1983) demonstrated that a part of bean phaseolin seed protein gene was transcribed in sunflower cells transformed by the tumor-inducing plant vector *Agrobacterium tumefaciens*, the first unequivocal demonstration of the transfer of a developmentally regulated plant gene from one plant species to another. In a similar development, bacterial genes were expressed in higher plants (Fraley et al. 1983), followed by the novel leaf disk *Agrobacterium*-mediated transformation method of Horsch et al. (1985) that combined gene transfer, plant regeneration, and an effective kanamycin-based selection to generate transgenic petunia, tobacco, and tomato. The most commonly used reporter gene to monitor transgene expression, the β-glucuronidase gene (*gusA* or *uidA*) (Jefferson et al. 1987), is one of the earliest examples of a successful molecular pharming product as its commercial production served as a model system for the production of proteins in transgenic corn plants (Kusnadi et al. 1998; Witcher et al. 1998). Prior to this, During (1988) demonstrated the wound-inducible expression and secretion of the T4 lysozyme and monoclonal antibodies in tobacco, the first report of human antibody expression in a transgenic plant. This was followed by the expression of secretory antibodies in transgenic plants (Hiatt et al. 1989), the secretion of biologically active blood substitutes, namely, human interleukin-2 and interleukin-4 in transgenic tobacco suspension cultures (Magnuson et al. 1998), the expression of nopaline synthase-human growth hormone chimeric gene in transgenic calli of tobacco and sunflower (Barta et al. 1986), human interferon (De Zoeten et al. 1989), and human serum albumin (Sijmons et al. 1990). The successes of molecular farming (extensively reviewed by Kumar et al. 2007) that involved the transfer of the desirable gene to an appropriate host system, optimization of the desirable pattern of gene expression, and optimal recovery of the recombinant protein in the form of a pharmaceutical product peaked with the achievement of oral immunization with a recombinant bacterial antigen produced in transgenic plants (Haq et al. 1995).

10.3 The Basic Goals and Strategies of Metabolic Engineering

10.3.1 Biochemical Pathways

The progress of metabolic engineering is closely related with the discovery and understanding of biochemical pathways. The more detailed and well-documented knowledge on the pathway of interest is known, the better chance exists that the engineering of such a pathway will be predictable and successful. Unfortunately, despite decades of elucidating pathways in various organisms, our knowledge is still far from being complete. The main problem is to identify all enzymes that catalyze individual reactions within the pathway. The analysis is still difficult due to their instability, low amount, and/or low activity among others. On the other hand, the isolation and identification of secondary metabolites is easier by using labeling techniques although many experimentally unconfirmed intermediates still exist. With the progress of molecular biology, other approaches were involved to help with the determination of involved enzymes. Most of them are based on "knocking out" the gene by various techniques such as transposon tagging, TILLING, RNAi (for details see Chap. 1-6 of this volume) or amiRNA, and then identify which enzyme was affected. This is usually easy when the knockout gene is manifested phenotypically, but it may be a daunting task if otherwise.

10.3.2 Functional Genomics

Using functional genomics is just another way of elucidating biochemical pathways starting from genes down to the proteins. Functional genomics is the usage of statistical methods and bioinformatics to determine the function of the genes (as the name suggests). It is obvious that the genomic sequence must be known in order to use this method. For time being there are not many sequenced plant genomes available but this will change in the near future; a number of nonmodel plant species are now being sequenced and more are planned in the near future. Functional genomics utilizes the "–omics" family tools, e.g., transcriptomics, proteomics, and metabolomics. The tool of metabolomics is particularly important because it qualitatively and quantitatively analyzes all metabolites in the organism. When this is combined with the transcriptomics and proteomics, the complete picture can be seen, e.g., involvement of regulatory and structural genes in the organism (for details see Chaps 2–10 of Volume 2 of this series).

10.3.3 Compartmentalization, Transport, and Storage

Engineering a metabolic pathway in plants needs to be calculated with intra- and/or intercellular compartments. The genes need to be expressed in the correct compartment and in the correct type of cell. If not, the expression system may not work, or only with low yield, or the product may even have a toxic effect. Thus production of metabolites occurs in intracellular compartments such as vacuoles, endoplasmic reticulum, cytosol, plastids, etc., whereas intercellular compartments cover various plant tissues. The existence of compartments requires the existence of various transport systems as the intermediate metabolite must be quickly moved between different intra- and intercellular compartments. A number of transport mechanisms were described in the literature. It is a rather complex process, and thus the biochemical reaction rate may be also limited on the level of transport. Intermediate metabolites are usually stored in vacuoles and thus transport mechanisms are required for an import.

10.3.4 Basic Strategies

Early experiments to increase the yield and productivity of secondary metabolite production relied on enzyme modification. These qualitative principles are based mainly on the view that control of pathways must reside in relatively few enzymes whose in vitro properties suggest that they could be controlling flux in vivo. However, manipulating enzymes considered to be "rate limiting" has rarely had the expected outcome. Metabolic pathways have evolved to exhibit control architectures that resist flux alterations at branch points. Stephanopoulos and Vallino (1991) therefore introduced the concept of flexible and rigid nodes. The rigidity of the biochemical network or its resistance to variations in metabolic change is due to control mechanisms established to ensure balanced growth. The more rigid the branch point, the harder it is to increase the flux through one of its branches. For an engineering strategy to be successful, a sound understanding of the host cell is necessary to determine the types of genetic modifications needed to achieve the final goal. Some of the physiological considerations that should be examined include the effects of genetic manipulation on growth and possible effects on unrelated systems. Traditional metabolic biochemistry did not provide the understanding needed to do this because it dealt with metabolic regulation in terms of a few qualitative principles and was thus sometimes called "reductionist approach." Recently, a fast-growing field was introduced in the biological research referred as "systems biology." As the name suggests systems biology aims at systems-level understanding, as distinct from understanding individual system components such as particular genes or enzymes (Kitano 2002). At the very core of systems biology is the goal of being able to model a living organism. The systems range from metabolic pathways and gene-regulatory networks all the way up to whole cells and organisms. Thus it offers the –omics integration together with phenotypic data for studying plant organisms and even their interaction within their ecosystems. As for metabolic engineering, the ultimate aim is to use the comprehensive experimental data sets describing changes in transcripts, proteins, metabolites, and flux to generate a complete mathematical description of the metabolism of a model plant species. It is envisaged that such a model would allow a truly predictive engineering of plant metabolism. This is an ambitious aim that will require a sustained commitment of resources and unprecedented technological developments to be achieved (Sweetlove et al. 2003). Within system biology it is possible to establish theoretical basis for determining which enzymes should be manipulated to achieve a desired outcome. One such theoretical basis is metabolic control analysis (MCA) which plays a central role in metabolic engineering.

Unlike traditional biochemistry, MCA is based upon the fact that a single rate-limiting step may not exist and several steps may share control of the metabolic network. Three commonly used normalized sensitivity measures, referred as Control Coefficients that quantify how the control of steady-state fluxes and concentrations is distributed between different reactions in a metabolic network, have been defined. Control Coefficients refer to the whole metabolic pathway (i.e., they are systemic or global properties). A Control Coefficient is a relative measure of how much a perturbation to, for example enzyme activity, affects a system variable, e.g., a flux or metabolite concentration. Flux Control Coefficient is the heart of the theory. It is a measure of how a change in the concentration of the enzyme affects the steady-state flux through that particular enzyme. That is, it is the measure of the degree of control exerted by enzyme on this steady-state flux. The Concentration Control Coefficient is a measure of the extent of control exerted by the enzyme on the steady-state concentration, while the Elasticity Coefficient is a measure of the response of the reaction rate upon changes in the concentration. Thus it refers to properties of individual enzymes in the pathway (they are local not systemic properties, and are related to classical enzyme kinetics).

Three main types of metabolic engineering based on MCA can be considered (Bailey 1991):

1. Extending an existing pathway to obtain a new product
2. Amplifying a flux-controlling step
3. Diverting flux at branch points ("nodes") to a desired product by:

(a) Circumventing a (feedback) control mechanism
(b) Amplifying the step initiating the desired branch (or the converse)
(c) Removing reaction products
(d) Manipulating levels of signal metabolites

These strategies have one major disadvantage: detailed knowledge of the network pathways and enzyme kinetics must be available. In contrast, the concept of inverse metabolic engineering does not require such knowledge. It is based on first obtaining the desired phenotype and later to determine environmental or genetic conditions that confer this phenotype, and finally to alter the phenotype of the selected host by genetic manipulation (Bailey et al. 1996; Delgado and Liao 1997).

10.4 Engineering Primary Metabolic Pathways

10.4.1 Carbohydrate Metabolism

In plants, the process of photosynthesis results in the production of sugars as direct products that undergo reversible conversion into storage carbohydrates such as starch and fructans, and the structural carbohydrate, cellulose (Fig. 10.1).

Starch as a storage carbohydrate accumulates transiently in leaves and stably in seeds, tubers, and roots. The importance of starch as a stable dietary carbohydrate and its many industrial uses render it as a favorite target for metabolic engineering in attempts to increase starch yields by changing the relative proportions of its structural components, amylose and amylopectin. Modulating the enzymes controlling starch synthesis and branching in potatoes resulted in the production of high-amylose starches that have important domestic uses such as improved frying and industrial uses as gelling agents and thickeners, and high-amylopectin starches notable in their use for improved freeze thaw characteristics, improved paper quality or adhesive manufacture (Capell and Christou 2004). Antisense expression of the *Waxy* gene in transgenic rice resulted in low-amylose rice grains of improved

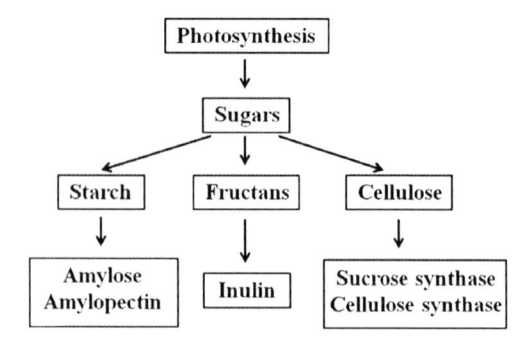

Fig. 10.1 Targets of carbohydrate metabolism

quality (Liu et al. 2003). Fujita et al. (2003) developed transgenic rice with modified amylopectin by antisense expression of the gene encoding isoamylase. Vincken et al. (2003) used a starch-binding domain from potato-granule bound starch synthase I to target luciferase to the inside of a starch granule, a promising method for directing recombinant enzymes into the starch granule for starch metabolic engineering. Novel starches were produced by changing the nature and frequency of branching directed toward a more versatile α-glucan synthesis by using bacterial enzymes (Kok-Jacon et al. 2003). Seed-specific overexpression of the potato sucrose transporter in transgenic pea was shown to increase sucrose uptake and growth rates of developing cotyledons (Rosche et al. 2002). Regierer et al. (2002) demonstrated that by increasing the activity of plasticidal adenylate kinase in transgenic potato, a larger pool of adenylates become available to fuel several metabolic pathways. This resulted in a 60% increase in overall starch levels, a two- to four-fold increase in amino acid levels combined with an increased tuber yield. Transgenic wheat and rice transformed with modified maize ADP-glucose pyrophosphorylase (*shrunken2*) targeting the enhancement of the enzyme activity in wheat endosperm and deregulation of the enzyme in rice endosperm resulted in a 40% and 20% increase in seed biomass, respectively, compared to wild-type controls (Smidansky et al. 2002, 2003).

The ability of fructans to substitute as low-calorie alternatives to fats due to their similar texture as fats attracted their attention in metabolic engineering. Inulin from chicory is a commercially available fructan. Bacterial and plant enzymes for fructan biosynthesis have been introduced into several crops to facilitate the large-scale extraction of fructans (reviewed in Ritsema and Smeekens 2003). The importance of fructans in protection of plants from abiotic stresses prompted their use in producing improved transgenic plant varieties by metabolic engineering (reviewed in Chen and Murata 2002).

The importance of the structural carbohydrate, cellulose, as pulp and fiber, and its role as a starting material for commercially important polymers make it an attractive target for metabolic engineering. Though the complete biosynthetic pathway of cellulose is not worked out, certain important enzymes involved in the process have been identified and exploited in metabolic engineering. Suppression of sucrose synthase gene expression was found to repress cotton fiber cell initiation, elongation, and seed development (Ruan et al. 2003). In *Arabidopsis*, the functional analysis of the cellulose synthase genes *CesA1*, *CesA2*, and *CesA3* revealed their role in primary and secondary cell wall formation (Burn et al. 2002), and the expression of a mutant form of cellulose synthase AtCes47 caused a dominant negative effect on cellulose biosynthesis (Zhong et al. 2003).

10.4.2   *Amino Acid Metabolism*

Metabolic engineering specifically targeted toward increasing the content of essential amino acids such as lysine, threonine, methionine, and tryptophan in food crops remains an exciting proposition (Galili and Hofgen 2002). More recently, metabolic engineering complemented by RNA interference (RNAi) resulted in effective protocols for multiple point intervention in well worked out amino acid pathways. A good illustration of this strategy can be seen in generating a novel opaque variant of maize by a single dominant RNAi-inducing transgene targeting the zein genes (Segal et al. 2003). This novel version of *opaque2* has increased lysine content. Increasing lysine content was also demonstrated in transgenic *Arabidopsis*, where multipoint engineering of lysine metabolism was achieved by combining the overexpression of a bacterial enzyme, dihydrodipicolinate synthase (DHPS) that is resistant to lysine inhibition together with knockout of lysine catabolism pathway (Zhu and Galili 2003). Whether either strategy was used alone, a 12-fold or five-fold increase in lysine content was observed, and when transgene expression and knockout were in combination, an 80-fold increase was observed. Yet another valuable amino acid for metabolic engineering is proline as its role in plant stress responses makes it an important target for modulation to generate transgenic stress tolerant plants (Chen and Murata 2002).

10.4.3 Polyamine Metabolism

Polyamines are low molecular weight polycationic molecules that are known to play an important role in plant defense and in the regulation of plant growth and development (Rajam 1997; Kumar et al. 2006). These small aliphatic amines are derived from the amino acids, ornithine and arginine, by a decarboxylation pathway (Bhattacharya and Rajam 2007). Three major polyamines found in plants are putrescine, spermidine, and spermine. Putrescine is produced by the decarboxylation of arginine catalyzed by the enzyme arginine decarboxylase (ADC) or as in fungi by the decarboxylation of ornithine catalyzed by ornithine decarboxylase (ODC). Putrescine then acts as a precursor for the higher polyamines, spermidine and spermine, the conversion catalyzed by spermidine and spermine synthases, respectively. The reactions proceed by the addition of propyl amino groups to the decarboxylated S-adenosylmethionine (SAM) that is generated from SAM by SAM decarboxylase (SAMDC). A diamine cadaverine that is a penta homolog of putrescine is mainly found in legumes and is produced by the decarboxylation of lysine catalyzed by lysine decarboxylase (LDC) (Rajam 1997). The polyamine biosynthetic pathway is thus very well worked out and offers an array of possibilities for metabolite manipulation (Fig. 10.2). The corresponding genes for all the enzymes have been cloned in plants, namely the *adc* gene for tomato (Rastogi et al. 1993), pea (Perez-Amadour and Carbonell 1995), *Arabidopsis* (Watson and Malmberg 1996), and rice (Chattopadhyay et al. 1997); the *odc* gene from *Datura* (Micheal et al. 1996), tobacco (Mallik et al. 1996), and tomato (Alabadi and Carbonell 1998); the *samdc* gene from *Arabidopsis*, *Datura*, potato (Taylor et al. 1992), *Catharanthus* (Schroeder and Schroeder 1995), tomato, tobacco (Kumar et al.

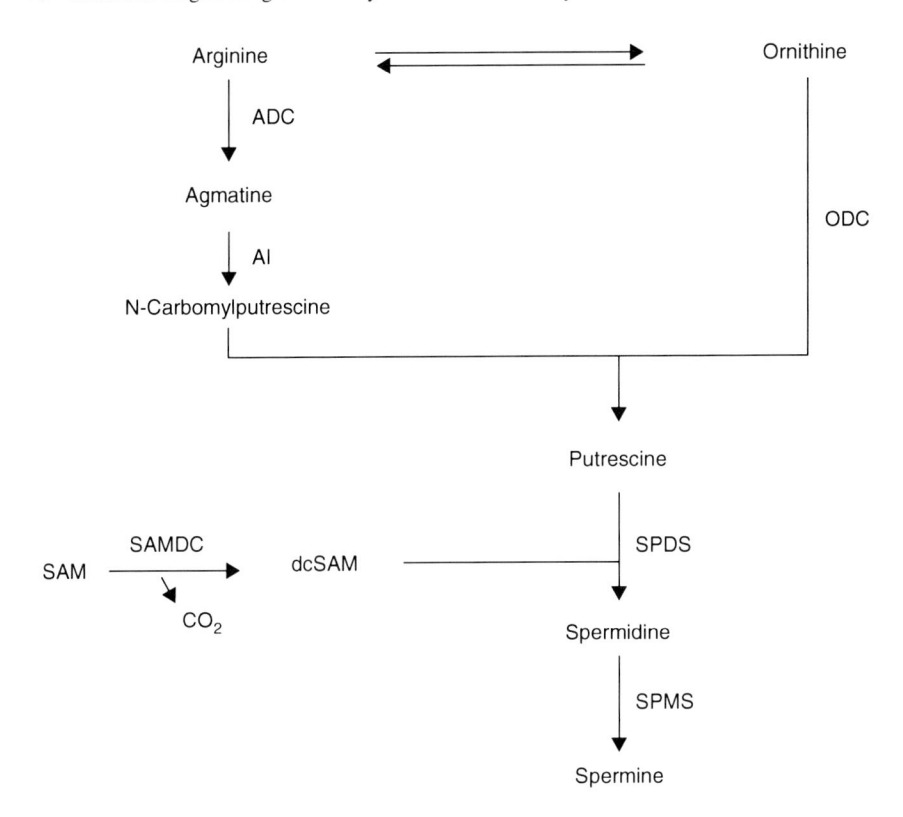

Fig. 10.2 Polyamine biosynthetic pathway in plants. ADC, Arginine decarboxylase; AI, Agmatine iminohydrolase; ODC, Ornithine decarboxylase; SAM, S-adenosyl methionine, SAMDC, SAM decarboxylase; dcSAM, decarboxylated SAM; SPDS, Spermidine synthase; SPMS, Spermine synthase. Reproduced from Bhattacharya and Rajam (2007)

1997; Park et al. 1998), and rice (Li and Chen 2000); and the *spd synthase* from tobacco, *Arabidopsis*, and *Hyoscyamus niger* (Hashimoto et al. 1998).

Metabolic engineering of polyamines has mainly utilized overexpression and antisense techniques. Overexpression of carrot *samdc* in rice showed increased levels of spermine and spermidine in seeds, and only spermine in leaves (Thu-Hang et al. 2002). The expression of yeast *samdc* in tomato under the control of a ripening-inducible E8 promoter increased spermidine and spermine levels in the fruit, and resulted in enhanced production of lycopene, a longer vine life, and nutritionally improved tomato juice (Mehta et al. 2002). The antisense suppression by an oat *adc* gene in rice reduced the putrescine and spermine levels, but no concomitant changes were observed in the downstream genes in the polyamine pathway (Trung-Nghia et al. 2003). Polyamines also act as precursors to many secondary metabolites and are thus important sources for engineering of secondary metabolic pathways.

10.4.4 Lipid Metabolism

The manipulation of oils and lipids in plants to change the quantity and nutritional quality of the plant fatty acids has far-reaching applications in food industry and human health as oils and fats are an important source of energy for the human body and form a vital component of many cell constituents. Since the main sources of fat in the human diet are vegetable oils, mostly soy, canola (oilseed rape), palm, peanut and sunflower, an attractive area of research has been the production of oilseed plants engineered to produce omega-3 long chain polyunsaturated fatty acids (LC-PUFAs) that have multiple health benefits in terms of cardiovascular and mental health. Attempts in this direction were deemed important in a quest to provide a successful alternative to that of the LC- PUFA-rich fish oils that proved to be undesirable food ingredients due to the increase in the vegetarian movement, the associated objectionable flavors, and more recently the rise in chemical and environmental contaminants in marine life that are difficult and cost prohibitive to remove from the fish oils. Improving the fatty acid content of plants has important industrial applications in production of detergents, fuels, lubricants, paints, and plastics. Our review touches on some of the important examples. For a more detailed exploration, the reader is directed to the many extensive reviews in this area published in recent years (Murphy 2002; Drexler et al. 2003; Singh et al. 2005; Damude and Kinney 2007).

Metabolic engineering of fatty acids is of great interest both at the laboratory and industrial research levels as even the most extensive modifications have no notable effect on the normal growth and development of the modified plant (Thelen and Ohlrogge 2002). Most higher plants have the ability to synthesize the main C18-PUFA, linoleic acid (LA), and α-linolenic acid (ALA), and to a lesser extent, γ-linolenic acid (GLA) and stearidonic acid (SDA). The inability of plants to elongate and desaturate these C18-PUFA acids into the beneficial LC-PUFA therefore makes it imperative to genetically engineer the genes that encode the required biosynthetic enzymes to convert the LA into an ω6 LC-PUFA like arachidonic acid (AA) or the ALA into an ω3 LC-PUFA like eicosapentaenoic acid (EPA) and docosahexaenoic acid (DHA) (for details of the steps in this pathways, see Singh et al. 2005 and Fig. 10.3).

Initial attempts to engineer plant fatty acid profiles focused on the redirection of fatty acid biosynthesis in the developing seed by blocking fatty acid desaturation resulting in a high oleic soybean (Kinney et al. 1998), and the introduction of enzymes that redirected fatty acid synthesis to new end products, such as medium chain fatty acid oils resulting in high laurate canola (Del Vecchio 1996). These technically successful experiments involving the introduction of one or two transgenes were aimed at improving the oxidative stability of the oil without hydrogenation. Liu et al. (2001) transformed the oilseed rape, *Brassica napus* with cDNAs encoding desaturation enzymes, 18:1 Δ12 desaturase alone or in combination with 18:2 Δ6 desaturase resulting in seeds producing 46% ALA and 43% GLA, respectively. Han et al. (2001) engineered a 60% erucic acid producing oilseed rape

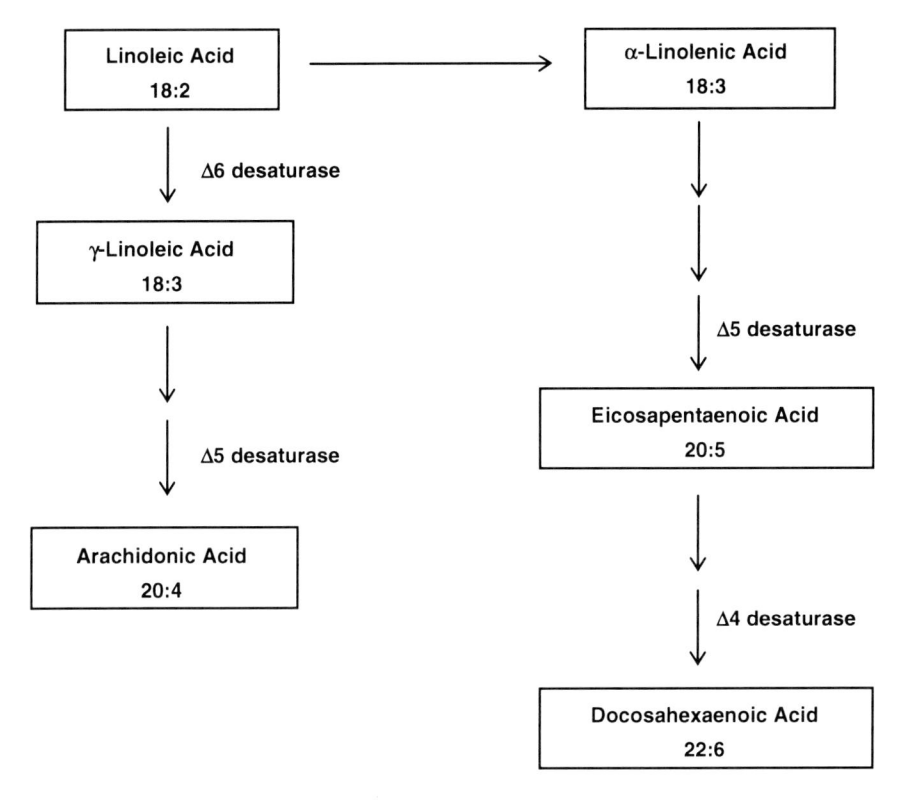

Fig. 10.3 Crucial steps in lipid metabolism of particular interest to plant metabolic engineers

by the combined expression of β-ketoacyl-CoA synthase and 22:1 acyl-CoA: lysophosphatidic acid acyltransferase. Anai et al. (2003) engineered rice with a soybean *FAD3* to increase the seed oil quality with a ten-fold increase of GLA. Sato et al. (2004) engineered marker-free soybean with levels of GLA as high as 50% by seed-specific expression of borage Δ6 desaturase gene.

With the advent of knowledge in gene discovery and corresponding genomic technologies, coupled with the spurt in elucidating gene expression pathways, the field of metabolic engineering forayed into technologies involving more complex manipulations of plant cell lipid metabolism involving engineering entire pathways in a single experiment. The first report of LC-PUFA production in higher plants as a "proof of concept" was by Qi et al. (2004) who demonstrated the increased synthesis of AA (6.6%) and EPA (3%) in *Arabidopsis* leaves by the transgenic expression of the individual genes in the Δ8 LC-PUFA pathway. The first success-ful reconstitution of Δ6 pathway for LC-PUFA production in plants and the first demonstration of LC-PUFA accumulation in seed oils were achieved in linseed (Abaddi et al. 2004). The most impressive demonstration of commercially signifi-cant concentrations of LC-PUFAs (19.5% of EPA) in plant seeds was achieved by placing the Δ6 pathway under the control of strong, seed-specific promoters in soybean (Kinney et al. 2004) where, apart from the minimal set of genes, coding the Δ6 pathway was engineered in combination with the Arabidopsis *Fad3* (Yadav et al. 1993) and the *Saprolegnia diclina* Δ17 desaturase (Pereira et al. 2004). In a further improvement of this technique, Damude and Yadav (2005) utilized the Δ15 desaturase from *Fusarium moniliforme* (Damude et al. 2004) instead of the *Fad3* to generate soybean plants with an overall 57% increase of ω3 LC-PUFA content.

10.4.5 Metabolic Engineering in Chloroplast Genome

The genetic engineering of plastids started in 1990, when Pal Maliga's lab demonstrated the first successful and stable transformation of plastids in higher plants (Svab et al. 1990). At that time, few scientists envisaged the use of plastids in metabolic engineering. However, 5 years later Maliga's lab published another paper describing the expression of *Bacillus thuringiensis* cry toxin in chloroplasts. Although this toxin was very difficult to express in a plant's nucleolar genome, expression in chloroplasts was shown to be at extraordinary levels (McBride et al. 1995). This article generated great interest among biotechnologists. Together with other advantages, such as lack of epigenetic processes and gene silencing, the possibility to use precise homologous recombination for transgene integration and lack of pollen transmission, plastids promise to become great tools in metabolic engineering. Moreover, plastids integrate and express foreign sequences as operons (Ruhlman et al. 2007) and the ability of plastid expression system to transcribe operons from a single promoter, and thus enabling the expression of multiple genes in a single recombination event, makes possible the expression of multienzyme pathways in the first transformed generation eliminating the need to cross lines recombinant for individual genes (Quesada-Vargas et al. 2005). These important properties make them an attractive alternative to nuclear genomic manipulations.

As mentioned earlier, the first use of plastids was in resistance engineering. Chloroplasts proved to be very suitable for expression of *B. thuringiensis cry* genes. Because of their different (prokaryotic) codon usage, expression from nuclear genome proved to be severely hampered. There is no need to adjust codon usage when expressed in chloroplasts (McBride et al. 1995; Kota et al. 1999). The expression efficiency was so high that cry protein crystals could be seen within chloroplasts (Daniell et al. 2001). Recently, the first insect-resistant soybean plants were generated, thus demonstrating that this technology works not only in tobacco model plant but also in important crops (Dufourmantel et al. 2005). Another example of resistance engineering is generation of glyphosate-tolerant plants. Glyphosate is a broad-spectrum herbicide, which blocks plant aromatic amino acids synthesis by competitively inhibiting the key enzyme 5-enol-pyruvyl shikimate-3-phosphate synthase (EPSPS). Thus, overexpression of EPSPS in plastids could block the inhibition effect of glyphosate. This presumption proved to be valid, as Ye et al. (2001) showed that chloroplast expression of an EPSPS gene yielded plants resistant to high doses of glyphosate.

Unlike resistance engineering, metabolic engineering does not require massive overexpression of the intermediate; to the contrary, this may even be disadvantageous. In plastids, therefore, expression mechanisms must be optimized by adjusting plastid expression signals. Unfortunately, this is likely to be a tedious process because the sequence of the coding region of the foreign gene itself often influences the accumulation of the expressed foreign protein and adjustment therefore relies on trial and error. On the other hand, the ability to engineer multiple genes (Daniell et al. 2005b), high levels of recombinant protein accumulation (Daniell et al. 1997), and the security of transgene containment due to maternal inheritance of plastid genomes in most crop species (Daniell 2002) are among the features that make the chloroplast system an efficient platform for metabolic engineering.

The first demonstration of the feasibility for engineering nutritionally important biochemical pathways in nongreen plastids was plastid expression of a bacterial lycopene β-cyclase gene in tomato chloroplasts (Wurbs et al. 2007). This resulted in herbicide resistance and triggered conversion of lycopene, the main storage carotenoid of tomatoes, to β-carotene (pro-vitamin A), an essential antioxidant. This yielded a four-fold enhancement in provitamin A content of the fruit. Thus far, the most complex and novel metabolic pathway introduced into tobacco plastids was that for the synthesis of the bioplastic polyhydroxybutyrate (PHB) from *Ralstonia eutropha* (Arai et al. 2001). Various techniques were used, of which the most promising appears to be plastid transgene expression using a nuclear-encoded and plastid-targeted ethanol-inducible T7 RNA polymerase promoter as this circumvents the toxic effect of constitutively expressed bacterial *phb* operon (Nakashita et al. 2001; Lossl et al. 2003).

Undoubtedly, the greatest importance for plastid metabolic engineering is the photosynthetic pathway where the *Rubisco* gene is their primary target. The work carried out in this field greatly exceeds the scope of this chapter; more than 5,000 manuscripts exist on this subject. Several extensive review articles were recently published covering this topic (Whitney and Andrews 2003; Bock and Khan 2004; Portis and Parry 2007).

Molecular pharming is the third category of plastid engineering. Extraordinary expression levels achieved in chloroplasts are undoubtedly the main reason for the high interest in the production of pharmaceuticals in plastids. For example, the tetanus toxin fragment was produced in tobacco chloroplasts with expression levels exceeding 25% of the total soluble protein (TSP) (Tregoning et al. 2003). Similarly, chloroplast expression of human serum albumin reached 11% of TSP (Fernandez-San Millan et al. 2003) and that for xylanase was 6% TSP (Leelavathi et al. 2004). Tobacco chloroplasts are able to correctly fold complex proteins with disulfide bridges, such as human somatotropin (Staub et al. 2000) and even full-size antibodies (Daniell et al. 2001). However, a significant increase in overall cost may arise if solubilization from inclusion bodies and refolding of these therapeutic proteins is necessary.

A disadvantage of using tobacco chloroplasts for protein production is that they are generally deficient in the capacity to glycosylate proteins since N- or O-glycosylation has a strong impact on the activity of several therapeutic proteins.

However, a recent discovery in *Arabidopsis* may remedy this situation. Villarejo et al. (2005) showed that a chloroplast-located protein in higher plants takes an alternative route through the secretory pathway and becomes N-glycosylated before entering the chloroplast. The other disadvantage of using tobacco plastids (and tobacco itself) is the high content of nicotine and other alkaloids that must be removed from the final product, increasing the overall cost. The choice of organism may therefore shift to edible plants since human proinsulin was shown to be produced in transgenic lettuce plastids (Ruhlman et al. 2007), and Daniell et al. (2005a) suggested that carrot appears to be ideal for oral delivery of therapeutic proteins. Commercialization of the expression of pharmaceutical proteins in chloroplasts is evidenced by an agreement made between Chlorogen (who has patented technology) and Sigma-Aldrich Fine Chemicals to produce four different proteins in tobacco plants.

10.5   Engineering Secondary Metabolic Pathways

Plant secondary metabolite pathways are the major target for metabolic engineering. Plant produces an enormous amount of secondary metabolites which play important roles in plant physiology. Some of plant secondary metabolites confer resistance against pests and diseases while some are the constituents of flower color, food flavor, and polymeric lignin for structural support and assorted medical agents such as phytoalexins, phytoestrogens (e.g., isoflavones and coumestrols) or chemopreventive anticancer agents (e.g., resveratrol), or regulate the development of fat cells (e.g., catechins), antimitotic, antimalarial, antioxidant, and antiasthmatic activities.

10.5.1   Transcription Factors as Tools for Metabolic Engineering

Transcription factors are regulatory proteins that can act as activators or repressors of gene expression through sequence-specific DNA binding and protein-protein interactions, mediating changes in the levels of mRNA accumulation. The molecular entities that are involved in such interactions are chromatin remodeling proteins other than the general transcription machinery (Latchman 2003). In recent years, a flurry in the knowledge of elucidating the functions of an array of transcription factors showed that many impact the flux through metabolic pathways and, since they tend to control multiple pathway steps, they are fast emerging as powerful tools to control complex metabolic pathways in plants (Broun 2004).

The potential of transcription factors as tools for manipulating metabolic pathways was recognized in the pioneering work of Goff et al. (1990) involving the maize flavonoid pathway regulators COLORLESS 1 and RED that were shown to induce flavonoid gene expression and anthocyanin accumulation in transgenic maize. Bovy et al. (2002) generated high-flavanol tomatoes by the heterologous expression of maize transcription factor genes *LC* and *C1* that resulted in an increased flux of flavonoid pathway throughout the fruit as opposed to expressing a chalcone isomerase gene that resulted in enhanced flavanol production only in the peel. This result proves that activating a pathway regulator as opposed to a pathway gene can induce metabolite accumulation in a tissue where most relevant enzymatic activities are insufficient. Transcription factors can also be valuable as discovery tools to identify enzymes and accessory proteins associated with complex pathways. This was effectively demonstrated by Broun et al. (2004) when they over-expressed an ETHYLENE RESPONSE FACTOR (ERF)-like transcription factor WAX INDUCER 1 (WIN1), that singly causes wax accumulation in *Arabidopsis*. When *WIN1* plants were examined by Northern and microarray analyses, an array of genes involved in wax biosynthesis, such as *CER1*, *KCS1* were seen to be up-regulated as were also other lipid biosynthetic genes and proteins involved in cellular trafficking. This shows that *WIN1* can be a useful tool to dissect molecular mechanisms underlying poorly understood, complex metabolic pathways. Transcription factors can also be utilized to downregulate pathway flux as shown by Kawaoka et al. (2000) who silenced a DNA-binding protein, *NTLIM1*, in transgenic tobacco that resulted in a dramatic reduction in lignin production due to a significant decrease in the expression of early phenylpropanoid pathway genes.

10.5.2 Flavonoids

Flavonoids are a large family of plant secondary metabolites synthesized from the phenylpropanoid pathway (Dixon and Steele 1999; Winkel-Shirley 2001). The biosynthetic pathway of flavonoids is the best characterized of plant secondary metabolites in terms of chemistry, biochemistry, genetics, and molecular biology (Harborne 1988, 1994; Stafford 1990; Winkel-Shirley 2001; Grotewold 2006). The knowledge of flavonoid biosynthesis and the important functions of flavonoid compounds in plants and in human nutrition have made flavonoids and isoflavonoids excellent targets for metabolic engineering.

In the phenylpropanoid pathway, phenylalanine ammonia lyase (PAL) catalyzes the conversion of phenylalanine to cinnamate. The cinnamate 4-hydroxylase (C4H) catalyzes the hydroxylation of cinnamate to *p*-coumarate that is converted by 4-coumarate: coenzyme A (CoA) ligase (4CL) to *p*-coumaroyl-CoA. The flavonoid biosynthesis starts with the condensation of one molecule of *p*-coumaroyl-CoA and three molecules of malonyl-CoA to produce tetrahydroxychalcone. This reaction is carried out by the enzyme chalcone synthase (CHS). Chalcone is isomerized to a flavanone by the enzyme chalcone isomerase (CHI) (Fig. 10.4).

Flavanones (e.g., naringenin) provide a central branch point in flavonoid biosynthesis. From these central intermediates, the pathway diverges into several side branches, each resulting in a different class of flavonoids as flavones, flavonols,

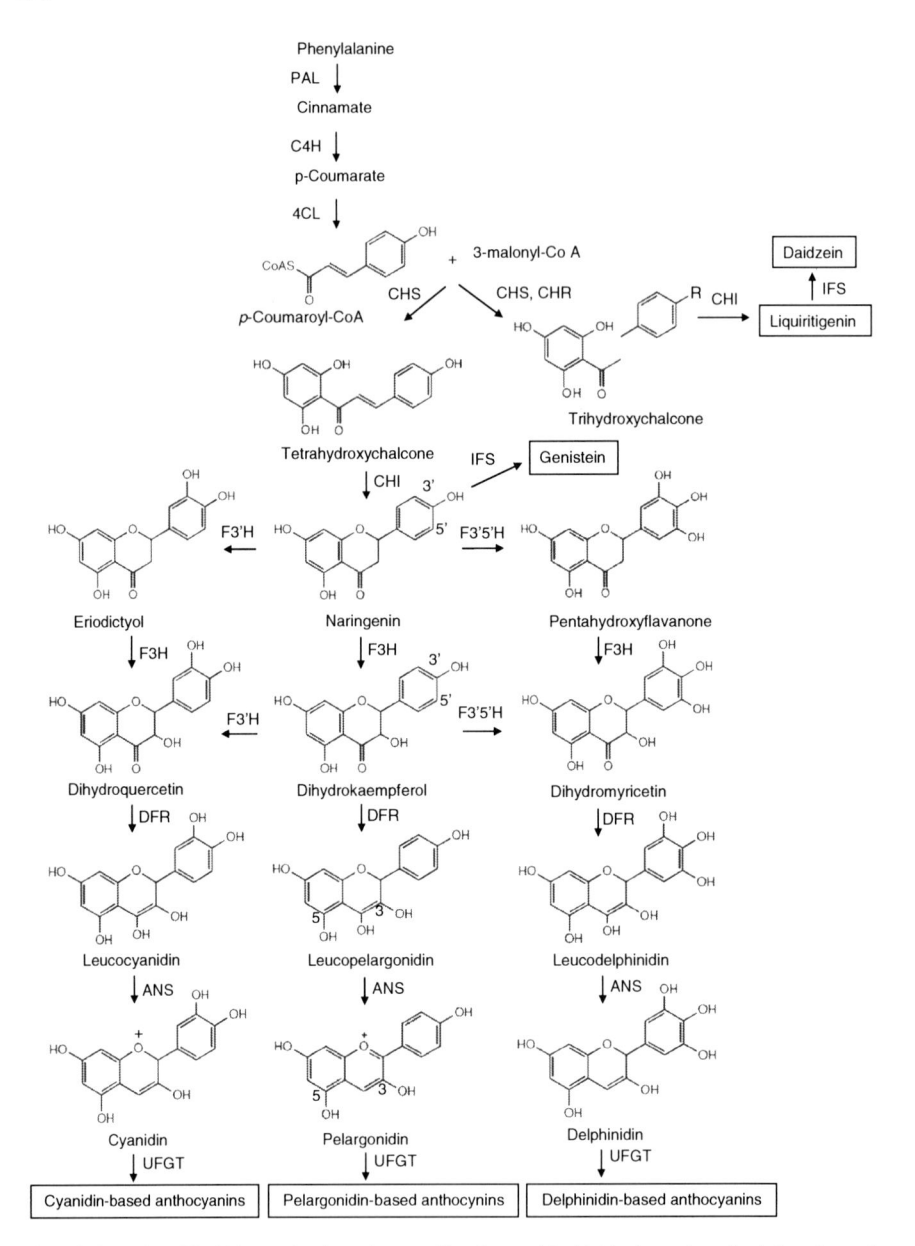

Fig. 10.4 A simplified biosynthesis pathway of isoflavonoids (daidzein and genistein) and cyanidin, pelargonidin and delphinidin-based anthocyanins. PAL, phenylalanine ammonia lyase; C4H, cinnamate 4-hydroxylase; 4CL, 4-coumarate:coenzyme A (CoA) ligase; CHS, chalcone synthase; CHI, chalcone isomerase; IFS, isoflavone synthase; F3′H, flavonoid 3′-hydroxylase; F3′5′H, flavonoid 3′, 5′-hydroxylase; F3H, flavanone 3-hydroxylase; DFR, dihydroflavonol reductase; ANS, anthocyanidin synthase and UFGT, UDP glucose:flavonoid 3-O- glucosyltransferase

isoflavones, anthocyanidins, and anthocyanins. Among these subclasses, isoflavones and anthocyanins are the main targets for metabolic engineering.

10.5.2.1 Isoflavone Biosynthesis and Metabolic Engineering

Isoflavones are mostly produced in the Papilionoideae subfamily of Leguminosae (Dewick 1994) such as soybean (*Glycine max*), green beans (*Phaseolus vulgaris*), and alfalfa (*Medicago sativa*). They are involved in plant defense mechanisms (Ebel et al. 1986; Rivera-Vargas et al. 1993; Graham and Graham 1996; Dixon and Sumner 2003) and symbiosis between the roots of leguminous plants and *Rhizobium* bacteria leading to the formation of nitrogen-fixing root nodules (Pueppke 1996; Spaink 2000; Ferguson and Mathesius 2003). Isoflavones have molecular structures similar to the human hormone estrogen and act as phytoestrogens. There are several reports of isoflavone activities important to human nutrition and medicine as anticancer and antioxidant compounds (for a review, see Ososki and Kennelly 2003; Cornwell et al. 2004).

The isoflavone phytoestrogens daidzein and genistein are synthesized from the phenylpropanoid pathway and stored in the vacuole as glucosyl- and malonyl-glucose conjugates (Graham and Graham 1996). The pathway to synthesize daidzein branches from the flavonoid biosynthesis catalyzed by chalcone synthase and a legume-specific enzyme, chalcone reductase (CHR) to generate trihydroxychalcone which is consequently converted to daidzein through reactions catalyzed by chalcone isomerase (CHI) and isoflavone synthase (IFS) (Fig. 10.4). Genistein synthesis from naringenin is mediated by IFS (Fig. 10.4). The soybean IFS is encoded by two genes, *IFS-1* and *IFS-2*, that have been cloned and examined in some detail by several groups (Akashi et al. 1999; Steele et al. 1999; Jung et al. 2000; Yu et al. 2000).

Metabolic engineering of isoflavones by increasing isoflavone levels in soybean and the introduction of isoflavone biosynthesis in nonlegume crops such as maize, wheat, or rice that do not naturally produce isoflavones has been a focus of research in recent years due to their significant roles in plant defense and in human medicine and nutrition. It has been shown that the level of genistein produced and accumulated in leaf and stem tissues of *Arabidopsis* transformed with soybean *IFS* (Jung et al. 2000) is enhanced when the phenylpropanoid pathway is activated by high UV-light (Yu et al. 2000). Genistein in IFS-transformed tobacco accumulates to higher levels in anthocyanin-producing flowers than in leaves. The production of genistein in maize black mexican sweet (BMS) cells required maize transcription factors C1 and R in addition to IFS. Further, BMS cells cotransformed with *IFS*, *CRC* (a chimeric transcription factor containing maize *C1* and *R* coding regions), and soybean *CHR* can produce the novel compound daidzein (Yu et al. 2000).

These results show that isoflavones can be synthesized in nonlegume plants, albeit at low levels compared to those in soybean. There are factors that limit the flow of intermediates toward isoflavone biosynthesis. These include flavanone

3b-hydroxylase (F3H), the major flavonoid enzyme that competes with IFS for the common substrate naringenin. Silencing of F3H reduced flavonoid biosynthesis and increased isoflavone accumulation. This was demonstrated in soybean lines transgenic for *CRC* as they accumulated isoflavones to much higher levels than in wild-type seed (Yu et al. 2003). Whereas the expression of *CRC* alone in soybean seeds gave only a small increase in isoflavone and flavonol levels, the coexpression of *CRC* together with a silencing construct targeting flavanone 3b-hydroxylase (F3H) resulted in increased total isoflavone content up four-fold higher than in wild-type seed. These high isoflavone soybeans would be useful for the production of soy foods providing potentially greater health benefits to consumers.

Another possibility for increasing the isoflavonoid content of nonlegume plants is by protein engineering. The expression of an IFS-CHI fusion protein in transgenic tobacco plants produced higher levels of the isoflavone genistein and genistein glycosides than plants transformed with IFS alone (Tian and Dixon 2006).

10.5.2.2 Anthocyanin Biosynthesis and Metabolic Engineering

Anthocyanins belong to the most important flavonoid class. They are major components of flower and fruit colors. The key enzymes to regulate the anthocyanin synthesis are flavonoid 3′-hydroxylase (F3′H) and flavonoid 3′, 5′-hydroxylase (F3′5′H) that catalyze hydroxylation at the 3′- or 3′5′-positions of the B-ring of flavonoid compounds. F3′H has a wide substrate range and can convert the naringenin to eriodictyol, the dihydrokaempferol (DHK) to dihydroquercetin (DHQ), and the flavonol kaempferol to quercetin. F3′5′H catalyzes the hydroxylation of both 3′ and 5′ positions of the B-ring leading to the conversion of naringenin and DHK to pentahydroxyflavanone and dihydromyricetin (DHM), respectively (Fig. 10.4). These compounds are catalyzed by flavanone 3-hydroxylase (F3H) to colorless dihydroflavonols, either DHK, DHQ, or DHM that will be reduced by dihydroflavonol reductase (DFR) to leucoanthocyanins. These compounds are converted to the corresponding leucoanthocyanidins by anthocyanidin synthase (ANS). Anthocyanidins serve as substrates for anthocyanin. They are unstable and will couple to sugar molecules by enzymes such as UDP-glucose:flavonoid 3-O- glucosyltransferase (UFGT) to yield the final relatively stable anthocyanins (Bohm 1998; Harborne 1994).

Cyanidin-, pelargonidin-, and delphinidin-based anthocyanins are responsible for flower colors. Cyanidin-based anthocyanin is the source of red and magenta colors, whereas pelargonidin-based anthocyanin is responsible for orange, pink, and bright red colors, and delphinidin-based anthocyanin for violet and blue colors. Limited ranges of flower color for individual plant species reflect the absence, mutation, or abundance of genes involved in anthocyanin biosynthesis, substrate-specificity of key enzymes, and/or the temporal and spatial regulation of the anthocyanin biosynthesis.

Some plants such as roses (*Rosa hybrida*), chrysanthemums (*Chrysanthemum morifolium*), and carnations (*Dianthus caryophyllus*) do not produce purple delphinidin-based anthocyanins because they lack the activity of F3′5′H (Elomaa and Holton 1994; Holton and Tanaka 1994; Tanaka et al. 1998; Mol et al. 1999). Petunia (*Petunia hybrida*), cymbidium (*Cymbidium hybrida*), tomato (*Solanum lycopersicum*), and cranberry (*Vaccinium macrocarpon*) do not produce brick red/ orange pelargonidin-based anthocyanins because their dihydroflavonol 4-reductases (DFRs) have strict substrate specificities and cannot utilize DHK as a substrate (Forkmann et al. 1980; Meyer et al. 1987; Johnson et al. 1999; Polashock et al. 2002).

Transformation of the maize *A* gene coding for the dihydroquercetin-4 reductase (DQR) into the petunia mutant, which shows no flower pigmentation (Meyer et al. 1987), demonstrated that the DQR in transgenic petunia can reduce the petunia DHK to leucopelargonidin, which leads to the production of red color pigmentation. This indicated that it is possible to generate a novel flower color in plants by introducing the gene involved in anthocyanin biosynthesis pathway.

However in some plants, the introduction of a foreign gene may not be sufficient to convert the metabolic flux of anthocyanin biosynthesis to obtain a plant with the desired flower color. It is necessary to select the suitable plant cultivars that have the appropriate genetic background and flavonoid composition and/or the artificial down-regulation of a competing endogenous pathway (Tanaka 2006; Tanaka and Brugliera 2006). Florigene Ltd. (Melbourne, Australia) generated violet carnations by transforming the petunia *F3'5'H* gene in combination with the *dfr* genes into white carnation cultivars that specifically lacked the *dfr* gene (Mol et al. 1999; Fukui et al. 2003).

Metabolic engineering of rose flower color is more complicated than in carnation as in nature there is no white rose lacking the *dfr* gene. To solve this problem, Florigene Ltd. (Melbourne, Australia) and Suntory Ltd. (Osaka, Japan) employed the gene-silencing technique to switch off the *dfr* gene that produces the red pigment in rose. The *F3'5'H* gene from iris was then inserted into *dfr*-silenced rose to produce the blue pigment of delphinidin-based anthocyanins (Katsumoto et al. 2007). However, the blue rose generated by Florigene and Suntory is not a "true" blue, but it is in fact of a pale violet color. In addition to anthocyanins that determine the flower color, other factors such as vacuolar pH, copigments, metal ions, and anthocyanin modifications (acylation, glycosylation, and methylation) that influence the shade and intensity of flower color must be taken into account (Yoshida, et al. 1995; Yabuya et al. 1997; Mol et al. 1998; Tanaka et al. 1998). Anthocyanins are bluer in weakly acidic and neutral pH and they are redder in acidic pH. Rose petals are moderately acidic with a pH around 4.0, which inhibits the blue pigment, while the carnation petals are less acidic with a pH of 5.5. Several genes such as *ph1-ph7* and *Pr*, encoding the proteins that control the vacuolar pH, have been identified from petunia by transposon tagging and transposon display (Mol et al. 1998; Fukada-Tanaka et al. 2000; Quattrocchio et al. 2006; Verweij et al. 2008). It may be feasible to engineer true blue roses by manipulation of the anthocyanin genes and the transcription factors regulating their spatial and temporal expression.

The intra- or intermolecular stacking of copigments such as flavones, flavonols, phenylpropanoids, organic acids, and aromatic acylated groups also leads to a shift in the visible absorption maximum of the complex toward longer wavelength (bathochromic shift) (Goto and Kondo 1991). Inhibition of DFR in torenia

(*Torenia fournieri*) by antisense silencing of DFR increases the level of flavones, which made wild-type violet flower color more intensely blue (Aida et al. 2000). Cosuppression of flavone synthase II (FNSII) in torenia decreased the amount of flavones and increased the amount of flavanones, and yielded paler flower color and increased that of flavanones, generating transgenic flowers paler than wild-type ones (Ueyama et al. 2002).

Extensively studied copigments of flavonoids are in blue-flowered plants (Harborne and Williams 2000; Grotewold 2006) and the effect of metal ions associated with anthocyanins in flower (Kondo et al. 1992; Yoshida et al. 2003; Shiono et al. 2005; Shoji et al. 2007).

Anthocyanins form complexes with copigments such as flavones and flavanols by aggregation, resulting in shift of the visible absorption maximum of the complex toward longer wavelength (bathochromic shift). This usually leads to darker flower colors (Forkmann 1991). Flavones are common copigments that form complexes with anthocyanins. Mixtures of various molar ratios of anthocyanin fraction and flavone fraction from flower extracts of torenia (*T. fournieri*) were prepared in a pH 5.4 buffer and their visible absorbance was monitored. Mixtures with molar ratios corresponding to the endogenous concentrations in the petal had absorbance values corresponding to the color of the petals (Aida et al. 2000). In addition, antisense silencing of DFR in Torenia caused a marked increase in flavones, resulting in transgenic plants with bluer flower color than CHS-silenced plants (Aida et al. 2000). FNS genes that are responsible for the biosynthesis of flavones have been isolated from *Torenia hybrida* and other species.

Inhibition of DFR in torenia (*T. fournieri*) by antisense silencing of DFR increases the level of flavones, yielding a more intense blue flower color than that of the wild type (Aida et al. 2000). Cosuppression of FNSII in torenia reduced the amount of flavones and increased that of flavanones, generating transgenic flowers paler than wild-type ones (Ueyama et al. 2002).

Many metal ions including Cu^{2+}, Ca^{2+}, Al^{3+}, Fe^{3+}, Mg^{2+}, and Mo^{2+} were found to coexist with anthocyanins (Ellestad 2006). Such associations usually have a significant impact on flower color. For instance, addition of metal ions (Mo^{2+}) in vitro to purified anthocyanins from *Brassica rapa* can result in a color change from pink to blue (Hale et al. 2001). Energy-dispersive X-ray analysis showed that metal ions accumulated predominantly in the vacuoles of the epidermal cells. In *Brassica juncea*, the X-ray absorption spectrum of plant-accumulated Mo^{2+} was different than that for molybdate, and correlated with the cellular and subcellular distribution of water soluble, pH-dependent anthocyanins (Hale et al. 2001).

10.6 Future Roadmap

The increased interest in plant metabolic engineering in recent years can be attributed to four main factors: plants as a major source for medicinal products; plants containing health-promoting secondary metabolites; plants resistant to pests and diseases; and plants with flowers of novel colors, patterns, scents. The rapid progress seen in successful and robust methods of plant genetic transformation (for details see Chaps 1-1, 1-2, 1-3, 1-4, 1-5, 1-7, 1-8 and 1-9 of this Volume) has revived the early promises of plant genetic engineering to provide many novel attributes to the world's flora.

10.6.1 Food for the World

The primary need of mankind for plants is to provide food security. As defined by Wikipedia, food security refers to the availability of food and its accessibility. A household is considered food secure when its occupants do not live in hunger or fear of starvation. It is hard for those of us living in well-developed nations that over 850 million of the world's 6.6 billion population people are chronically hungry and up to two billion people lack food security intermittently due to varying degrees of poverty (FAO 2003).

10.6.2 Biofortification of Plant-Based Foods

The requirement of a minimal daily intake of essential micronutrients, vitamins, and minerals for the maintenance of optimal human health has long driven the focus of plant science research toward combating micronutrient malnutrition by developing superior plant varieties with improved nutritional value. The approach evolved into what is now known as biofortification where efforts are on to deliver the daily micronutrients directly into the staple crops consumed by mankind. This approach was intended to alleviate the increased industrial costs incurred by the fortification of processed foods with the micronutrients. The dissection of plant metabolic pathways involved in synthesis of essential dietary micronutrients showed that all plants have the biochemical activities necessary to synthesize and accumulate a near full complement of essential dietary micronutrients with the exception of vitamins D and B_{12} (Dellapenna 2007). If we observe the dietary habits of populations with the maximum risk of micronutrient malnutrition, they consume foods like rice, wheat, cassava, and maize that contain insufficient daily intake levels of essential dietary micronutrients. These specific staple crops are being targeted by plant metabolic researchers for biofortification so that the levels of the limiting micronutrients in these crops can be increased by an effective combination of breeding and genetic engineering. The first successful and also the most popular example in this area included the biofortification of provitamin A in rice that resulted in the nutritionally enhanced "golden rice" that will help toward combating malnutrition-induced eye defects (Ye et al. 2000; Datta et al. 2003). The historical and scientific details of the steps involved in the golden rice technology are given in

Emani et al. (2008). Biofortification of the vitamin E family of lipophilic antioxidants called tocochromanols that protect against effects of free radicals, reactive oxidation species, and lipid oxidation was achieved in barley (Cahoon et al. 2003), soybean (Van Eenennaam et al. 2003), and oilseeds (Karunanandaa et al. 2005). Biofortification of another important B-vitamin, folate, was achieved in tomato fruit (Diaz de la Garza et al. 2007).

Initially, plant metabolic engineering involving manipulating nutritional levels relied on expression or silencing of a single gene in well-studied metabolic pathways that proved effective when the engineered step was at a potential metabolic branch point. Future research especially in the area of biofortification should consider expressing heterologous enzymes at such steps that can potentially create novel substrates for already existing enzymes that may lead to the creation of an entirely new branch in the pathway and formation of novel products (Kinney 2006).

10.6.3 *Biofuel From Plants*

Biofuels have come of age as attractive sources of energy around the world as finite petroleum reserves, increasing demands of energy in both industrially developed and rapidly industrializing countries combined with negative environmental effects of petroleum undermines both economic strength and threatens national security (Bordetsky et al. 2005). The biofuel that has the potential for extensive usage around the globe is ethanol due to its environmentally friendly nature owing to low toxicity and ready biodegradability, and the ability to be produced from the abundant biomass of land plants. Ethanol production from biomass also reduces the levels of greenhouse gasses. The usage of plants for cellulosic ethanol production as compared to other sources like starch and sugar-derived ethanol is because of the lower costs and abundance of biomass as compared to the limited supplies and the food supply competition related to starch and sugar. Food crops such as corn, rice, sugarcane, perennial grasses such as switchgrass and giant miscanthus and woody crops such as polar and shrub willow are potential sources for ethanol production (Sticklen 2008). Plant cell wall is the source for the lignocellulosic biomass, and the secondary cell wall contains cellulose, hemicellulose, and lignin (Sticklen 2008). Enzymatic hydrolysis utilizing cellulases and hemicellulases can convert the cell wall polysaccharides to fermentable sugars, the main barrier to overcome being the lignin that prevents accessibility of the enzymes to the polysaccharides. Lignin breakdown by chemical and heat treatments combined with microbial production of cellulases was the starting point in developing efficient processes to produce fermentable sugars for biofuels. Metabolic engineers can play a vital role in research aimed at characterizing the cell-wall deconstruction enzymes, especially in isolating enzymes that can resist higher conversion temperatures and a range of pHs during the pretreatments aimed at lignin hydrolysis that is one of the important challenges in cellulosic ethanol production (Sticklen 2008). Presently, the successes seen in plant genetic transformation can be exploited to design strategies to express plant cell wall deconstructing enzymes in transgenic plants to enable cheaper processes for producing cellulosic hydrolysis enzymes (Sticklen et al. 2006). A comprehensive characterization of all the steps in cellulose biosynthesis (Kawagoe and Delmer 1997) has been the focus of plant molecular biologists (Arioli et al. 1998) now being complemented by the latest advances of genomics and microarray technology to identify the relevant useful genes (Persson et al. 2005; Andersson-Gunneras et al. 2006). This would enable the efforts to increase the plant cellulosic biomass in terms of increased cell-wall polysaccharide content by genetic manipulations. The increase seen in the overall plant biomass in rice by the elevated expression of ADP-glucose pyrophosphorylase by an endosperm-specific promoter (Smidansky et al. 2003) throws open the doors for metabolic engineers to explore manipulations of other enzymes of the starch biosynthetic pathway to aid in a shift to increasing biomass for biofuel production. A better understanding of the lignin biosynthesis aimed at down-regulation of the involved enzymes to modify structural components of lignin or reduce the lignin content itself is the need of the hour to avoid the need for the expensive pretreatments (Sticklen 2008). For a more exhaustive review of transgenic technology related to biofuels, refer to Chaps 2–6 of the Volume 2 of this series.

10.7 Conclusion: Factories of the Future

Plant metabolic engineering has had a fairly successful run in the academic and industrial circles, but a fact that cannot be ignored is that it was punctuated by several failures and limitations. Several "proof of concept" experiments successful in model plants failed to live up to expectations in the cultivars. The key to successfully overcoming such challenges is to fully exploit the advent of applied genomics, proteomics, and metabolomics to comprehensively understand poorly characterized metabolic pathways. The newly emerging discipline of systems biology should be used to see beyond the boundaries of metabolic pathways that are subject to engineering to create and understand the complete metabolite profiles in the plant world. The acquired knowledge will enable researchers worldwide to successfully dissect and understand metabolic pathways, and successfully increase their ability to both model and implement multipoint metabolic manipulations. This would in turn result in an avalanche of desirable products in transgenic plants that would be rightly called the "factories of the future." The advent of "molecular pharming" has readily identifiable benefits for mankind in a cost-effective, nutritionally wholesome, and environmentally sustainable manner. Together with enhancing the esthetic nature of the world through the development of novel ornamental plants, it can supplement the development of the still irreplaceable traditional agriculture to meet the rising food security in the centuries to come.

References

Abaddi A, Domergue F, Bauer J, Napier J, Welti R, Zahringer U, Cirpus P, Heinz E (2004) Biosynthesis of very long chain polyunsaturated fatty acids in transgenic oilseeds: constraints on their accumulation. Plant cell 16:2734–2748

Aida R, Kishimoto S, Tanaka Y, Shibata M (2000) Modification of flower color in torenia (*Torenia fournieri* Lind.) by genetic transformation. Plant Sci 153:33–42

Akashi T, Fukuchi-Mizutani M, Aoki T, Ueyama Y, Yonekura-Sakakibara K, Tanaka Y, Kusumi T, Ayabe S (1999) Molecular cloning and biochemical characterization of a novel cytochrome P450, flavone synthase II, that catalyzes direct conversion of flavanones to flavones. Plant Cell Physiol 40:1182–1186

Alabadi D, Carbonell J (1998) Expression of ornithine decarboxylase is transiently increased bypollination, 2, 4-Dichlorophenoxyacetic acid and gibberellic acid in tomato ovaries. Plant Physiol 118:323–328

Anai T, Koga M, Tanaka H, Kinoshita T, Rahman S, Takagi Y (2003) Improvement of rice (*Oryza sativa* L.) seed oil quality through introduction of a soybean microsomal omega-3 fatty acid desaturase gene. Plant Cell Rep 21:988–992

Andersson-Gunneras S, Mellerowicz EJ, Love J, Segerman B, Ohmiya Y, Coutinho PM, Nillson P, Henrissat B, Moritz T, Sundberg B (2006) Biosynthesis of cellulose-enriched tension wood in *Populus*: global analysis of transcripts and metabolites identifies biochemical and developmental regulators in secondary wall biosynthesis. Plant J 45:144–165

Arai Y, Nakashita H, Doi Y, Yamaguchi I (2001) Plastid targeting of polyhydroxybutyrate biosynthetic pathway in tobacco. Plant Biotechnol 18:289–293

Arioli T, Peng L, Betzner A, Burn J, Wittke W, Herth W, Camillerie C, Hoffe H, Plazinski J, Birch R, Cork A, Glover J, Redmond J, Williamson R (1998) Molecular analysis of cellulose biosynthesis in *Arabidopsis*. Science 279:717–720

Bailey JE (1991) Toward a science of metabolic engineering. Science 252:1668–1675

Bailey JE, Sburlati A, Hatzimanikatis V, Lee K, Renner WA, Tsai PS (1996) Inverse metabolic engineering: a strategy for directed genetic engineering of useful phenotypes. Biotechnol Bioeng 52:109–121

Barta A, Sommergruber K, Thomson D (1986) The expression of a nopaline synthase- human growth hormone chimeric gene in transformed tobacco and sunflower callus tissue. Plant Mol Biol 6:347–357

Bhattacharya E, Rajam M (2007) Polyamine biosynthetic pathway: a potential target for alkaloid production. In: Verpoorte R, Alfermann A, Johnson T (eds) Applications of plant metabolic engineering. Springer, Dordrecht, pp 129–143

Bock R, Khan MS (2004) Taming plastids for a green future. Trends Biotechnol 22:311–318

Bohm BA (1998) Introduction to Flavonoids, 1st edn. Taylor and Francis, London

Bordetsky A, Hwang R, Korin A, Lovaas D, Tonachel L (2005) Securing America: solving our oil dependence through innovation. Natural Resources Defense Council, New York

Bovy A, de Vos R, Kemper M, Schiljen E, Almenar Pertejo M, Muir S, Collins G, Robinson S, Verhoyen M, Hughes S (2002) High-flavanol tomatoes resulting from the heterologous expression of the maize transcription factor genes LC and C1. Plant cell 14:2509–2526

Broun P (2004) Transcription factors as tools for metabolic engineering in plants. Curr Opin Plant Biol 7:202–209

Broun P, Poindexter P, Osborne E, Jiang C-Z, Riechmann J (2004) WIN1, a transcriptional activator of epidermal wax accumulation in *Arabidopsis*. Proc Natl Acad Sci USA 101:4706–4711

Burn J, Hocart C, Birch R, Cork A, Williamson R (2002) Functional analysis of the cellulose synthase genes *CesA1, CesA2, and CesA3* in *Arabidopsis*. Plant Physiol 129:797–807

Cahoon E, Hall S, Ripp K, Ganzke T, Hitz W, Couglan S (2003) Metabolic redesign of vitamin E biosynthesis in plants for tocotrienol production and increased antioxidant content. Nat Biotechnol 21:1082–1087

Capell T, Christou T (2004) Progress in plant metabolic engineering. Curr Opin Biotechnol 15:148–154

Chattopadhyay M, Gupta S, Sengupta D (1997) Expression of arginine decarboxylase in seedlings of *indica* rice (*Oryza sativa* L.) cultivars as affected by salinity stress. Plant Mol Biol 34:477–483

Chen T, Murata N (2002) Enhancement of tolerance of abiotic stress by metabolic engineering of betaines and other compatible solutes. Current Opin Plant Biol 5:250–257

Cornwell T, Cohick W, Raskin I (2004) Dietary phytoestrogens and health. Phytochemistry 65:995–1016

Damude H, Kinney A (2007) Mertabolic engineering of seed oil biosynthetic pathways for human health. In: Verpoorte R, Alfermann A, Johnson T (eds) Applications of Plant Metabolic Engineering. Springer, Dordrecht, The Netherlands, pp 237–247

Damude H, Yadav N (2005) Cloning and sequences of fungal $\delta-15$ desaturases suitable for production of polyunsaturated fatty acids in oilseed plants for food or industrial uses. In: PCT Int ApplWO2005047479

Damude H, Zhang H, Farrall L (2004) Identification of bifunctional $\delta12/\omega3$ fatty acid desaturases for improving the ratio of $\omega3$ to $\omega6$ fatty acids in microbes and plants. Proc Natl Acad Sci USA 103:9446–9451

Daniell H (2002) Molecular strategies for gene containment in transgenic crops. Nat Biotechnol 20:581–586

Daniell H, Guda C, McPherson DT, Zhang X, Xu J, Urry DW (1997) Hyperexpression of a synthetic protein-based polymer gene. Methods Mol Biol 63:359–371

Daniell H, Streatfield SJ, Wycoff K (2001a) Medical molecular farming: production of antibodies, biopharmaceuticals and edible vaccines in plants. Trends Plant Sci 6:219–226

Daniell H, Wiebe PO, Millan AF (2001b) Antibiotic-free chloroplast genetic engineering – an environmentally friendly approach. Trends Plant Sci 6:237–239

Daniell H, Chebolu S, Kumar S, Singleton M, Falconer R (2005a) Chloroplast-derived vaccine antigens and other therapeutic proteins. Vaccine 23:1779–1783

Daniell H, Kumar S, Dufourmantel N (2005b) Breakthrough in chloroplast genetic engineering of agronomically important crops. Trends Biotechnol 23:238–245

Datta K, Baisakh N, Oliva N, Torrizo L, Abrigo E, Tan J, Rai M, Rehana S, A-B S, Beyer P, Potrykus I, Datta S (2003) Bioengineered golden *Indica* rice cultivars with beta-carotene metabolism in the endosperm with hygromycin and mannose selection systems. Plant Biotechnol J 1:81–90

De Zoeten G, Penwick J, Horisberger M (1989) The expression, localization and effect of a human interferon in plants. Virology 172:213–222

Del Vecchio A (1996) High laurate canola. Inform 7:230–243

Delgado J, Liao JC (1997) Inverse flux analysis for reduction of acetate excretion in *Escherichia coli*. Biotechnol Prog 13:361–367

Dellapenna D (2007) Biofortification of plant-based food: enhancing folate levels by metabolic engineering. Proc Natl Acad Sci USA 104:3675–3676

Dewick PM (1994) The Flavonoids: Advances in Research since 1986. In: Harborne JB (ed) The Flavonoids: Advances in Research. Chapman and Hall, London, New York, pp 117–232

Diaz de la Garza R, Gregory J, Hanson A (2007) Folate biofortification of tomato fruit. Proc Natl Acad Sci USA 104:4218–4222

Dixon RA, Steele CL (1999) Flavonoids and isoflavonoids – a gold mine for metabolic engineering. Trends Plant Sci 4:394–400

Dixon RA, Sumner LW (2003) Legume natural products: understanding and manipulating complex pathways for human and animal health. Plant Physiol 131:878–885

Drexler H, Spiekermann P, Meyer A, Domergue F, Zank T, Sperling P, Abaddi A, Heinz E (2003) Metabolic engineering of fatty acids for breeding of new oilseed crops: strategies, problems and first results. J Plant Physiol 160:779–802

Dufourmantel N, Tissot G, Goutorbe F, Garcon F, Muhr C, Jansens S, Pelissier B, Peltier G, Dubald M (2005) Generation and analysis of soybean plastid transformants expressing *Bacillus thuringiensis* Cry1Ab protoxin. Plant Mol Biol 58:659–668

During K (1988) Wound-inducible expression and secretion of T4 lysozyme and monoclonal antibodies in *Nicotiana tabacum*. PhD thesis. Mathematish-Naturwissenschaftlichen Fakultaet der Universitaet Zu Koeln

Ebel J, Schmidt WE, Loyal R (1986) Phytoalexin synthesis: the biochemical analysis of the induction process. Annu Rev Phytopathol 24:235–264

Ellestad G (2006) Structure and chiroptical properties of supramolecular flower pigments. Chirality 18:134–144

Elomaa P, Holton T (1994) Modification of flower color using genetic engineering. Biotechnol Genet Eng Rev 12:63–88

Emani C, Jiang Y, Miro B, Hall T, Kohli A (2008) Rice. In: Kole C, Hall T (eds) Compendium of transgenic crop plants, vol 1, Cereals and forage grasses. Wiley-Blackwell, Oxford, pp 1–47

Ferguson BJ, Mathesius U (2003) Signaling interactions during nodule development. J Plant Growth Reg 22:47–72

Fernandez-San Millan A, Mingo-Castel A, Miller M, Daniell H (2003) A chloroplast transgenic approach to hyper-express and purify Human Serum Albumin, a protein highly susceptible to proteolytic degradation. Plant Biotechnol J 1:71–79

Fischer R, Emans N (2000) Molecular farming of pharmaceutical proteins. Transgenic Res 9:279–299

Forkmann G (1991) Flavonoids as flower pigments: the formation of the natural spectrum and its extension by genetic engineering. Plant Breed 106:1–26

Forkmann G, Heller W, Grisebach H (1980) Anthocyanin biosynthesis in flowers of *Matthiola incana* flavanone 3- and flavonoid 3'-hydroxylases. Z Naturforsch C 35:691–695

Fraley R, Rogers S, Horsch R (1983) Expression of bacterial genes in plant cells. Proc Natl Acad Sci USA 80:4803–4807

Fujita N, Kubo A, Suh D, Wong K, Jane J, Ozawa K, Takaiwa F, Inaba Y, Nakamura Y (2003) Antisense inhibition of isoamylase alters the structure of amylopectin and the physicochemical properties of starch in rice endosperm. Plant Cell Physiol 44:607–618

Fukada-Tanaka S, Inagaki Y, Yamaguchi T, Saito N, Iida S (2000) Color-enhancing protein in blue petals. Nature 407:581

Fukui Y, Tanaka Y, Kusumi T, Iwashita T, Nomoto K (2003) A rationale for the shift in color towards blue in transgenic carnation flowers expressing the flavonoid 3', 5'-hydroxylase gene. Phytochemistry 63:15–23

Galili G, Hofgen R (2002) Metabolic engineering of amino acids and storage proteins in plants. Metab Eng 4:3–11

Goff S, Klein T, Roth B, Fromm M, Cone K, Radicella J, Chandler V (1990) Transactivation of anthocyanin in biosynthetic genes following transfer of B regulatory genes into maize tissues. EMBO J 9:2517–2522

Goto T, Kondo T (1991) Structure and molecular stacking of anthocyanins – flower Color variation. Angew Chem 30:17–33

Graham TL, Graham MY (1996) Signaling in soybean phenylpropanoid responses (dissection of primary, secondary, and conditioning effects of light, wounding, and elicitor treatments). Plant Physiol 110:1123–1133

Grotewold E (2006) The genetics and biochemistry of floral pigments. Annu Rev Plant Biol 57:761–780

Hale K, McGrath S, Lombi E, Stack S, Terry N, Pickering I, George G, Pilon-Smits E (2001) Molybdenum sequestration in *Brassica* species. A role for anthocyanins? Plant Physiol 126:1391–1402

Han J, Luhs W, Sonntag K, Zahringer U, Borchardt D, Wolter F, Heinz E, Frentzen M (2001) Functional characterization of β-ketoacyl-CoA synthase genes from *Brassica napus* L. Plant Mol Biol 46:229–239

Haq T, Mason H, Clements J (1995) Oral immunization with a recombinant bacterial antigen produced in transgenic plants. Science 268:714–716

Harborne JB (ed) (1988) The Flavonoids: advances in research since 1980. Chapman and Hall, London

Harborne JB (ed) (1994) The Flavonoids: advances in research since 1980, 1st edn. Chapman and Hall, London

Harborne JB, Williams CA (2000) Advances in flavonoid research since 1992. Phytochemistry 55:481–504

Hashimoto T, Tamaki K, Suzuki K (1998) Molecular cloning of plant spermidine synthases. Plant Cell Physiol 39:73–79

Hiatt A, Cafferkey R, Bowdish K (1989) Production of antibodies in transgenic plants. Nature 342:76–78

Holton TA, Tanaka Y (1994) Blue roses – a pigment of our imagination? Trends Biotechnol 12:40–42

Horsch RB, Fry JE, Hoffman NL, Eicholtz D, Rogers SG, Fraley RT (1985) A simple and general method for transferring genes into plants. Science 227:1229–1231

Jefferson R, Kavanaugh T, Bevan M (1987) GUS fusion: β-glucuronidase as a sensitive and versatile gene fusion marker in higher plants. EMBO J 6:3901–3907

Johnson ET, Yi H, Shin B, Oh BJ, Cheong H, Choi G (1999) *Cymbidium hybrida* dihydroflavonol 4-reductase does not efficiently reduce dihydrokaempferol to produce orange pelargonidin-type anthocyanins. Plant J 19:81–85

Jung W, Yu O, Lau SMC, O'Keefe DP, Odell J, Fader G, McGonigle B (2000) Identification and expression of isoflavone synthase, the key enzyme for biosynthesis of isoflavones in legumes. Nat Biotechnol 18:208–212

Karunanandaa B, Qi Q, Hao M, Baszis S, Jensen P, Wong Y, Jiang J, Venkatramesh M, Gruys K, Moshiri F (2005) Metabolically engineered oilseed crops with enhanced seed tocopherol. Metab Eng 7:384–400

Katsumoto Y, Fukuchi-Mizutani M, Fukui Y, Brugliera F, Holton T, Mirko K, Noriko N, Yonekura-Sakakibara K, Togami J, Pigeaire A, Tao G, Nehra N, Lu CY, Dyson B, Tsuda S, Ashikari T, Kusumi T, Mason J, Tanaka Y (2007) Engineering of the rose flavanoid biosynthetic pathway succesfully generated blue-hued flowers accumulating delphinidin. Plant Cell Physiol 48:1589–1600

Kawagoe Y, Delmer D (1997) Pathways and genes involved in cellulose biosynthesis. Genet Eng 19:63–87

Kawaoka A, Kaothien P, Yoshida K, Endo S, Yamada K, Ebinuma H (2000) Functional analysis of tobacco LIM protein Ntlim1 involved in lignin biosynthesis. Plant J 22:289–301

Kinney A (2006) Metabolic engineering in plants for human health and nutrition. Curr Opin Biotechnol 17:130–138

Kinney A, Knowlton S, Blackie L (1998) Designer oils: the high oleic soybean. In: Roller S, Harlander S (eds) Genetic modification in the food industry. Blackie, London, pp 193–213

Kinney A, Cahoon E, Damude H (2004) Production of very long chain polyunsaturated fatty acids in oilseed plants. PCT Int ApplWO2004071467

Kitano H (2002) Systems biology: a brief overview. Science 295:1662–1664

Kok-Jacon G, Ji Q, Vincken J-P, Visser R (2003) Towards a more versatile α-glucan synthesis in plants. J Plant Physiol 160:765–777

Kondo T, Yoshida K, Nakagawa A, Kawai T, Tamura H, Goto T (1992) Structural basis of blue-color development in flower petals from *Commelina communis*. Nature 358:515–518

Kota M, Daniell H, Varma S, Garczynski SF, Gould F, Moar WJ (1999) Overexpression of the *Bacillus thuringiensis* (*Bt*) Cry2Aa2 protein in chloroplasts confers resistance to plants against susceptible and Bt-resistant insects. Proc Natl Acad Sci USA 96:1840–1845

Kumar A, Altabella T, Taylor M (1997) Recent advances in polyamine research. Trends Plant Sci 2:124–130

Kumar S, Sharma M, Rajam M (2006) Polyamine biosynthetic pathway as a novel target for potential applications in plant biotechnology. Physiol Mol Biol Plant 12:53–58

Kumar G, Ganapathi T, Srinivas L, Bapat V (2007) Plant molecular farming: host systems, technology and products. In: Verpoorte R, Alfermann A, Johnson T (eds) Applications of plant metabolic engineering. Springer, Dordrecht, pp 45–77

Kurnaz I (2005) Biochemical modelling tools and applications to metabolic engineering. Turk J Biochem 30:200–207

Kusnadi A, Hood E, Witcher D (1998) Production and purification of two recombinant proteins from transgenic corn. Biotechnol Prog 14:149–155

Latchman D (2003) Eukaryotic transcription factors, 4th edn. Academic, San Diego, CA

Leelavathi S, Sunnichan VG, Kumria R, Vijaykanth GP, Bhatnagar RK, Reddy VS (2004) A simple and rapid *Agrobacterium*-mediated transformation protocol for cotton (*Gossypium hirsutum* L.): embryogenic calli as a source to generate large numbers of transgenic plants. Plant Cell Rep 22:465–470

Li Z, Chen S (2000) Differential accumulation of the S-adenosylmethionine decarboxylase transcript in rice seedlings in response to salt and drought stress. Theor Appl Genet 100:782–788

Liu J-W, Huang Y-S, DeMichele S, Bergana M, Bobik EJ, Hastilow C, Chuang L-T, Mukerji P, Knutzon D (2001) Evaluation of the seed oils from a canola plant genetically transformed to produce high levels of γ-linoleic acid. In: Hunag Y-S, Ziboh V (eds) γ-linoleic acid: recent advances in biotechnology and clinical applications. AOCS Press, Champaign, IL, pp 61–71

Liu Q, Wang Z, Chen X, XL C, Tang S, Yu H, Zhang J, Hong M, Gu M (2003) Stable inheritance of the antisense *Waxy* gene in transgenic rice with reduced amylose level and improved quality. Transgenic Res 12:71–82

Lossl A, Eibl C, Harloff HJ, Jung C, Koop HU (2003) Polyester synthesis in transplastomic tobacco (*Nicotiana tabacum* L.): significant contents of polyhydroxybutyrate are associated with growth reduction. Plant Cell Rep 21:891–899

Magnuson N, Linzmaier P, Reeves R (1998) Secretion of biologically active human interleukin-2 and interleukin-4 from genetically modified tobacco cells in suspension culture. Protein Expr Purif 13:45–52

Mallik V, Watson M, Malmberg R (1996) A tobacco ornithine decarboxylase partial cDNA clone. J Plant Biochem Biotechnol 5:109–112

McBride KE, Svab Z, Schaaf DJ, Hogan PS, Stalker DM, Maliga P (1995) Amplification of a chimeric Bacillus gene in chloroplasts leads to an extraordinary level of an insecticidal protein in tobacco. BioTechnology 13:362–365

Mehta R, Cassol T, Li N, Ali N, Handa A, Mattoo A (2002) Engineered polyamine accumulation in tomato enhances phytonutrient content, juice quality, and vine life. Nat Biotechnol 20:613–618

Meyer P, Heidmann I, Forkmann G, Saedler H (1987) A new petunia flower color generated by transformation of a mutant with a maize gene. Nature 330:677–678

Micheal A, Furze J, Rhodes M (1996) Molecular cloning and functional identification of a plant ornithine decarboxylase cDNA. Biochem J 314:241–248

Mol J, Grotewold E, Koes R (1998) How genes paint flowers and seeds. Trends Plant Sci 3:212–217

Mol J, Cornish E, Mason J, Koes R (1999) Novel colored flowers. Curr Opin Biotechnol 10:198–201

Murai N, Kemp JD, Sutton DW, Murray MG, Slightom JL, Merlo DJ, Reichert NA, Sengupta-Gopalan C, Stock CA, Barker RF (1983) Phaseolin gene from bean is expressed after transfer to sunflower via tumor-inducing plasmid vectors. Science 222:476–482

Murphy D (2002) Biotechnology and the improvement of oil crops-genes, dreams and realities. Phytochem Rev 1:67–77

Nakashita H, Arai Y, Shikanai T, Doi Y, Yamaguchi I (2001) Introduction of bacterial metabolism into higher plants by polycistronic transgene expression. Biosci Biotechnol Biochem 65:1688–1691

Ososki AL, Kennelly EJ (2003) Phytoestrogens: a review of the present state of research. Phytother Res 17:845–869

Park W, Lee S, Park K (1998) Cloning and characterization of genome clone (Accession no. U64927) encoding S-adenosyl-L-methionine decarboxylase whose gene expression was regulated by light in morning glory (*Ipomea nil* L.). Plant Physiol 116:867–872

Pereira S, Leonard A, Huang Y-S (2004) Identification of two novel microalgal enzymes involved in the conversion of the ω3-fatty acid, eicosapentaenoic acid, into docosahexaenoic acid. Biochem J 384:357–366

Perez-Amadour M, Carbonell J (1995) Arginine decarboxylase and putrescine oxidase in ovaries of *Pissum sativum* L. changes during ovary senescence and early fruit development. Plant Physiol 107:865–872

Persson S, Wei H, Milne J, Page GP, Somerville CR (2005) Identification of genes required for cellulose synthesis by regression analysis of public microarray data sets. Proc Natl Acad Sci USA 102:8633–8638

Polashock JJ, Griesbach RJ, Sullivan RF, Vorsa N (2002) Cloning of a cDNA encoding the cranberry dihydroflavonol-4-reductase (DFR) and expression in transgenic tobacco. Plant Sci 163:241–251

Portis AR Jr, Parry MA (2007) Discoveries in Rubisco (Ribulose 1, 5-bisphosphate carboxylase/oxygenase): a historical perspective. Photosynth Res 94:121–143

Pueppke SG (1996) The genetic and biochemical basis for nodulation of legumes by rhizobia. Crit Rev Biotechnol 16:1–51

Qi B, Fraser T, Mugford S, Dobson G, Sayanova O, Butler J, Napier J, Stobart A, Lazarus C (2004) Production of very long chain polyunsaturated omega-3 and omega-6 fatty acids in plants. Nat Biotechnol 22:739–745

Quattrocchio F, Verweij W, Kroon A, Spelt C, Mol J, Koes R (2006) PH4 of Petunia is an R2R3 MYB protein that activates vacuolar acidification through interactions with basic-helix-loop-helix transcription factors of the anthocyanin pathway. Plant Cell 18:1274–1291

Quesada-Vargas T, Ruiz ON, Daniell H (2005) Characterization of heterologous multigene operons in transgenic chloroplasts: transcription, processing, and translation. Plant Physiol 138:1746–1762

Rajam M (1997) Polyamines. In: Prasad M (ed) Plant ecophysiology. Wiley, New York, pp 343–374

Rastogi R, Dulson J, Rothstein S (1993) Cloning of tomato (*Lycopersicum esculentum* Mill.) arginine decarboxylase gene and its expression during fruit ripening. Plant Physiol 103:829–834

Regierer B, Fernie A, Springer F, Perez-Melis A, Leisse A, Koehl K, Willmitzer L, Geigenberger P, Kossmann J (2002) Starch content and yield increase as a result of altering adenylate poolsin transgenic plants. Nat Biotechnol 20:1256–1260

Ritsema T, Smeekens S (2003) Fructans: beneficial for plants and humans. Curr Opin Plant Biol 6:223–230

Rivera-Vargas LI, Schmitthenner AF, Graham TL (1993) Soybean flavonoid effects on and metabolism by *Phytophthora sojae*. Phytochemistry 32:851–857

Rosche E, Blackmore D, Tegeder M, Richardson T, Schroeder H, Higgins T, Frommer W, Offler C, Patrick J (2002) Seed-specific overexpression of the potato sucrose transporter increases sucrose uptake and growth rates of developing pea cotyledons. Plant J 30:165–175

Ruan Y, Llewellyn D, Furbank R (2003) Suppression of sucrose synthase gene expression represses cotton fiber cell initiation, elongation, and seed development. Plant Cell 15:952–964

Ruhlman T, Ahangari R, Devine A, Samsam M, Daniell H (2007) Expression of cholera toxin B-proinsulin fusion protein in lettuce and tobacco chloroplasts—oral administration protects against development of insulitis in non-obese diabetic mice. Plant Biotechnol J 5:495–510

Sato S, Xing A, Ye X, Schweiger B, Kinney A, Graef G, Clemente T (2004) Production of γ-linoleic acid and stearidonic acid in seeds of marker-free transgenic soybean. Crop Sci 44:646–652

Schroeder G, Schroeder J (1995) cDNA for S-adenosyl-L-methionine decarboxylase from *Catharanthus roseus*, heterologous expression, identification of the proenzyme processing site, evidence for the presence of both subunits in the active enzyme and a conserved region in the 5′ messenger RNA leader. Eur J Biochem 228:74–78

Segal G, Song R, Messing J (2003) A new opaque variant of maize by a single dominant RNA-interference -inducing transgene. Genetics 165:387–397

Shiono M, Matsugaki N, Takeda K (2005) Phytochemistry: structure of the blue cornflower pigment. Nature 436:791

Shoji K, Miki N, Nakajima N, Momonoi K, Kato C, Yoshida K (2007) Perianth bottom-specific blue color development in Tulip cv. Murasakizuisho requires ferric ions. Plant Cell Physiol 48:243–251

Sijmons P, Dekker B, Schrammeijer B (1990) Production of correctly processed human serum albumin in transgenic plants. Biotechnology 8:217–221

Singh S, Zhou X, Liu Q, Stymne S, Green A (2005) Metabolic engineering of new fatty acids in plants. Curr Opin Plant Biol 8:197–203

Smidansky E, Clancy M, Meyer F, Lanning S, Blake N, Talbert L, Giroux M (2002) Enhanced ADP-glucose pyrophosphorylase activity in wheat endosperm increases seed yield. Proc Natl Acad Sci USA 99:1724–1729

Smidansky E, Martin J, Hannah L, Fischer A, Giroux M (2003) Seed yieldand plant biomass increases in rice are conferred by deregulation of endosperm ADP-glucose pyrophosphorylase. Planta 216:656–664

Spaink HP (2000) Root nodulation and infection factors produced by rhizobial bacteria. Annu Rev Microbiol 54:257–288

Stafford HA (1990) Flavonoid metabolism. CRC Press, Boca Raton FL

Staub JM, Garcia B, Graves J, Hajdukiewicz PT, Hunter P, Nehra N, Paradkar V, Schlittler M, Carroll JA, Spatola L, Ward D, Ye G, Russell DA (2000) High-yield production of a human therapeutic protein in tobacco chloroplasts. Nat Biotechnol 18:333–338

Steele CL, Gijzen M, Qutob D, Dixon RA (1999) Molecular characterization of the enzyme catalyzing the aryl migration reaction of isoflavonoid biosynthesis in Soybean. Arch Biochem Biophys 367:146–150

Stephanopoulos G, Vallino JJ (1991) Network rigidity and metabolic engineering in metabolite overproduction. Science 252:1675–1681

Sticklen M (2008) Plant genetic engineering for biofuel production: towards affordable cellulosic ethanol. Nat Rev Genet 9:433–443

Sticklen M, Dale B, Maqbool S (2006) Transgenic plants containing ligninase and cellulase which degrade lignin and cellulose to fermentable sugars. US Patent 7,049,485

Svab Z, Hajdukiewicz P, Maliga P (1990) Stable transformation of plastids in higher plants. Proc Natl Acad Sci USA 87:8526–8530

Sweetlove LJ, Last RL, Fernie AR (2003) Predictive metabolic engineering: a goal for systems biology. Plant Physiol 132:420–425

Tanaka Y (2006) Flower color and cytochromes P450. Phytochem Rev 5:283–291

Tanaka Y, Brugliera F (2006) Flower color. In: Ainsworth C (ed) Flowering and its manipulation. Blackwell, Oxford, pp 201–239

Tanaka Y, Tsuda S, Kusumi T (1998) Metabolic engineering to modify flower color. Plant Cell Physiol 39:1119–1126

Taylor M, Mad Arif S, Kumar A (1992) Expression and sequence analysis of cDNAs induced during the early stages of tuberization in different organs of potato plant (*Solanum tuberosum* L.). Plant Mol Biol 20:641–651

Thelen J, Ohlrogge J (2002) Metabolic engineering of fatty acid biosynthesis in plants. Metabol Eng 4:12–21

Thu-Hang P, Bassie L, Safwat G, Trung-Nghia P, Christou P, Capell T (2002) Expression of a heterologous S-adenosylmethionine decarboxylase cDNA in plants demonstrates that changes in S-adenosyl-L-methionine decarboxylase activity determine levels of the higher polyamines spermidine and spermine. Plant Physiol 129:1744–1754

Tian L, Dixon RA (2006) Engineering isoflavone metabolism with an artificial bifunctional enzyme. Planta 224:496–507

Tregoning JS, Nixon P, Kuroda H, Svab Z, Clare S, Bowe F, Fairweather N, Ytterberg J, van Wijk KJ, Dougan G, Maliga P (2003) Expression of tetanus toxin Fragment C in tobacco chloroplasts. Nucl Acids Res 31:1174–1179

Trung-Nghia P, Bassie L, Safwat G, Thu-Hang P, Lepri O, Rocha P, Christou P, Capell T (2003) Reduction in the endogenous arginine decarboxylase transcript levels in rice leads to depletion of the putrescine and spermidine pools with no concomitant changes in the expression of downstream genes in the polyamine biosynthetic pathway. Planta 218:125–134

Ueyama Y, Suzuki K, Fukuchi-Mizutani M, Fukui Y, Miyazaki K, Ohkawa H, Kusumi T, Tanaka Y (2002) Molecular and biochemical characterization of torenia flavonoid 3′-hydroxylase and flavone synthase II and modification of flower color by modulating the expression of these genes. Plant Sci 163:253–263

Van Eenennaam A, Lincoln K, Durrett T, Valentin H, Shewmaker C, Thome G, Jiang J, Baszis S, Levering C, Aasen E (2003) Engineering vitamin E content: from *Arabidopsis* mutant to soy oil. Plant cell 15:3007–3019

Verweij W, Spelt C, Di Sansebastiano G-P, Vermeer J, Reale L, Ferranti F, Koes R, Quattrocchio F (2008) An H+ P-ATPase on the tonoplast determines vacuolar pH and flower color. Nat Cell Biol 10:1456–1462

Villarejo A, Buren S, Larsson S, Dejardin A, Monne M, Rudhe C, Karlsson J, Jansson S, Lerouge P, Rolland N, von Heijne G, Grebe M, Bako L, Samuelsson G (2005) Evidence for a protein transported through the secretory pathway en route to the higher plant chloroplast. Nat Cell Biol 7:1224–1231

Vincken Q, Suurs L, Visser R (2003) Microbial starch-binding domains as a tool for targeting proteins to granules during starch biosynthesis. Plant Mol Biol 51:789–801

Watson M, Malmberg R (1996) Regulation of *Arabidopsis thaliana* (L.) arginine decarboxylase by potassium deficiency stress. Plant Physiol 111:1077–1083

Whitney SM, Andrews TJ (2003) Photosynthesis and growth of tobacco with a substituted bacterial Rubisco mirror the properties of the introduced enzyme. Plant Physiol 133:287–294

Winkel-Shirley B (2001) Flavonoid biosynthesis. A colorful model for genetics, biochemistry, cell biology, and biotechnology. Plant Physiol 126:485–493

Witcher D, Hood E, Petersen D (1998) Commercial production of b-glucuronidase (GUS): a model system for the production of proteins in plants. Mol Breed 4:301–312

Wurbs D, Ruf S, Bock R (2007) Contained metabolic engineering in tomatoes by expression of carotenoid biosynthesis genes from the plastid genome. Plant J 49:276–288

Yabuya T, Nakamura M, Iwashina T, Yamaguchi M, Takehara T (1997) Anthocyanin-flavone copigmentation in bluish purple flowers of Japanese garden iris (*Iris ensata Thunb.*). Euphytica 98:163–167

Yadav N, Wierzbicki A, Aegerter M (1993) Cloning of higher plant ω-3 fatty acid desaturases. Plant Physiol 103:467–476

Ye X, Al-Babili S, Kloti A, Zhang J, Lucca P, Beyer P, Potrykus I (2000) Engineering the provitamin A (beta-carotene) biosynthetic pathway into (carotenoid-free) rice endosperm. Science 287:303–305

Ye GN, Hajdukiewicz PT, Broyles D, Rodriguez D, Xu CW, Nehra N, Staub JM (2001) Plastid-expressed 5-enolpyruvylshikimate-3-phosphate synthase genes provide high level glyphosate tolerance in tobacco. Plant J 25:261–270

Yoshida K, Kondo T, Okazaki Y, Katou K (1995) Cause of blue petal color. Nature 373:291

Yoshida K, Toyama-Kato Y, Kameda K, Kondo T (2003) Sepal color variation of Hydrangea macrophylla and vacuolar pH measured with a proton-selective microelectrode. Plant Cell Physiol 44:262–268

Yu O, Jung W, Shi J, Croes RA, Fader GM, McGonigle B, Odell JT (2000) Production of the isoflavones genistein and daidzein in non-legume dicot and monocot tissues. Plant Physiol 124:781–793

Yu O, Shi J, Hession AO, Maxwell CA, McGonigle B, Odell JT (2003) Metabolic engineering to increase isoflavone biosynthesis in soybean seed. Phytochemistry 63:753–763

Zhong R, Morrison W, Freshour G, Hahn M, Ye Z (2003) Expression of a mutant formof cellulose synthase AtCesA7 causes dominant negative effect on cellulose biosynthesis. Plant Physiol 129:797–807

Zhu X, Galili G (2003) Increased lysine synthesis coupled with a knockout of its catabolism synergistically boosts lysine content and also transregulates the metabolism of other amino acids in *Arabidopsis* seeds. Plant cell 15:845–853

Index